Exotic Vegetables

The Book

In recent past, exotic vegetables like broccoli, red cabbage, Savoy cabbage, Chinese cabbage, celery, gherkin etc. have been introduced and their adaptability studies have been undertaken in different parts of the country. These vegetables are getting increasing demand in the five star cultures of the cosmopolitan cities. They generally require cool temperature, however, agro climatic conditions prevailing in several parts of north-western Himalayas of J&K, Himachal Pradesh and Kumaon and Garwhal division of Uttarakhand are very favourable for growing these exotic vegetables. As the day by demand is increasing, the cultivation of exotic vegetables has started picking up in different parts of the country. Earlier these vegetables were not in list of our traditional vegetables; hence, we require special attention to understand their agronomic practices as well as other scientific facts related to nutritional, medicinal attributes, and issues related to improvement of the crop to address the problem of biotic, abiotic stresses etc. Keeping these facts in view, efforts have been made to bring the publication on exotic vegetables. This current publication deals about 37 exotic vegetables namely, cherry tomato, paprika, snow pea, gherkin, baby corn, sweet corn, broccoli, Romanesco cauliflower, red cabbage, Savoy cabbage, Chinese cabbage, Brussels sprouts, Welsh onion, leek, chives, Chinese chives, Rakkyo, parsnip, horse radish, Jerusalem artichoke, endive, lettuce, sugar loaf chicory, red chicory, globe artichoke, parsley, Florence fennel, celery, celeriac, Swiss chard, kale, collard, water cress, winter cress, rhubarb, asparagus and winter squash. Further, it is expected that in near future more research attention will be paid towards advance production and protection technology, genetic improvement and application of biotechnological tools to address the several complex issues of different exotic vegetables grown across the country. I assure to the readers that new and enlarged edition of this title will be brought for their reference.

Exotic Vegetables

Dr. A. K. Panday
Dean
College of Horticulture and Forestry, CAU
Pasighat-791102, Arunachal Pradesh, India

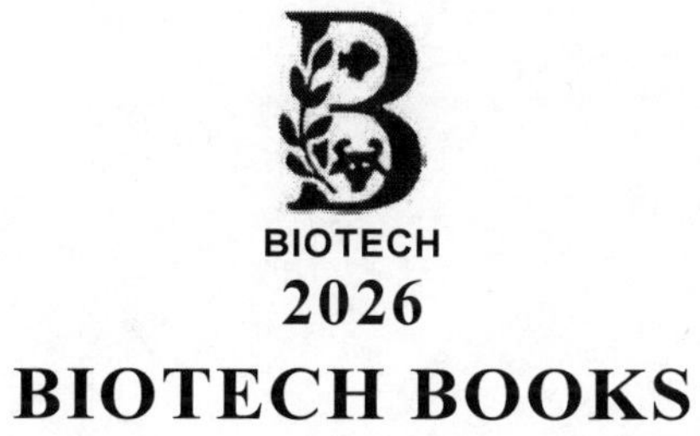

2026
BIOTECH BOOKS

ISBN: 978-81-7622-359-1

Published by : **BIOTECH BOOKS**
4762-63/23, Ansari Road,
Darya Ganj, New Delhi - 110 002
Phone: +91-011-23262132
E-mail: biotechbooks@yahoo.co.in

Laser Typesetting : **Sushil Kumar**

Digitally Printed at : **Replika Press Pvt. Ltd.**

PRINTED IN INDIA

Prof. S.N. Puri, FNAAS

Vice-Chancellor
Central Agricultural University
Imphal-795 004, Manipur

FOREWORD

Vegetable consumption has always been an integral part of every diet and is an important component of a healthy diet. They provide vitamins, minerals and fibre, some energy, as well as certain micro elements often referred to as phytochemicals or secondary plant products; potentially beneficial for our health. Epidemiological studies have shown that high intake of fruits and vegetables are directly associated with a low risk of chronic diseases; particularly, cardiovascular disease.

Exotic vegetables, a status symbol, food of the effluent and as the name suggests has always been preference for the rich people. Further, the restaurant industry is booming because of the country's young population, their growing disposable income, and a trend towards eating out. There is greater awareness about international cuisine too. According to National Restaurant Association of India and Technopak the food services market in India is estimated to be about 48 billion US dollar and in coming five years, it could be worth of 78 billion US dollar; that is nearly what the Indian IT industry currently exports. Further, present changing trend in the eating habit of exotic vegetables are not only restricted among the effluent society of the people but it has now reached to the common men both in urban and rural areas. Hence, growers have taken initiative for its cultivation as well as crop diversification in much bigger way having known its market demands and good returns.

I congratulate the author for successfully putting all efforts in bringing out this valuable publication on Exotic Vegetables which someone with a deep and practical knowledge about these vegetables has put it in together. I am sure this book will cater not only to the inquisitive minds of research scholars, university but would also be extremely helpful to progressive growers and exporters.

(S.N.Puri)

PREFACE

Vegetables are considered one of the most integral parts of human diets, providing some of the most precious and necessary nutritional components, vitamins, minerals, fibers, antioxidants and other micronutrients. Apart from these, vegetables contain several phyto- nutrients, playing a beneficial role in improving human health. Further, an intricate synergy and convergence exists among various nutrients which promotes the sustainable health. Thus promoting any one source of nutrient and attributing to it all health consequences is to ignore the accumulated knowledge of much inter-relationship among different micro-macro nutrients and their role in human nutrition. The immense need of numerous phytonutrients necessitates enlarging the food baskets utilized for human nutrition. According to the Recommended Dietary Allowances (RDA) of the Indian Council of Medical Research (ICMR), per caput consumption of vegetable must be 280 g, whereas availability of vegetable per caput per day is hardly 241 g. Recently it has been observed that along with production of traditional vegetables, consumers are allured for exotic vegetables which are quite attractive in shape, size and colour, and claimed with the enormous research facts that they are immensely rich in several health related factors.

'Exotic Vegetables' the word exotic conjures up images of something strange, exciting, unusual, or simply something different. This book does just that. It brings to you the exciting and strange world of exotic vegetables grown/found in the any part of the country. By definition, exotic vegetables are those that have been introduced into the country from outside, something that is not native to the country. Of course, given the fastness of the country, something found in one region of the country could be exotic in some other regions of the same country. In this context, India has got unique gift blessed with diverse climatic condition for growing an array of these exotic vegetables and all efforts can be taken to trap the country's potential in growing these vegetables and cater to the needs of the ever growing population and thereby reducing the cost on imports. In recent past, exotic vegetables have entered in to the kitchens of the country in a very big way, particularly so in bigger cities and towns and any hypermarket of repute today has a plenty of stock of these vegetables, catering to the need of dedicated and growing clienteles. This is supported by the hard facts and figures that the demands for such vegetables have been growing at the rate of 15-20 per cent over the past few years. How many of us have not heard of broccoli, baby corn, lettuce, Brussels sprouts, endive, chive, chicory, Chinese cabbage etc. But not many do actually know about the several important facts like origin, the nutrient value, benefits, methods of growing etc. of these vegetables. This book seeks to present these and numerous more facts about these vegetables in a very interesting and easy to understand manner. The author is well certain that this book will be of good use to people of all sections of the society particularly to researchers, students, growers, exporters, extension worker, planners and policy makers.

It is expected that in near future more attention will be paid towards unrevealing the various facts related to nutritional, medicinal importance, advance production and protection technology, genetic improvement and application of biotechnological tools to address the several complex issues of various exotic vegetables grown across the country. I assure to the readers that new and enlarged edition of this title will be brought for their reference, however, in this context suggestion and observations of readers are welcome.

In bringing out this publication, I express my profound gratitude to Prof. S.N.Puri, Hon'ble Vice-Chancellor, Central Agricultural University, Imphal, Manipur, for his constant encouragement, admirable suggestion and appreciation throughout the preparation of manuscript. I am grateful to eminent vegetable Scientist Prof. G. Kalloo, Ex-Vice Chancellor, Jawaharlal Nehru Agricultural University, Jabalpur, (M. P.) and Dr. Kirti Singh, Ex. Chairman, Agricultural Scientists' Recruitment Board, New Delhi for their continuous guidance and support.

I express my heartfelt appreciation to Dr. S. Ayyappan, Secretary (DARE) & DG, ICAR, for his unstinted support and co-operation. I extend my deep appreciation to Dr. N. Krishna Kumar, Deputy Director General (Horticulture), I.C.A.R., New Delhi whose incisive inputs contributed much to the quality of the book. I am thankful to Dr. Premjit Singh, Director, Extension Education, Central Agricultural University, Imphal, Manipur; Dr. A.N. Maurya, Ex-Director, Institute of Agricultural Sciences, BHU, Varanasi and Dr. B. Singh, Director, Indian Institute of Vegetable Research, Varanasi for giving their support and valuable suggestions for dwelling upon the various aspects of exotic vegetables.

I would like to put on record my appreciation to Prof. S. D.Warade, Professor and Head , Dr. R .K. Dubey and Dr. Vikas Singh, Asstt. Professors, Department of Vegetable Crops, College of Horticulture and Forestry, CAU, Pasighat for their critical inputs. My thanks are also due to Ms. Th. EloniVida ,Ms.NeetaLongjam for their patience in proof reading of the manuscript. I thank Mr. Ripon Dam and Shri Deepak Vishwakarma for word processing and beautifully arranging the photographs.

Finally, my great appreciation to my wife, Radhika , daughter Pratbha and son Saurabh who over the years has always been very patient while I have been busy in preparing this manuscript.

Author

Contents

5
Gherkin 43

6
Baby Corn 54

1

EXOTIC VEGETABLES - AN INTRODUCTORY CONSIDERATION

About 3 billion people in the world are malnourished due to imbalanced diets or a lack of food. Those without sufficient food are undernourished and have weakened immune systems; undernourished children may be stunted with impaired cognitive development. More than 70 per cent of malnourished children live in Asia. People unable to consume a balanced diet suffer from micronutrient malnutrition; they may be over nourished, and prone to develop chronic diseases such as anemia, blindness, cardiovascular disease, diabetes and cancer. In Asia and the Pacific, almost 600 million malnourished people add a considerable burden to already overstretched healthcare systems. Presently our country's population is increasing at the rate of 1.548 per cent and it is projected that it will be around 1256 million up to 2015 and 1331 million in 2020. In our country where problem of malnutrition is prevailing in general and micronutrient malnutrition in particular, addressing the household nutritional security is *sine qua non*. Recent studies indicate that intake of micronutrients in daily diet is far from satisfactory and largely less than 50 per cent RDA is consumed by over 70 per cent of Indian population. The loss due to micronutrient deficiency costs India 1 per cent of its GDP. This amounts to a loss of Rs. 27,720 crore per annum in terms of productivity, illness, increased health care costs and death. Every day, more than 6,000 children below the age of five die in India. More than half of these deaths are caused by malnutrition mainly the lack of vitamin A, iron, iodine, zinc and folic acid. About 57 per cent of preschoolers and their mothers have sub- clinical vitamin A deficiency. Anemia prevalence among children under five years is 69 per cent and among women it is over 55 per cent. The consequences of micronutrient malnutrition are unacceptably high morbidity and mortality. Vitamin A, iron and zinc deficiency when combined constitute the second largest risk factor in the global burden of diseases; 330,000 child deaths are precipitated every year in India due to vitamin A deficiency; 22,000 people, mainly pregnant women, die every year in India from severe anemia; 6.6 million children are born mentally impaired every year in India due to iodine deficiency; intellectual capacity is reduced by 15 per cent across India due to iodine deficiency; and 200,000 babies are born every year with neural tube defects in India due to folic acid deficiency.

Owing to the vicious circle of malnutrition managing global malnutrition is complex, especially with the potential consequences of climate uncertainty and social unrest. For addressing the problem of malnutrition different nations/international agencies launched several programmes like bio-fortification, food fortification, dietary supplements etc. however, results were not upto the satisfactory levels. In this context, vegetables are a key source of micronutrients to improve the health of the malnourished poor, particularly vulnerable women and children. With locally available, high nutrient content vegetables, communities can ensure good nutrition through balanced diets and also increase household incomes. According to the Recommended Dietary Allowances (RDA) of the Indian Council of Medical Research (ICMR), per caput consumption of vegetable must be 280 g, whereas availability of vegetable per caput per day is hardly 241 g. In recent past, along with production of traditional vegetables, consumers are allured for exotic vegetables which are quite attractive in shape, size and colour, further rich in several health related factors (Pandey and Rai, 2006).

NUTRITIONAL AND MEDICINAL IMPORTANCE OF EXOTIC VEGETABLES

Nutritional Importance of Exotic *Brassica*

Recently, enormous attention has been paid towards harnessing the amazing nutritional and medicinal properties of broccoli. In fact now days broccoli is one of the most preferred *Brassica* vegetables. It is also a good source of zeaxanthin and beta-carotene. All three of these carotenoids function as key antioxidants. Other antioxidants provided by broccoli in beneficial amounts include vitamin E and the minerals manganese and zinc. The unique combination of antioxidant, anti-inflammatory, and pro-detoxification components in broccoli make it a unique food in terms of cancer prevention. Relationship between cancer development and oxidative stress, chronic inflammation, and inadequate detoxification are so well-documented in the research that any food improving all three of these metabolic problems would be highly likely to lower our risk of cancer. In the case of regular consumption of broccoli avoids the risk of prostate cancer, colon cancer, breast cancer, bladder cancer and ovarian cancer. The B-complex vitamins in broccoli can also make a major contribution to our cardiovascular health especially with respect to excessive formation of homocystein which raises our risk of atherosclerosis, stroke, and heart attack. Broccoli being a rich source of vitamin B_6, vitamin B_{12}, and folate helps in reducing the risk of excessive homocysteine formation and cardiovascular problems that are related to excess homocysteine.

While green cabbage is the most commonly eaten variety of cabbage, but our familiarity to red cabbage is restricted. Red cabbage is nutritionally rich and besides possessing all required nutrition, it is having added nutritional benefits of rich red color with concentration of anthocyanin polyphenols, which is considered one of the most protective phytonutrients. Anthocyanin pigments acts as, apart from these, it is a potentially protective, preventative and plays therapeutic roles in a number of human diseases. The antioxidant richness of cabbage is partly responsible for its cancer prevention benefits. Without sufficient intake of antioxidants, our oxygen metabolism can become compromised, and we can experience a metabolic problem called oxidative stress. Chronic oxidative stress in and of itself can be a risk factor for development of cancer. In *Brassica* group of vegetables, Brussels sprouts are grown in very restricted area and it is a very interesting vegetable where cabbage likes small heads commonly known as sprouts are formed in the axils of leaves. It is a rich source of vitamins C, E, and A (beta-carotene). The anti-oxidant mineral manganese is also found in Brussels sprouts. Brussels sprouts also contain important antioxidant like isorhamnetin, quercitin, and kaempferol, caffeic acid and ferulic acid. Some of the antioxidant compounds found in Brussels sprouts may be somewhat rare in foods overall. One such compound is a sulfur-containing compound called D3T (3H-1, 2-dithiole-3-thione).

Nutritional Importance of Exotic Greens

Leafy greens are important constituent of our diets. Demand of edible greens varies from region to region depending upon preference, growing conditions and availability etc. In recent years, people have paid good interest towards exotic greens. Among leafy greens, kale is a good source of vitamin C, beta-carotene, and manganese. Kale contains flavonoids, including kaempferol and quercetin. Many of the flavonoids in kale are also now known to function not only as antioxidants, but also as anti-inflammatory compounds (Pandey and Rai, 2007; Pandey, 2008). It is a good source of cancer-preventing glucosinolates. Kale contains nearly twice the amount of vitamin K as compared to other *Brassica* vegetables. Besides above nutrients, kale is an excellent source of copper, tryptophan, calcium, vitamin B_6 and potassium. It is considered a good source of iron, magnesium, vitamin E, vitamin B_2, protein, vitamin B_1, folate, phosphorous and vitamin B_3.

Among the exotic greens, Swiss chard is another leafy vegetable tinged with the multiple colors. Swiss chard is an excellent source of bone-building vitamin K, manganese, and magnesium; antioxidant vitamin A, vitamin C, and vitamin E; potassium and iron. It is a very good source of copper, calcium, vitamin B_2 and vitamin B_6. In addition, Swiss chard is a good source of energy-producing phosphorus, vitamin B_1, vitamin B_5, biotin, and niacin; immune supportive zinc; and folate. Swiss chard has unique benefits for blood sugar regulation. In addition, chard may provide special benefits in the diets of individuals diagnosed with diabetes. In addition to its syringic acid, chard contains an excellent amount of fiber and protein. Fiber and protein-rich foods are an excellent way to help stabilize blood sugar levels. Among the edible greens, asparagus is one of the most important vegetables. The fleshy green spears of asparagus are both succulent and tender and have been considered a delicacy since ancient times. Asparagus is anti-inflammatory nutrients like saponins, including asparanin A, sarsasapogenin, protodioscin, and diosgenin. Among these saponins, sarsasapogenin has been of special interest in relationship to amyotrophic lateral sclerosis (ALS), also known as "Lou Gehrig's Disease."

Among the salad vegetables, lettuce is having significant edge over other greens owing to its high vitamin C and beta-carotene content. Vitamin C and beta-carotene functions together to prevent the oxidation of cholesterol. When cholesterol becomes oxidized, it becomes sticky and starts to build up in the artery walls forming plaques. If these plaques become too large, they can block off blood flow or break, causing a clot that triggers a heart attack or stroke. Fiber rich lettuce adds another plus in its column of heart-healthy effects. In the colon, fiber binds to bile salts and removes them from the body. This forces the body to make more bile, which is helpful because it must break down cholesterol to do so. This is just one way in which fiber is able to lower high cholesterol levels. Equally beneficial to heart health is lettuce's folic acid content. In addition, lettuce is a very good source of potassium, which has been shown in numerous studies to be useful in lowering high blood pressure, another risk factor for heart disease. With its folic acid, vitamin C, beta-carotene, potassium and fiber content, lettuce can significantly contribute to a heart-healthy diet.

Production of Exotic Vegetables in India

Recently few exotic vegetables including broccoli, red cabbage, Savoy cabbage, Chinese cabbage, celery, leek, parsley have been introduced and their adaptability studies have been undertaken in different parts of the country. These vegetables are getting increasing demand in the five star cultures of the cosmopolitan cities (Pandey, 2003; Pandey and Singh, 2011). These vegetables generally require cool temperature. The agro climatic conditions prevailing in several parts of north-western Himalayas of J & K, Himachal Pradesh and Kumaon and Garwhal division of Uttarakhand are very favourable for growing these exotic vegetables. The cultivation of exotic vegetables has started picking up in Himachal Pradesh with farmers opting for diversification in agriculture. Having the most of climatic and soil conditions of the region conducive for exotic vegetables, farmers in Theog, Matyana, Narkanda, Sainj valley, Sproon valley in Solan, Nauradhar in Sirmaur, Katrain and Manali towns have taken to growing exotic vegetables like asparagus, broccoli, lettuce, coloured capsicum, celery, Chinese cabbage, Brussels sprouts, European carrot, parsley, leek and snow peas. Himachal Pradesh's climatic as well as soil conditions which is at an altitude of 2,000 to 2,500 meters above sea level, are most suitable for growing these exotic vegetables. The Theog, Matyana belt which was earlier growing traditional vegetables has switched over to exotic vegetables and presently broccoli, savoy cabbage and Chinese cabbage have got sizable areas. These exotic vegetables are marketed in Chandigarh and Delhi as they have a huge demand for these vegetables keeping in view of the foreign tourists and five star hotels in these cities. Even in local markets in Shimla there is a growing demand and growers have also tied up with some hotels to cater to them in the winters. Many areas in the Himachal Pradesh are giving good production of exotic vegetables, like snow pea is picking up fast with farmers in Nuradhar in Sirmaur. Similarly there is good production of European carrot in Kinnaur district and coloured capsicum is growing well in Solan. A part from hill states of the country, these cold season exotic vegetables grow well during winter season in plains of north India including Punjab, Uttar Pradesh and Haryana. Cultivation of celery and parsley were started during 1990s under controlled conditions at very small extent in Chatarpur near Mahrauli in Delhi and Kairana in Muzaffarnagar by progressive vegetable growers to supply Delhi, Mumbai and Kolkata markets.

Recently in Ahmedabad, Gujarat a motley group of farmers in Bharuch district is attempting to grow colourful Israeli varieties of cherry tomato and capsicum. This produce has now developed national and international retail chains for their marketing. This clearly shows transformation in the global food system are causing changes in food production and marketing. In our country as the purchasing power of people is increasing there is a growing demand of exotic vegetables. Further, production and marketing arrangements are responding to changing demand driven by urbanization and diet change (Priya *et al.*, 2003).

Global Attention towards promoting the Exotic Vegetables

The first symposium on Diversification of Vegetable Crops was held in Angers, France in September, 1988. In this symposium, vegetable scientists expressed their opinion towards breeding, cultural management, and marketing of new vegetable crops. At the conclusion of the first Symposium it was decided to hold the second one in Miami, Florida, USA in 1992. With this background, second International Symposium was organized on Specialty and Exotic Vegetable Crops. The scope of this symposium was expanded to include not only new vegetable crops, but also those crops that are grown on a small scale to satisfy special market demands.

Priority on Organic Production of Exotic Vegetables.

According to IFOAM (International Federation of Organic Agriculture Movements) Organic agriculture is a production system that sustains the health of soils, ecosystems and people. It relies on ecological processes, biodiversity and cycles adapted to local conditions, rather than the use of inputs with adverse effects. Organic agriculture combines tradition, innovation and science to benefit the shared environment and promote fair relationships and a good quality of life for all involved. According an estimate 35 million hectares of agricultural land are managed organically by almost 1.4 million producers. The regions with the largest areas of organically managed agricultural land are Oceania (12.1 million hectares), Europe (8.2 million hectares) and Latin America (8.1 million hectares). The countries with the most organic agricultural land are Australia, Argentina and China. The highest shares of organically managed agricultural land are in the Falkland Islands (36.9 per cent), Liechtenstein (29.8 per cent) and Austria (15.9 per cent). The countries with the highest numbers of producers are India (3, 40,000 producers), Uganda (1,80, 000) and Mexico (1,30,000). More than one third of organic producers are in Africa. On a global level, the organic agricultural land area increased in all regions, in total by almost three million hectares, or nine per cent, compared to the data from 2007. It has been estimated that twenty-six per cent (or 1.65 million hectares) moreland under organic management was reported for Latin America, mainly due to strong growth in Argentina. In Europe the organic land increased by more than half a million hectares, in Asia by 0.4 million. About one-third of the worlds organically managed agricultural land - 12 million hectares is located in developing countries. Most of this land is in Latin America, with Asia and Africa in second and third place. The countries with the largest area under organic management are Argentina, China and Brazil. Further, it has been recorded that 31 million hectares are organic wild collection areas and land for bee keeping. The majority of this land is in developing countries - in stark contrast to agricultural land, of which two-thirds is in developed countries. Further organic areas include aquaculture areas (0.43 million hectares), forest (0.01 million hectares) and grazed non-agricultural land (0.32 million hectares). The global market of organic produce is given in Table (1.1).

Table (1.1). Global Market of Organic Produce (US dollar)

Year	Value
2001	20.0 billion
2002	23.0 billion
2007	46.0 billion.
2008	50.8 billion
2009	55.0 billion
2010	59.0 billion

Source: IFOAM

Continent wise growth of organic agriculture is mentioned in the table (1.2).

Indian organic - a need of integrating traditional knowledge and wisdom

In India approximately 2.6 m/ha is under organic certification. Out of which cultivated area is only 1 lakh ha while remaining 2.5 million ha is wild forest area. North East accounts for nearly 4400 ha under certified organic farming (including under conservation). Nation steering committee on Accreditation is the apex body for accreditation and has authorized 11 certification agencies. Ministry of Agriculture has initiated a National Project on Organic Farming for promotion of organic agriculture and for ensuring effective implementation of regulatory mechanism at affordable cost. The project is operated through a National Centre of Organic

Table (1.2). Continent wise growth of organic agriculture

Continent	Total area under organic (lack ha)	Per cent area under organic	Organic Rich Countries
Africa	9	2.5	Uganda Tunisia, Ethiopia
Asia	33	9.0	India, China
Europe	82	23.0	
Latin America	81	23.0	Argentina, Brazil and Uruguay
North America	25	7.0	U.S.
Oceania (Australia, New Zealand, Fiji, Papua New Guinea, Tonga and Vanuatu)	121	35	Australia, New Zealand, Vanuatu

Farming at Ghaziabad and six Regional Centres of Organic Farming at Bangalore, Bhubaneswar, Hisar, Imphal, Jabalpur and Nagpur.

Demand for Indian organic products in domestic and export markets

The products available in the domestic markets in the organic quality are vegetables, fruits, tea, coffee, cereals and pulses. Major markets for organic products are in metropolitan cities like Mumbai, Delhi, Kolkata, Chennai, Bangalore and Hyderabad. According to an estimate domestic sales of organic products are 7.5 per cent of the total organic production in the country. According to an estimate made by Org- Marg 2002, demand of domestic organic produce during 2005-2006 and 2006-2007 was around 1457 and 1568 tonnes, respectively which is getting momentum.

Components of organic farming

Various components required for addressing the complete protocols of organic production has been described as under:

Nutrient Management: There are a number of organic sources of nutrition and among them green manuring, composting, bio-fertilizer, cakes, vermicompost and bio-dynamics are important.

Green Manuring: Green manuring is the main source of nutrition for organic production. Experimentally it has been confirmed that a number of green manures are major source of nitrogen, carbon and for the improvement of physical condition of soil. Generally, dhaincha (*Sesbenia esculenta*) and *S. rostrata*/ other species) and sun hemp (*Crotolaria juncea*) have proved important and successful green manure crops which are ploughed in the soil after about 6 to 8 weeks of sowing when adequate vegetative growth and nodule formation is attained. Green manure increases water holding capacity of soil as well as increase the population of micro-organism also.

Compost: Compost is the traditional source of nutrients for the crops. Through it is a low nutrient source, its special merit lies in its capacity to supply in addition to NPK a large number of essential micro-nutrients which are becoming deficient in the intensively cultivated areas. The supply of micronutrients particularly satisfies the hidden hunger in plants and safeguards against toxicity/injury which can be caused at times when improvement of soil physio-chemical and biological properties like soil structure, water holding capacity, nutrient supply and the benefit to the soil microbial population have been well established. Benefits of compost application are given as under:

a) In order to encourage soil biological properties, regular use of Cow Pat Pit (CPP) Cow Horn Manure (BD-500) are helpful.

b) Use of liquid manure prepared from cow dung, cow urine, leguminous leaves or Vermiwash are also effective in promotion of growth and fruiting.

c) Bio-dynamic compost is an effective soil conditioner and is an immediate source of nutrient for a crop. Biodynamic compost can be prepared by using green leaves (nitrogenous material) and dry leaves (carbonaceous material) in 8-12 weeks.

d) Integrating with cow dung slurry is always good in the decomposition process. The enrich compost is ready in 75-100 days depending upon the prevailing temperature.

Nadep compost: Nadep compost has been found very effective due to an aerobic respiration. Composting is fast and nutritional status of the compost is better than the ordinary compost. In this method of composting, farm wastes (cow dung, green/dry grasses, wheat/paddy straw and weeds) are used. Wide variations in nutrient status of composts and CPP have been observed. This can be further enriched through incorporation of rock phosphate, bone meal, slacked lime blood and fishmeal. Various combination of green vs dry, leguminous non-leguminous may be helpful. These need to be worked out for meeting the nutritional requirement of various horticultural commodities.

Role of cow dung: Since ancient times, human beings have recognized the importance of cows and their contribution to the society. Through her calm, loving nature, the cow teaches humanity how to behave with each other and with the environment. Cow dung and urine are the important component in bio farming. Their role is summarized below:

a) Cow dung cake and ghee are basic component of Agnihotra (Homa farming).

b) Japan uses cow dung to get protection from atomic emissions.

c) Cow dung is the best soil conditioner.

d) *Actinomycetes* are helpful in control of many plant diseases like gummosis, die back, anthracnose etc.

e) Cow dung has been a regular component for plastering houses in the Indian villages.

Cow urine: Cow urine contains copper, which transforms into Gold in human body. Gold has power to destroy all diseases and is considered an antidote.

(i) Besides, copper, cow urine contains iron, calcium, phosphorus, carbonic acid, potash and lactase.

(ii) It contains 24 types of salts and the medicines made from cow urine are used to cure several diseases.

(iii) Cow urine is disinfectant and prophylactic and these purifies and improves the fertility of land.

Bio-fertilizers: Bio-fertilizers are bacterial cultures of appropriate species which have the capability of fixing atmospheric nitrogen such as *Rhizobium* species in leguminous crop and *Azotobacter* and *Azospirillum* in non-leguminous crops. The Phosphate Solubilizing Bacteria (PSB) have proven utility in making soil unavailable phosphorus in form of available phosphorus to the crop. The renewable source of nutrient supply to the crops has become popular during the last decade or so among the farmers. A great potential exists for its widespread use in all crops under all situations. *Azospirillum* has been recommended in a number of crops for increasing the production and productivity of crops. *Rhizobium* is now commercially being used in almost all the leguminous crops for the nitrogen fixation. *Rhizobium* cultures against specific crop plants need to be applied in order to boost the nodulation which will help in enriching the soil fertility. The free living bacteria such as *Azotobacter*, *Azospirillum* and other species can also be applied for enhancing the nitrogen fixation in the soil. Use of blue green algae and azola also helps in fixing the atmospheric nitrogen and enhancing the soil fertility. Antagonistic bacteria which are capable of suppressing harmful bacteria and fungi in the rhizosphere of legumes are capable of enhancing the performance of *Rhizobium* for nodulation, nitrogen fixation and improving yield. Co-inoculation of *Rhizobium* and antagonistic bacteria along with *Azospirillum* and/*Azotobacter* can enhance the nodulation and yield. Following steps may be initiated for regular incorporation of organic waste through NADEP, Vermi, Biodynamic Compost (BD) or Microbe Mediated Compost (MM compost):

a) Use of cakes (neem, mahuwa, *Pongamia*, castor, ground nut etc.) as per availability need to be promoted.
b) Promotion of green manuring and legumes as inter and cover crops whenever and wherever possible.
c) Promotion of mulching with organic wastes which can be further promoted by spread of 5-20 vermi / BD kg compost, or 100g of CPP and incorporation of 50-100 earthworms.

Vermi compost: Vermi-culture is an aspect involving the use of earthworms as versatile natural bioreactors for effective recycling of non-toxic organic wastes to the soil. They effectively harness the beneficial soil micro flora, destroy soil pathogens, and convert organic wastes into valuable products such as bio fertilizers, bio pesticides, vitamins, enzymes, antibiotics, growth hormones and proteinous biomass. Earthworm functions in soil farming system in following ways:

a) Through their influence on soil pH.
b) As agents of physical decomposition.
c) Promoting humus formation.
d) Improving soil structure.
e) Enriching biological properties of soil.

Vermi-wash: Vermi-wash is prepared from the population of earthworms reared in earthen pots or plastic drums. The extract contains major, micronutrients, vitamins (such as B_{12}) and hormones (gibberellins) secreted by the earthworms. Earthworms produce bacteriostatic substances and its use in the form vermin- wash can protect the bacterial infections. Vermi-wash can be sprayed on vegetable crops for better growth, yield and quality. Continuous pouring of water in the pot/drum drop wise and the organic waste in the pot/drum should be changed regularly, after its full conversion in to the compost.

Biodynamic approach: It has been designed as a biodynamic system of farming and so is an enhanced state of organic agriculture. It views the farm as an individual organism and in the addition to the normal organic practices; this system emphasizes animal husbandry, a set of eight fermented minerals and herbal preparations, and the use of astronomical influence on growth and development of the plants. The accelerated aerobic composting uses a mixture of cow manure and urine with organic waste from the farm, which is then formed into heaps and allowed to heat above 80 ^{0}C killing all weed seeds, thus preventing weed infestation. A special variety of worm is also introduced increasing the potency, as it accelerates the bioprocesses that naturally maintain soil fertility.

IPM strategies for organic farming of vegetables

Routine agronomic practices like selection of varieties, intercultural operation, sowing/planting date, trap cropping/intercropping should be suitably modified to make uncongenial condition for pests and to enhance the activity of natural enemies. Keeping in mind the diversity and intensity of pests in particular place, selection of resistant / less susceptible varieties, holds good in pest management. Plant resistance provides a built in ability to allow less pest and cut off the extra load of insecticides in the crop. Being completely safe, host plant resistance fits well with all other components. Unlike cereals, in vegetables very less number of resistant/ less susceptible varieties has been developed (Table1.3). The pests of sucking nature can be combated to greater extent with the adoption of resistant varieties.

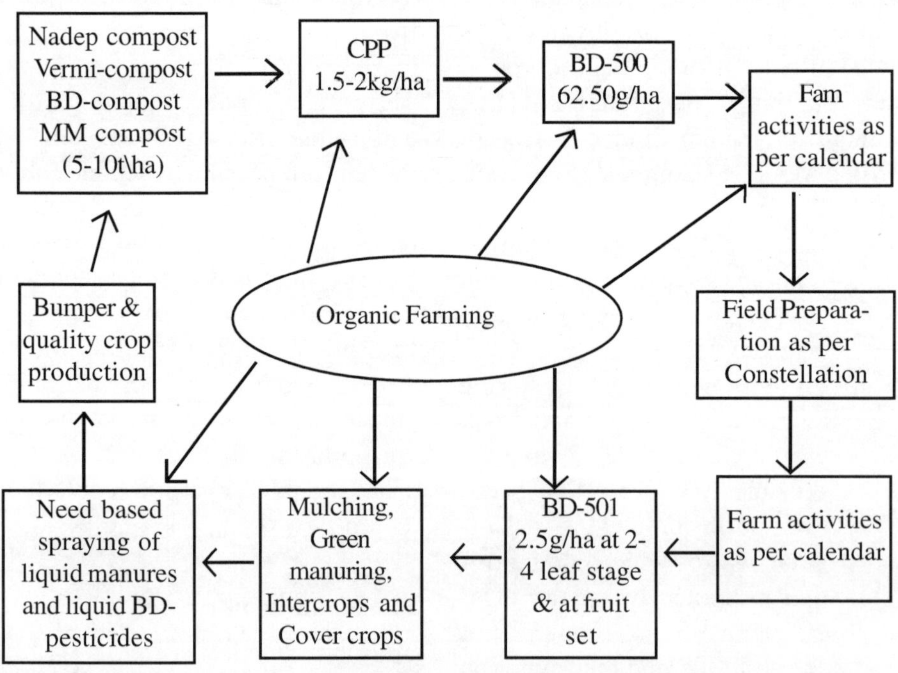

Fig. (1.1): Schematic Presentation of organic farming

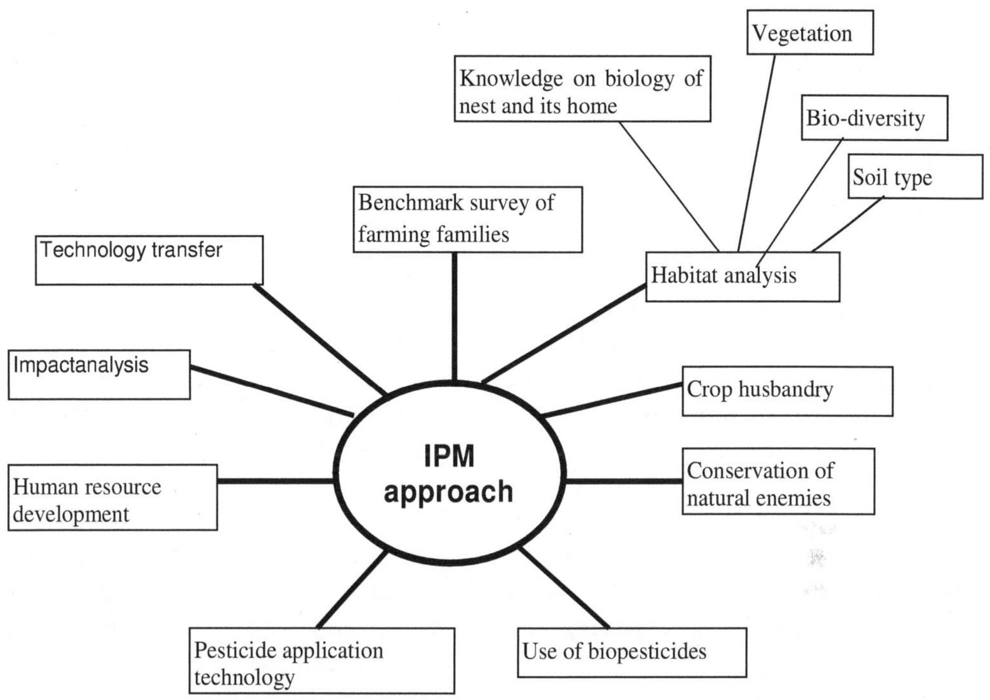

Fig. (1.2): Approaches for the successful IPM

Table (1.3): Less susceptible varieties of some vegetable crops

Crop	Pest	Varieties
Brinjal	Shoot and fruit borer (*Leucinodes orbonalis*)	SM 17-4, PBR 129-5, Punjab Barsati, ARU 2C, Pusa Purple Round, Punjab Neelam
	Aphid, jassid, thrips whitefly	Kalyanpur-2, Punjab Chamkila, Gote-2, PBR-91, GB-1, GB-6
Cabbage	Aphid (*Brevicoryne brassicae*)	All season, Red Drum Head, Sure Head, Express Mail
Cauliflower	Stem borer (*Hellula undalis*)	Early Patna, EMS-3, KW-5, KW-8, Kathmandu Local
Okra	Jassid (*Amrasca biguttula*)	IC-7194, IC-13999, New Selection, Punjab Padmini
Onion	Thrips (*Thrips tabaci*)	PBR-2, PBR-6, Arka Niketan, Pusa Ratnar, PBR-4, PBR-5, PBR-6
Round gourd	Fruit fly (*Dacus cucurbitae*)	Akra Tinda
Pumpkin	Fruit fly (*Dacus cucurbitae*)	Arka Suryamukhi
Bitter gourd	Fruit fly (*Dacus cucurbitae*)	Hisar-II

Adjustment of sowing and planting date keeping in mind the peak activity of the damaging stage of the pest helps in avoiding the critical stage of the crop from pest attack. Careful consideration of sowing / planting date in vegetable reduced the attack of red pumpkin beetle, fruit fly, and shoot and fruit borer. Early planting of cucurbits in November escapes the attack of red pumpkin beetle whereas flowering beyond October in bitter gourd suffers less from fruit fly. Sowing of okra during second week of June retain less population of borers, there by incurred maximum healthy yield whereas July planted brinjal face the ravages of shoot and fruit borer. Thus, synchronization of most susceptible stage of the crop with the inactive period of insect reduces the infestation and chemical intervention. Intercropping of crops with diverse plant geometry and insect pest breaks the standard mono-cropping and limits the infestation from insect (Table 1.4). Diverse nature of plant not only obstructs the adults from egg laying but also the release of volatile allelochemicals from a particular crop deters the adult insect from damaging the other. These effects did not provide suitable condition for development of micro-climate favouring the multiplication of particular type of pest. Above all such planting combination enhances the activity of predators and parasites also.

Table (1.4). Intercrop combination effective in vegetable pest management

Crop combination	Target pest
Cabbage + Carrot	Diamond back moth
Broccoli + Faba bean	Flea beetle
Okra + Cowpea	Yellow vein mosaic
Cabbage + French bean	Root fly
Cabbage + Tomato	Diamond back moth

Trap crop must be distinctly attractive to the pest than the main crop. It provides protection either by preventing the pest from reaching the main crop or by concentrating them in certain part of the field where they can be economically destroyed. Mustard as a trap crop along with cabbage has been successfully utilized for the control of diamond back moth, aphid and leaf webber. African marigold in tight bud stage functions as good source to trap the adults of *Helicoverpa armigera* besides it also attracts the adults of leaf miner for egg laying on the leaves. Maize plants applied with bait spray trap may kill fruit fly adults when sown in combination with bitter gourd.

Natural enemies- an integral part of IPM module in vegetables

Vegetable ecosystem is endowed with a large number of natural enemies consisting of predators, parasites and parasitoids causing damage to different stages of the insect to varying extent (Table1.5). Till now except a few, most of these natural enemies are under exploited for pest management in vegetables. Among the egg parasites *Trichogramma* spp. has been utilized to some extent for control of tomato fruit borer. Release of 2,50,000 parasitized *Corcyra* eggs has been found to control the fruit borer convincingly. Another predator which needs immediate attention is *Chrysoperla carnea*. This is an effective predator for control of aphid, jassid and eggs of some lepidopterist borers, when the first instar larvae are released @ 50,000 /ha. Mass rearing of this predator needs to be streamlined for economizing its use as bio-control agent.

The larval parasites of diamond back moth, *Apanteles plutellae* and *Diadegma semiclausum* can be incorporated into biological pest management because of their potential or damaging the host larvae. These will prove effective in areas where diamond back moth possesses problem because of insecticidal resistance.

Export Potential of Exotic Vegetables

Diversification in horticulture resulted in multiple increase in total export of horticultural produce over last decades. India is exporting 2.43mMT of fresh fruits and vegetables including 16.67lakh metric tones of onions worth Rs. 1966.62crores (2012-13). In addition, 68520 MT dried and preserved vegetables worth Rs. 637.95 crores are also exported to different countries. During 2012-13, country exported 2386424 MT cucumber and Gherkin worth Rs.637.95 crores. Agricultural and Processed Food Products Export Development Authority (APEDA) is the nodal organization to access the market, guide and promote the export of agricultural commodities. APEDA apart from tadtional vegetables, it has non-traditional exotic vegetables like asparagus, celery, sweet pepper, sweet corn, baby corn, snow peas, cucumber, gherkins, broccoli and cherry tomato which are getting momentum in export. Cultivation of new vegetables like

Table (1.5). Natural enemies of insect pests of some important vegetable crops

Pest	Number recorded	Parasite / Predator	Parasitisation (%)
Crucifers			
Plutella xylostella	29	*Apanteles plutellae*	3.5
		Diadegme semiclausum	20.35
		Brachymeria excarinata	59
Crocidolomia binotalis	10	*Alanteles crocidolomia*	-
Hellula undalis	4	*Bracon* spp.	16.7
Tomato			
Helicoverp aarmigera	120	*Trichogramma* spp.	80
Okra			
Earias spp.	27	*Trichogramma* spp.	0.7-56
Spodoptera litura	42	*Telenomusr owani*	10-20
		Peribae aorbata	24
Aphis spp.	34	*Coccinella septempuctata*	-
		Chrysoperla carnea	-

Table (1.6). Export of vegetables (quantity in MT and value in lakh)

Products	2011-12		2012-13	
	Qty.	Value	Qty.	value
Floriculture				
Floriculture	30926.02	36532.15	27121.86	42344.6
Fruits and Vegetable seeds	15205.81	28776.35	17168	34772.39
TOTAL	**46131.83**	**65308.5**	**44289.86**	**77116.99**
Fresh Fruits & Vegetables				
Fresh onion	1309924.82	172299.8	1666872.6	196662.66
OTHER FRESH VEGETABLES	734178.83	131048.2	768627.2	151633.56
TOTAL	**2044103.65**	**303348.0**	**2435499.8**	**348296.22**
Processed Vegetables				
Dried and preserved vegetables	64794.09	52678.47	68520.25	63795.76
Other processed fruits and vegetables	274807.05	157759.82	269217.26	173305.54
SUB TOTAL	**339601.1**	**210438.3**	**337737.5**	**237101.3**

Source: NHB Data Base, 2013-14.

gherkin, baby corn, sweet corn, broccoli, Brussels sprouts, zucchini, Chinese cabbage, red cabbage, asparagus, celery and parsley is on the rise trends and provide better returns. The area and production of gherkins is increasing in Karnataka, Andhra Pradesh and Maharashtra for export. Export of vegetables is given in the table (1.6).

Initiatives for Promotion of Export

The level of processing of fruits & vegetables in the country is estimated as 2.20 per cent. The low level of fruit and vegetable processing is due to non-availability of processable varieties of raw materials at right quantity and quality, seasonable nature of industry, lack of adequate post-harvest infrastructure such as lack of cold chain facilities, transportation, proper storage facilities etc. However, over the last few years, there has been a positive growth in fruit and vegetable processed products such as ready-to-serve beverages, fruit juices and pulps, dehydrated and frozen fruits and vegetable products, tomato products, pickle, convenience veg-spice pastes, processed mushrooms and curried vegetables. In order to increase level of processing and to promote food processing industries to exploit both the domestic and international market potential for processed food products, Vision 2015 Document has been finalized by the Ministry of Food Processing Industries (MFPI), which envisages tripling the size of the processed food sector by increasing the level of processing of perishables from 6% to 20%, value addition from 20% to 35% and share in global food trade from 1.5% to 3% by 2015. An integrated strategy for promotion of agribusiness Vision, Strategy and Action Plan for the Food Processing Sector has also been approved by the Government.

Ministry of Food Processing Industries (MFPI) has been implementing various schemes for promotion and development of food processing industries in the country. MFPI through its various schemes for financial assistance and other promotional measures facilitates creation of food related infrastructure including processing facilities aimed at reducing wastages, enhancing value addition and increasing shelf life. The Scheme for Technology Upgradation / Establishment/ Modernization of Food Processing Industries is aimed at creation of new processing capacity and up-gradation of existing processing capabilities, modernization of Food Processing Sector including fruit & vegetables processing units. Under the Scheme for Technology Upgradation/ Modernization/Establishment of Food Processing Industries, MFPI extends financial assistance to food processing industries including fruit and vegetable processing units in the form of grant-in-aid @ 25% of the cost of plant and machinery and technical civil works subject to a maximum of Rs. 50.00 lakh in general area or 33.33% subject to maximum of Rs. 75.00 lakh in difficult areas such as Jammu & Kashmir, Himachal Pradesh, Uttarakhand, Sikkim and North Eastern States, A & N Islands, Lakshadweep and ITDP areas. In addition, under the Technology Mission for Integrated Development of Horticulture in North Eastern and Himalayan States, higher levels of assistance @ 50% up to maximum of Rs. 4.00 crore for setting up and Rs. 1.00 crore for upgradation of fruit and vegetables processing is available. National Horticulture Mission

(NHM) has been launched with an objective to boost the horticulture sector.

APEDA's role in Export Promotion

The Agricultural and Processed Food Products Export Development Authority (APEDA) was established by the Government of India under the Agricultural and Processed Food Products Export Development Authority Act passed by the parliament in December, 1985. The Act (2 of 1986) came into effect from 13th February, 1986 by a notification issued in the Gazette of India: Extraordinary: Part-II [Sec. 3] (ii): 13.2.1986). The Authority replaced the Processed Food Export Promotion Council (PFEPC). It has major mandate of promotion of export oriented production and development of the scheduled products.

Establishment of Agri-Export Zones

Govt.'s decision to establish the agri-export zone in specific areas of particular horticultural/export oriented commodities have paved significant role in enhancing the crop specific export. In recent past, country's vast export of gherkin has been only possible due to catering the export need by exporting zones. Presently, 60 agri-export zones have been established in different parts of the country. Vegetable specific export zone

Table (1.7). Vegetable based Agri-Export Zones

S. No.	Product	State	District/Area
1.	Gherkins	Karnataka	Tumkur, Bangalore Urban, Bangalore Rural, Hassan, Kolar, Chtradurga, Dharwad and Bagalkot
2.	Vegetables	Punjab	Fatehgarh Sahib, Patiala, Sangrur, Ropar & Ludhiana
3.	Potatoes	Uttar Pradesh	Agra, Hathras, Farrukhabad, Kannoj, Meerut, Aligarh and Bagpat
4.	Mangoes and Vegetables	Uttar Pradesh	Lucknow, Unnao, Hardoi, Sitapur&Barabanki
5.	Potatoes	Punjab	Singhpura, Zirakpur (Patiala), RampuraPhul, Muktsar, Ludhiana, Jallandhar
6.	Mango Pulp & Fresh Vegetables	Andhra Pradesh	Chittor District
7.	Potatoes, Onion and Garlic	Madhya Pradesh	Malwa, Ujjain, Indore, Dewas, Dhar, Shajapur, Ratlam, Neemuch and Mandsaur
8.	Mangoes & Vegetables	Gujarat	Districts of Ahmadabad, Khaida, Anand, Vadodra, Surat, Navsari, Valsad, Bharuch and Narmada
9.	Potatoes	West Bengal	Districts of Hoogly, Burdwan, Midnapore (W), UdayNarayanpur and Howrah
10.	Rose Onion	Karnataka	Bangalore (Urban), Bangalore (Rural), Kolar
11.	Onion	Maharashtra	Districts of Nasik, Ahmednagar, Pune, Satara and Solapur
12.	Vegetables	Jharkhand	Ranchi, Hazaribagh and Lohardaga
13.	Vegetables	West Bengal	Nadia, Murshidabad and North 24 Parganas
14.	Dehydrated Onion	Gujarat	Districts of Bhavnagar, Surendranagar, Amreli, Rajkot, Junagadh and Jamnagar.
15.	Gherkins	Andhra Pradesh	Districts of Mahboobnagar, Rangareddy, Medak, Karinagar, Warangal, Ananthpur and Nalgonda
16.	Chilli	Andhra Pradesh	Guntur

has been listed in the Table (1.7).

Contract Farming of Exotic Vegetables

Contract farming can be defined as an agreement between a farmer and processing and/or marketing firm for the production and supply of agricultural products under forward agreements, frequently at predetermined prices (Charles and Shepherd, 2001). The arrangement also invariably involves the purchaser in providing a degree of production support through, for example, supply of inputs and provision of technical advice. The basis of such arrangements is a commitment by the farmer to provide a specific commodity in quantities and on quality standards determined by the purchaser and a commitment by the company to support the farmer's production and to purchase the commodity. Contract farming is an intermediate production and marketing system that spreads production and marketing risks between agribusiness and smallholders. Similarly, it also provides agribusiness companies with the opportunity to guarantee a reliable source of supplies of required quantity and quality. It can be regarded as a means of reducing high transaction costs that result from the failure of the market and/or government to provide required inputs and market institutions. Contract farming has been found very effective in enhancing the production as well as export of snow pea in Jharkhand. Success of gherkin production in India has been only due to very effective and farmers friendly programme of contract farming.

Export of Organically Grown Vegetables

The Government of India has implemented the National Program for Organic Production (NPOP). The national program involves the accreditation program for certification bodies, norms for organic production, promotion of organic farming etc. The NPOP standards for production and accreditation system have been recognized by European Commission and Switzerland as equivalent to their country standards. Similarly, USDA has recognized NPOP conformity assessment procedures of accreditation as equivalent to that of US. With these recognitions, Indian organic products duly certified by the accredited certification bodies of India are accepted by the importing countries. India produced around 3,96,997 MT of certified organic products which includes all varieties of food products namely Basmati rice, Pulses, Honey, Tea, Spices, Coffee, Oil Seeds, Fruits, Processed food, Cereals, Herbal medicines and there value added products. The production is not limited to the edible sector but also produces organic cotton fiber, garments, cosmetics, functional food products, body care products, etc. India exported 86 items last year (2007-08) with the total volume of 37533 MT. The export realization was around 100.4 million US $ registering a 30% growth over the previous year. Organic products are mainly exported to EU, US, Australia, Canada, Japan, Switzerland, South Africa and Middle East. Cotton leads among the products exported (16,503 MT).

References

Charles, E. and Shepherd, A.W. (2001). Contract Farming: Partnerships for growth, FAO Agricultural Services Bulletin, 145.

Pandey, A.K. (2003). *Grih Vatika avam Hamara Swasth.* First Edition- 2003, page IX + 286, A.K. Prakashan, Varanasi.

Pandey, A.K. (2008). Underutilized Vegetable Crops. First Edition-2008, page XIII+ 366, ISBN:81-89304-53-4, Satish Serial Publishing House, New Delhi-10033.

Pandey, A.K. and Rai, Mathura (2006). Under Utilized Exotic Vegetables. IIVR, Technical Bulletin. No. 31, pp. 1-35.

Pandey, A.K. and Rai, Mathura (2007). *Hari Pattedar Sabjiyan* (in Hindi). Directorate of Extension, Department of Agriculture and Cooperation, Ministry of Agriculture, Govt. of India, New Delhi, pp 1-48.

Pandey, A.K. and Singh, B. (2011). *Niryat yogy sabziy ki Kheti*, Director, IIVR, Varanasi, Technical Bulletin. No. 40, pp. 1-70.

Pandey, A.K. and Singh, B. (2011). Production Technology of Underutilized Vegetables. Director, IIVR, Varanasi, Technical Bulletin. No. 39, pp. 1-87.

Priya, D., Kulkarni, U., Rao, L. and Rao, S. (2003). Changing food systems in India: resource sharing and marketing arrangements for vegetable production in Andhra Pradesh. Development Policy Review, 21(5-6):627-639.

●●●

2

CHERRY TOMATO

Cherry tomato is a small variety of tomato that has been cultivated since at least the early 1800s and thought to have originated in Peru and Northern Chile (Andrew, 1994). Cherry tomatoes range in size from 1.25 cm to 2.25 cm in diameter and either red or yellow in color, two-celled with numerous comparatively small, kidney-shaped seeds. The more oblong ones often share characteristics with plum tomatoes, and are known as grape tomatoes. The cherry tomato is regarded as a botanical variety of the cultivated berry, *Solanum lycopersicum* var. cerasiforme. Cherry tomato have a higher flavonoid content than larger fruited types (Crozier *et al.*, 1997). Rune and Michel (2011) reported that fresh cherry tomatoes contained more of the two flavonoids chalconaringenin (CN) and rutin than lycopene. The two flavonoids were extracted from peel and isolated by use of different chromatographic methods. Molecular absorbtivities were found to be 26907 for CN and 20328 abs M^{-1} cm^{-1} for rutin. Both compounds exhibited properties as antioxidants. The red color of tomatoes is produced by lycopene which is beneficial in heart diseases, lowers blood pressure, protects bones and is beneficial for the body in general. Hayman (1999) reported that cherry tomatoes contains higher lycopene levels when fully ripe. The cherry tomato is low in sodium, and very low in saturated fat and cholesterol. It is also a good source of vitamin E (alpha tocopherol), thiamin, niacin, vitamin B_6, folate, magnesium, phosphorus and copper, and a rich source of dietary fiber, vitamin A, vitamin C, vitamin K, potassium and manganese. Nutritive value of cherry tomato is given in table (2.1).

Table (2.1): Nutritive value of cherry tomato (per 100 g of edible portion)

Constitutents	Amount
Calories	18 kcal
Total Fat	0.2g
Saturated Fat	0.046g
Polyunsaturated Fat	0.135g
Monounsaturated Fat	0.051g
Sodium	5 mg
Potassium	237mg
Total Carbohydrate	3.92g
Dietary Fiber	1.2g
Sugars	2.63g
Protein	0.88g
Vitamin A	17μg
Vitamin C	21 mg
Calcium	1 mg
Iron	2 mg

Botanical Description

Cherry tomato belongs to family Solanaceae, there are 42 genera in family Solanaceae, 103 species under the genus *Solanum*, 2 varieties in *Solanum lycopersicum* L. Taxonomic position of the genus *Solanum* is given as under:

Taxonomic position of the genus *Solanum*

Kingdom	Plantae
Subkingdom	Tracheobionta
Superdivision	Spermatophyta
Division	Magnoliophyta
Class	Magnoliopsida
Subclass	Asteridae
Order	Solanales
Family	Solanaceae
Genus	*Solanum* L.
Species	*Solanum lycopersicum* L.
Variety	*Solanum lycopersicum* L. var. cerasiforme (Dunal) Spooner, G.J. Anderson & R.K. Jansen

Morphological Description of Plant

Plant vigorous with stout branches which are distinctly trailing in habit. Leaves flat or but slightly curled. Fruit very abundant, borne in short, branched clusters, globular, perfectly smooth, with no apparent sutures. The cherry tomato is regarded as a botanical variety of the cultivated berry, *Solanum lycopersicum* var. cerasiforme

Cultivars

Santorini cherry tomato : It was developed from seeds brought to Santorini (Greece) in 1818. It is known for its characteristic flavor.

Tomaccio : It was developed in Israel, using wild Peruvian tomato species to create a sweet taste of tomato.

The Super Sweet 100 : It is a common hybrid cultivar possesses resistance to both *Fusarium* and *Verticillium* wilt.

Climate and Soil

Cherry Tomato is a relative warm season crop. Plants grow well at the temperature range of 19°C to 30°C. It also requires plenty of sunshine but low humidity. It has been observed that continuous rain in the hot weather promotes disease problems such as bacterial wilt, blight, rot and fruit cracking. Tomato grows best on deep sandy loam or clay loam soil with pH 6-7, good water holding ability, rich organic matters and good drainage. The location should be free from nematodes and other soil borne diseases. Succession cropping should be avoided because of bacterial wilt and other soil borne disease problems. It is best to rotate tomato with rice and legume crops for 3 years before plant it again on the same ground. It does best when grown on deep loam soil with good fertility, light irrigation, adequate drainage and plenty of sunshine.

Seed germination

There was a positive correlation ($r = 0.68$, $P \leq 0.05$) between germination and α-amylase activity during seed develo-pment. Further, placenta water potential decreased with fruit and seed maturation, and correlated negatively with improved seed germination and α-amylase activity (Modi and White, 2004). As the seed of cherry tomato is small, it is better to raise seedling using tray and cocopeat media; for this 104 cup trays can be used. Raise the seedling in proper nursery condition. Irrigate the seedlings to keep enough moisture. High temperature and high humidity results in diseases like color rot, root rot. It takes 20-30 days from sowing to transplanting.

Transplanting and Spacing

Transplant the seedlings at 5-6 true leaves stage. Space them 60 cm apart in double rows on each bed spaced at 1.5-2.0 m wide. The seedlings must be sufficiently watered several hours before transplanting to make it easier to remove the plant from the pot or seedling bed. Thus the roots of seedlings may hold as much soil as possible, so as to prevent wilting during transpla-nting. Irrigation must be started immediately after the transplanting.

Manures and Fertilizers

As the crop of cherry tomato is heavy nutrient feeder, fertilizer application must be based on soil analysis. In general, at the time of field preparation apply well rotten cow dung or FYM @ 25 t/ha. Among the chemical fertilizers, NPK @ 150:60:80 kg/ha should be given. Full dose of P and K and one third of N may be given at the time of planting and rest of two third of N should be given in three split dozes at the interval of 30 days from the date of planting.

Water Management, Staking and Pruning

Tomato is a shallow rooted crop and requires constant and mild moisture throughout the crop period. It is especially sensitive to water stress during fruit setting period. When the weather is hot and dry flowers and fruits drop easily. Therefore irrigation should be applied frequently to maintain a steady growth. Water management thruogh drip has been found effective in respect of plant growth, yield and quality attributes. As plant lodges from fruit setting, it requires proper staking with help of bamboo sticks and wires. Depending on the type of variety keep the height of bamboo and wire support. Pruning is recommended to reduce the plant load and quality fruits. Allow only 3-4 stems to grow and pruning the other side branches. Under protected cultivation, production was better with a spacing of 0.5 m between plants and the use of two stems per plant (Charlo *et al.*, 2006).

Ontogenic stages of fruit development in cherry tomato

While studying the ontogenic stages of fruit development in cherry tomato, Ishida *et al.* (1994) observed that in small green fruits, large amounts of water were located in the epidermis and neighbouring cells, and in seed envelopes and seeds. High mobility water with long spin-lattice relaxation times (T_1) in the seeds of small green fruits disappeared as the fruit grew. The content of high mobility water in locular cavities, including seed envelopes, increased as ontogeny advanced. Low mobility water (short T_1) in the pericarp and dissepiments of the small and large green fruits increased in quantity and mobility during maturation. In some of the several water compartments in large green fruits water content declined during maturation. Changes in the physical state of water appeared to reflect changes in the physiology of the fruits. The mobility of sugars increased during fruit ripening.

Polyhouse Production of Cherry tomato

For quality production of cherry tomato, preference may be given for poly house production. First of all, seedlings are raised and when seedlings attain proper size they are planted in green house at specified intervals spacing 45 cm apart. Water the growing tomatoes regularly to keep the soil moist. Irregular watering can cause the fruit to split or to develop blossom end rot. Feed with a soluble fertilizer about once per week. The tomato plant is a vine so the stem will not remain upright on its own. Push a stake - I use a 2.5m bamboo cane into the growing bag adjacent to the stem. At intervals loosely tie the stem to the stake as the plant grows. Alternatively, wind the stem up a well anchored but slack length of twine. A very little pruning is given to maintain the canopy. Where the leaves grow out from the main stem a side shoot develops and it should be pinched out from each leaf joint as they develop in size of about 5 cm long. When the plants are about 1.2 m remove the leaves from below the first truss. Old and yellowing leaves should be removed as the plants develop and season progresses. Picking of the fruits should be made when they are ripe and fully coloured.

Harvesting and Yield

Discard the first fruit setting if the size of fruit is bigger. Fruits are harvested early morning (when temperature is low) by keeping calyx if possible (it can give attractive look to the fruit). Grading is performed as per size and quality. Discard the abnormal shape/size, diseased, cracked fruits. Pack the fruits in ventilated boxes and end to the market. Cherry tomatoes harvested at full ripeness exhibited the highest level of carotenoids and antioxidant activity in the water-insoluble fraction. On the other hand, no significant differences in ascorbic acid content were observed at different ripening stages, whereas the main phenolics content and the antioxidant activity of water-soluble fraction showed slight, but significant, decreases at later stages of ripeness (Antonio *et al.*, 2002).

Major Diseases

1. **Damping off** (*Rhizoctonia solani, Phytophthora capsici,* or *Pyhthium aphanidermatum*).

Pre-emergence damping- off occurs when the seed or seedling decay before it emerges. Post-emergence damping- off occurs above ground, and young seedlings eventually wilt and falling over. The stem usually becomes soft, water soaked, dark and shriveled at the soil line before the plant collapse. Roots may also be affected resulting in seedling death. The disease may occur under warm and high soil moisture condition. Seedlings during the first to second weeks after emergency are particularly susceptible. Usually, it occurs in over watering, overcrowding, poor ventilation and cloudy in the seedbeds.

Control:

1) Use high quality seed with good vigor.
2) Soil sterilization of seed beds or the use of sterile growing media.
3) Proper drainage of nursery areas.
4) Drenching with effective fungicides after planting.

*2. **Sclerotium*** **wilt**

It is caused by *Sclerotium rolfsii.* The first symptoms are general wilting of the plant. Wilting progress without a change in foliage color until the plant finally dies. The base of the plant is girdled with a white growth covering the dead tissue. Embedded in this white mat are light brown bodies about the size of mustard seed, which are characteristic of this disease. Fruits also infected when they are touching the soil. The disease occurs under high temperature and high moisture levels.

The fungus can survive in soil for several years and can be spread in running water, in infested soil, in infested plants, on tools and implements, and as sclerotia among the seed.

Control:

1) Removal and burning of all infected plants.
2) Crop rotation at least three years with crops such as corn or sorghum.
3) Follow good agricultural practices.

3. Late blight (*Phytophthora infestans*)

The first symptoms of the disease are a bending down of the leaf petiole. Lesions produced on the leaves and stems are large, irregular, greenish, water soaked patches. As these patches enlarge, they turn brown and paper-like. During wet weather, the lesions on the leaf under surface have a fine, white mold ring around them. Blighting of the entire foliage may occur during moist, warm periods. Entries fields can have extensive foliar and fruit damage. Fruit lesions are firm, large, irregular, brownish-green blotches. The surface of the fruit lesion has a greasy, rough appearance. The fungus survives on weeds of solanace-ous crops. Spores of the fungus can be carried long distances by storms. Cool (15-20°C), wet weather is required for late blight to develop. Under this condition the disease progresses rapidly and can ruin a mature tomato field in a few day.

Control:

1) Avoid field previously to crop Solanaceous vegetables.
2) Apply chlorothalonil in every 7-10 days.

4. Early blight (*Alternaria solani*)

It occurs on foliage, stem, and fruit of plants. The symptom appears first as small brownish, black lesion on the older leaves. The lesions area are about 6 to 16 mm diameter. These leaf spots typically have concentric, black rings giving an oyster shell, or target-board appearance to the lesions. A yellow area surrounds the spots and of there are many lesions the whole leaf turns yellow and quickly dries up. Lesions may occur as brown elongated, sunken areas anywhere on the stem or petioles. The lesions appear leathery, sunken with concentric ringing. The fungus can be survived between crops on infected debris in the soil and on seed. It also survives from season to season on volunteer tomato plants and other solanaceous hosts. First infection occurs during periods of warm (24-30°C), rainy or humid weather. The fungus progrsses rapidly by wind and rain.

Control:

1) Apply benlate (benomyl 0.14 kg a.i./ha) with Manzate D (0.6 kg a.i./ha).
2) Destroy affected plants after harvest.

5. Black leaf mold (*Pseudocercospora fuligena*)

Small lesions develop on young leaflets as indistinct discolorations with no definite margins. As the size of a lesion increases, a faint halo appears at the lesion margin and surrounds a brownish margin of collapsed tissue on both the upper and lower sides of the leaflet. Under humid conditions, heavy conidial production may be observed, on lower leaf surfaces. The spores are disseminated by splashing rain, running water, or machinery. Infection occurs rapidly, but symptoms develop slowly over 2 weeks. Sporulation occurs on the lower leaf surfaces under high humidity. Disease development is favored by warm (27°C) and wet weather.

Control :

1) Burn the debris after harvest of the crop.
2) Follow staking and pruning to reduce disease severity.
3) Use tolerant and resistant cultivars.

6. Leaf mold (*Cladosporium fulvum*)

First symptoms appear on the upper side of older leaves as yellowish or pale green spots. Purplish, or olive green mold appears on the lower side which coincides with a yellowing on the upper leaf surface. Later the infected lower leaves of the plant turn yellow and drop off. The disease develops quickly on high relative humidity (90%) and temperature (between 22°C and 24°C). The conidia, which are readily disseminated by rain or wind, can survive at least one year.

Control:

1) Crop residue should be removed and destroyed once the tomato crop is over.
2) The greenhouse production areas should be steamed (55°C) for at least 6 hrs once the crop residue is removed out.
3) Staking and pruning to avoid the spread of disease.
4) Grow resistant cultivars.

7. Powdery mildew (*Leveillua taurica*)

Its symptoms appear as light green and bright yellow lesions on the upper leaf surface. Necrotic spots, sometimes with concentric rings similar to those of early blight lesion, may develop in their centers. A light powdery covering of these lesions may occur on the lower leaf surface. Heavily infected leaves die but seldom drop from the plant. The spores of the fungus can travel long distances in air currents and are able to germinate under low relative humidities. The disease develops quickly in warm (near 25°C) and relative humidity of 70 to 100 per cent

Control:

1) Apply sulfur fungicides.
2) Maintain adequate row and plant spacing.
3) Maintain relative humidity levels below 85 per cent.

8. *Fusarium* wilt (*Fusariunt oxysporum* Schlechtend: Frfsp.)

The earliest symptom is the yellowing of the older leaves. This often develops on only one side of a leaf or branch. The affected leaves wilt and die though the remain attached to the stem. The whole plant appears stunted. A characte-ristic red-brown to brick red discoloration of the outer portion of the vascular tissue is evident, which extends far up the plants. But the pith remains healthy. It is a warm-weather (soil and air temperatures of 28°C) disease, most prevalent on acid sandy soils with low N and P and high K, short day length, and low light intensity. The pathogen is soil borne and remains in infested soils for several years. This disease can be dissemination by seed, tomato stakes, soil, infected plants, farm machinery, windborne, and waterborne infested soil.

Control:

1) Follow long term crop rotation.
2) Remove and destroy infested plant material after harvest.

9. ***Verticillium* wilt (*Verticillium dahliae*)**

Often the first indication of *Verticillium* wilt is a diurnal wilting pattern. Plants show mild to moderate wilting during warmest part of the day but recover at night. As the disease advances, some marginal and interveinal chlorosis develops on lower leaflets. These lesions show characteristic N-shaped. When the base of the main stem is cut a light tan discoloration with scattered, darker brown spots can be seen, which extends across the vascular system. The discoloration usually does not extend far up the plant, so there is no vascular browning in petioles or pith. This disease is a cool-weather disease the day time high temperature averaged 20-24°C. It seems to be more severe in neutral to alkaline soils. The infection is usually through wounds on roots.

Control:

1) Use resistant cultivars/hybrids.
2) Fumigation using methyl bromide-chloropicrin.
3) Solar sterilization in hot, acid areas is also effective.

***10. Phytophthora* blight (*Phytophthora capsici* and *P. parasitica*)**

Above or below the soil line brown lesion develops which may eventually girdle the stem or root. These brown lesions become large and sunken. A chocolate-brown internal discoloration of the vascular system extends above and below these sunken, brown lesions a short distance. Eventually the stem or roots may rot and plant wilts and dies. The fungus is favored by high soil moisture and hot weather (30-35°C). High moisture conditions favor the spread of this disease.

Control:

1) Avoid soil compaction.
2) Follow raised beds technique to avoid the water logging.
3) Use drip irrigation to avoid excess moisture.

11. Anthracnose (*Colletotrichum coccodes*)

The fruit may be infected when green and small but symptom do not appear until it begins to ripen. Symptoms of the disease appear on the ripe fruit as small, slightly depressed, circular lesions. A lesion may enlarge to 12 mm in diameter and becoming more sunken with concentric ring markings. The flesh beneath a lesion may be of a lighter color than the surrounding tissue and granular in texture. The center of a lesion is usually tan, and as the lesion matures it becomes dotted with small black specks. The surface of a mature lesion generally remains smooth and intact. Leaf infection is characterized by small, circular, brown lesions surrounded by yellow holes. The fungus survives on decayed plant material in the soil. Splashing rain carries the fungus from the soil to the fruit.

Control:

1) Rotation with non-solanaceous crops.
2) Weed management to reduce hosts.
3) Staking plants and mulching can be to reduce infection.

Virus Diseases

Tomato mosaic (Tomato mosaic virus)

The symptoms are mottled areas of light and dark green on the leaves which caused by common strains. Some other strains may cause a striking yellow mottling. Plant infected in early stage of growth are usually stunted and have a yellowish cast leaves may also be curled, reduce in size, and malformed (fern leaf). Sometimes, the necrosis of leaves stems or fruit result from infection by some strains of TMV, while chlorosis of leaves results from other. The virus is easily transmitted mechanically by machinery or workers from infected to healthy plants any time during the handling of the tomato plants. Infected debris from a previous crop can lead to infection through the roots of the new tomato plants. Chewing insects, such as grasshoppers and beetles, can transmit the virus, but are not considered a major source of infection.

Control:

1) Seed treatment with a 10% solution of trisodium phosphate for 20 min, or heat treatment of dry seed two days at 78°C.
2) Washing hands with soap during the handling of plants.
3) Avoid planting in soil which contains the virus from previous crops of tabacco, tomato, pepper or eggplant.
4) The soil media of seedbed should be steam sterilized for 60 min at 100°C All pruning tools should be dipped in mild or 10% trisodium phosphate solution.
5) Grow resistant cultivars.

Cucumber mosaic

The systems in early stages are yellow, bushy and stunted. The leaves may show a mottle similar to TMV. The most characteristic symptom of cucumber mosaic is shoestring-like leaf blades. In fern leaf, the blade of the leaflet is not as completely suppressed as in a shoestring leaflet, but it is abnormally long and narrow. Severely affected plants produce small, few fruit and irregular shape. Aphids spread CMV into a tomato crop from weeds or crop plants in adjacent fields. Secondary infection by aphids may occur, or less often. Workers may spread the virus from infected to healthy plants.

Control:

1) Check the population of aphids to reduce spread of disease.
2) Raising seedlings under insect proof net cover.
3) Follow clean cultivation practice.
4) Remove the infected plants and burn them as soon as possible.

Tomato yellow leaf curl

The plants infected at an early age are severely stunted and bushy; their terminal and axially shoots are erect, and their leaflets are reduced in size and abnormally shaped. Foliages are vein clearings, reduced leaflet size, downward curving and upward curling of leaflets, interveinal chlorosis; a rosette like growth habit and a profusion of axillary branch formation. Later leaf veins become thickened and puckered with enations. Usually flowers are sterile.

Control:

1) Managing the population of white fly.
2) Grow resistant varieties.

Tomato spotted wilt (TOSPO)

Symptoms of spotted wilt vary, but young leaves usually turn bronze and later develop numerous small, dark spots. Growing tips may dieback, and stem of terminals may be streaked. Affected plants may have a one-sided growth habit or may be entirely stunted and have dropping leaves, suggesting a wilt. Plants infected early in the season may produce no fruit, and those infected after fruit-set produce fruit with chlorotic ring spots. Green fruit has slightly raised areas with faint concentric rings; on ripen fruit these turn into obvious rings, which become red-and-white or red-and-yellow. It is transmitted by thrips.

Control:

1) Use reflective mulches.
2) Managing the population of thrips and host plants.

Root-knot nematode (*Meloidogyne incognita*)

The root-knot nematode causes galls on tomato roots. These galls may vary in size from pinhead to over 3 cm in diameter. The first above ground symptoms are a stunting, wilting and generally off-colored appearance of the plant. Often, whole areas in a field are affected. When diseased plants are pulled up, the irregular swellings of the root and gall formation can easily be seen. Nematodes are more severe in lighter and sandier soils. The host range of the root knot nematode is very wide with many crops and weeds begin susceptible. Moderate soil temperatures of 16°C to 27°C favor development of the nematode. Spread of the nematode may occur from infected plants, farm machinery and irrigation water.

Control:

1. Using resistant cultivars provides the most effective control.
2. Follow long term crop rotation.

Major Pests

1. Fruit borer (*Helicoverpa armigera*)

It attacks fruit in all stages of development. The fruits are spoiled by feeding marks (small or deep holes). The female is active at night and lays her eggs singly on the sepals of flowers or young fruits. The newly hatched larvae are dark-gray and feed on the flowers or fruits. Older leaves vary in color from brown to green and have a characteristic yellow strip on either side of the body.

Control:

1. Remove weeds and crop residues.
2. Crop residue should not be left out at the end of the cycle.
3. Follow crop rotations. Remove and destroy the pruning. In greenhouses, install a double door at the entrance.
4. Do not remove the fruits without taking appropriated measure to prevent pest dispersion.

2. Leaf miners

A few leaf-mining flies are common pests of tomato plants, including *Liriomyza sativae*, *L. trifolii* and *L. huidobrensis*. These small yellow-and-black flies lay their eggs inside the leaves of tomatoes, where larvae hatch and proceed to green pigment of the leaves.

Control:

1. Apply Profenophos 40% + Cypermethrin 4% (Rai *et al.*, 2013).
2. Avoid use of high levels of nitrogen fertilization as it promotes leaf mining.

3. White flies (*Bemisia tabacilar-gentifolii*)

The adult white fly is small (1-2 mm long) and winged. Its body colour is yellow but appears white due to the waxy dust covering the body and the wings. The forewings are slightly longer than the hind wings. At rest, the wings cover the abdomen like a roof. The eggs are elliptical and elongated, attached vertically to the leaf surface by a short stalk which is inserted into the leaf tissue. They are normally laid in a circle comprising 20-40 eggs, on the underside of the leaf. The main damage is indirect - by transmitting tomato yellow leaf curl virus diseases.

Control:

1. Reflective plastic mulches can repel white flies.

Genetic Resources and Improvement

Under the tomato improvement program of AVRDC, 27seven cherry tomato lines were evaluated in 2000 and 2001 (Table 2.2). In 2000, lines CLN1555-106-4 and CLN1561-124-2 gave the highest fruit yields, exceeding 55 t/ha. In 2001, lines CLN155- 104-4 and CLN1558-100-10 gave the highest yields, approximately 45 t/ha. line CLN1561- 124-2 produced a relatively high number of fruits/plant at 251 and 149 during 2000 and 2001, respectively. Fruit yields were generally higher in 2000 compared to 2001. Because of their good yields under Arusha conditions, lines CLN1555-106-4, CLN1561-124-2, CLN155-104-4 and CLN1558-100- 10 have good potential for intensive cropping (Chadha, 2002). While evaluating a total of 14 cultivars of cherry tomatoes and four cultivars of high-pigment tomato hybrids their content in different classes of antioxidants and for their antioxidant activity like lycopene, beta-carotene, alpha-tocopherol, vitamin C (ascorbic acid and dehydro ascorbic acid), and total phenolic and flavonoid contents, Lenucci *et al.* (2006) observed that LS203 and Corbus appeared to be the cultivars with the highest content of lipophilic and hydrophilic antioxidants among cherry tomatoes, respectively. All cultivars of high-pigment tomato hybrids showed an expected exceptionally high lycopene content. Among them, the highest content of lipophilic and hydrophilic antioxidants was found in cv. HLY 13. Hydrophilic and lipophilic antioxidant activities were both significantly influenced by genotype. Soluble sugars and cell wall polysaccharides are well known for contributing to a range of 'quality' characteristics of fresh vegetables such as flavour, texture and healthy properties. Red-ripe berries of 14 cultivars of cherry tomatoes and four cultivars of high-pigment tomato hybrids, cultivated in the south of Italy, were analyzed for their content of these important qualitative parameters. Sakura appeared to be the cultivar with the highest amount of soluble sugars (53 g kg^{-1} fresh weight, mainly glucose and fructose, and, hence the 'Sweetest' among cherry tomatoes. High pigment tomatoes, especially HLY02 and HLY13, showed a soluble sugar content much lower than cherry tomatoes, as expected for industrial, normal-size tomatoes. Cellulose to matrix polysaccharide ratio was highly variable and ranged between 0.06-1.48 and 0.17-0.77 in cherry and high-pigment tomato cultivars, respectively (Marcello *et al.*, 2008).

LePR-5, a putative PR5 like protein gene was amplified from a cherry tomato which encodes a precursor protein of 250 amino acid residues, and shares high degrees of homology with a number of other PR5 genes. Expression of LePR-5 in different tomato organs was analyzed with Semi-quantitative RT-PCR, showing that LePR-5 expressed at different levels in leaves, stems, roots, flowers and fruits. In addition,

Table (2.2): Evaluation of cherry tomato lines grown at AVRDC-RCA, Fall 2000 and 2001

Sl. no.	Lines	Flowering (50%) period	Maturity (50%) period	Maturity of ist fruit (Days)	Fruits / plant		Yield / plant(kg)		Yield (t/ha)	
					2000	2001	2000	2001	2000	2001
1.	CLN1561-64	27	72.1	63	228.6	138	1.52	0.61	40.5	20.4
2.	CLN1561-124-2	29	73.3	66	251.0	149	2.04	0.88	55.2	29.2
3.	CLN1558-2-2	-	-	-	274.1	-	1.89	-	50.4	
4.	CLN1561-124-25	22	74.3	46	106.6	119	1.83	1.11	48.9	37.0
5.	CLN1555-106-4	-	-	-	177.8	-	2.11	-	56.4	-
6.	CLN1555-35-1	27	76.7	66	129.7	93	1.60	0.92	42.6	27.4
7.	CLN1561-7-5	22	73.0	63	120.7	96	1.49	1.09	39.9	36.4
8.	CH155	31	74.7	67	105.6	80	1.76	0.87	47.0	29.0
9.	CLN1558-100-10	25	71.3	65	164.9	100	1.84	1.35	49.2	45.1
10.	CLN1561-7-13	-	-	-	182.4	-	1.70	-	45.3	-
11.	CH157 47	21	66.6	57	144.6	110	1.76	1.18	47.0	39.2
12.	CH152	-	-	-	135.7	-	1.94	1.09	51.8	
13.	CH154	24	69.7	64	132.3	103	1.68	-	44.8	36.4
14.	CLN1559-94-8	-	-	-	155.5	-	1.85	-	49.5	-
15.	CH-7C	-	-	-	161.0	-	1.97	-	52.5	-
16.	CH33-7C-9-8-0	27	75.7	68		109	-	1.17	-	38.9
17.	CLN152 – 33.5 a-c	22	70.7	64		77	-	1.01	-	33.5
18.	CLN2070A – 26.4 bc	24	73.7	62		43	-	0.79	-	26.4
19.	CLN2071C – 27.6 a-c	26	73.7	72		56	-	0.83	-	27.6
20.	CLN1561 – 26.7 a-c	26	70.7	64		126	-	0.80	-	26.7
21.	CLN155-104-4 – 45.9 a	26	74.0	67		123	-	1.38	-	45.9
22.	CLN1558 – 24.4 c – 0.73 ae	22	65.3	53		144	-	0.73	-	24.4
23.	CLN1859-94-8 – 25.7 c	27	70.0	64		84	-	0.77	-	25.7
	CV (%)	5.0	5.1	7	20.6	28	16.90	29	16.9	16.9

Source : Chadha (2002)

expression of LePR-5 under different abiotic stresses was carried out at different time points. Three of the four tested abiotic stimuli, ethophen, salicylic acid and methyl jasmonate, triggered a significant induction of LePR-5 after treatment. However, LePR-5 was weaker induced by abscisic acid than by others. The positive responses of LePR-5 to the three abiotic stimuli suggested that LePR-5 may play an important role in response to abiotic stresses, and it may also be involved in plant defense system against pathogens. In addition, different expression patterns between tomato fruit and seedling suggested that LePR-5 may play a distinctive role in the defensive system protecting tomato fruit and seedling (Ren, *et al.*, 2011).

References

Andrew Smith, F. (1994). The tomato in America: early history, culture, and cookery. ISBN 978-157003000.

Antonio Raffo, Cherubino Leonardi, Vincenzo Fogliano, Patrizia Ambrosino, Monica Salucci (2002). Nutritional value of cherry tomatoes (*Lycopersicon esculentum* cv. Naomi F_1) harvested at different ripening stages. J. Agric. Food Chem. 50(22):6550-6556.

Chadha, M.L. (2002). Open pollinated cherry tomato lines for Africa. AVRDC Progress Report for 2001.

Charlo, H.C.O., Castoldi, R. Ito, L.A., Fernandes, C. and Braz, L.T. (2006). Productivity of cherry tomato under protected cultivation carried out with different types of pruning and spacings. Acta Horticulturae 761: XXVII International Horticultural Congress - IHC2006: International Symposium on Advances in Environmental Control, Automation and Cultivation Systems for Sustainable, High Quality Crop Production under Protected Cultivation.

Crozier A., *et al.* (1997). Quantitative analysis of the flavonoid content of commercial tomato, onion, lettuce and celery. J.Agric. Food Chem. 45:599-595.

Hayman, G. (1999). Tomatoes preliminary investigation on the effect of cultivars, location of production, harvesting stage on fruit lycopene content. Horticulture Development Council Research Report PC 16 .

Ishida, N., Koizumi, M. and Kano, H. (1994). Ontogenetic changes in water in cherry tomato fruits measured by nuclear magnetic resonance imaging. Scientia Horticulturae, 57(4):336-346.

Lenucci, M.S., Cadinu, D., Taurino, M., Piro, G. and Dalessandro, G. (2006). Antioxidant composition in cherry and high-pigment tomato cultivars. J Agric Food Chem. 54(7):2606-26013.

Marcello Salvatore Lenucci, Maria Rosaria Leucci, Gabriella Piro and Giuseppe Dalessandro (2008). Variability in the content of soluble sugars and cell wall polysaccharides in red-ripe cherry and high-pigment tomato cultivars. Journal of the Science of Food and Agriculture, 88(10):1837-1844.

Modi, A.T. and White, B. J. (2004). Water potential of cherry tomato (*Lycopersicon esculentum* Mill.) placenta and seed germination in response to desiccation during fruit development. Seed Science Research, 14: 249-257 .

Rai, D., Singh, A.K., Sushil, S.N., Rai, M.K., Gupta,. J.P. and Tyagi, M.P. (2013). Efficacy of insecticides against American serpentine leaf miner, *Liriomyza trifolii* (Burgess) on tomato crop in N-W region of Uttar Pradesh, India. International Journal of Horticulture, 3(5):19-21.

Ren, X., Kong, Q., Wang, P., Jiang, F., Wang, H., Yu, T. and Zheng, X. (2011). Molecular cloning of a PR-5 like protein gene from cherry tomato and analysis of the response of this gene to abiotic stresses. Molecular Biology Reports, 38:801-807.

Rune, S. and Michel, V. (2011). Properties of chalconaringenin and rutin isolated from cherry tomatoes. J. Agric. Food Chem. 59(7):3180-3185.

3

PAPRIKA

Paprika or Hungarian Paprika also called sweet pepper or Spanish pimento is the mild or non-pungent variety of chilli or capsicum. The dried ripe red paprikas are valued chiefly for their excellent red colour and mild flavour. The European paprikas are different from their cousin red chillies which are grown extensively in India. Presently there is huge demand for paprika powder in the Western world. It is desirable to extend the area under paprika in India with the ultimate object of diversification of exports. Some success has already been attained both at the IARI, New Delhi and CFTRI Mysore where several varieties of paprika have been successfully grown indicating good scope for expansion of area under paprika. The pungent principle, capsaicin, is contained only in small amounts, as low as 0.001 to 0.005% in mild and 0.1% in hot cultivars. Apart from capsaicin, the taste of paprika is mostly due to essential oil (<1%; with long-chain aliphatic hydrocarbons, fatty acids and their methyl esters); paprika flavour is mostly due to a range of alkylmethoxypyrazines (e.g., 3-isobutyl 2-methoxy pyrazine, earthy flavour). Ripe paprika contains up to 6% sugar. Furthermore, paprika contains sizable amounts (0.1%) of vitamin C; this substance was first isolated from ripe paprika pods by the Hungarian chemist Albert Szent-György, who later won the Nobel Prize for this work. Paprikas derive their colour in the ripe state mainly from carotenoid pigments, which range from bright red (capsanthrine, capsorubin and more) to yellow (cucubitene); total carotenoid content in dried paprika is 0.1 to 0.5 per cent. Chile cultivars which produce yellow but no or little red pigments appear yellow to orange when ripe. A small number of cultivars do not produce significant amounts of carotenoids; when chlorophyll levels decrease in the last stages of ripening; these chilies develop a pale hue often referred to as white. Due to small amounts of chlorophyll and/or yellow carotenoids, the white is, however, more precisely described as a pale greenish-yellow. Some varieties of paprika contain pigments of anthocyanin type and develop dark purple, aubergine-coloured or almost black pods; in the last stage of ripening, however, the anthocyanins get decomposed, and the unusual darkness thus gives way to normal orange or red colours. The same anthocyanins cause the dark spots which are sometimes seen on unripe fruits or particularly the stems of paprika plants and which almost all paprika varieties can develop.

In other *Capsicum species*, anthocyanin production is a rare phenomenon (e.g., scarlet lantern, an Andean cultivar of *C. chinense*). The volatile compounds extracted from both sweet and hot samples of commercial Spanish paprika were analyzed and substances identified belonged to several chemical classes: phenols, aldehydes, acids, ketones, alcohols, ethers, nitrogen compounds, aromatic hydrocarbons, alkanes, esters, and lactones. Acetic acid was by far the most abundant compound followed by 1,3- and 2,3-butanediol, acetoine, 3-methylbutanal, ethyl acetate, and dimethoxyphenol. The presence of 2-methoxy-3-isobutyl-pyrazine, the characteristic aroma compound of fresh capsicum, was not detected. Of the volatiles isolated, acetic acid, phenols, ethyl acetate, methyl-branched aldehydes and acids, and other carbonyls would be expected to contribute to the overall paprika flavor. Pungency is generally expressed in 'Scoville units' a somewhat subjective measurement based on the so-called 'Scoville Organoleptic Test'. To determine the 'heat level' volunteers are given samples of Chillies, which are subsequently diluted with water until pungency can no longer be detected. The scale ranges from 0 to about 300000. Unfortunately such tests are not very reliable as a degree of tolerance is quickly developed. More recently a scientific method has been devised, known as the 'high-performance liquid chromatography test (HPLC)', which measures the type and quantity of capsaicinoids present in the sample.

Table (3.1): The approximate chemical composition of Hungarian paprika powder (in mass per cent)

Moisture	7.0-9.5
Carbohydrate	58-60
Protein	13.8-17.5
Crude fibre	14-20
Total ash	5.6-7.6
Water soluble ash	4.7-5.5
Acid insoluble ash	0.0-0.35
Volatile ether extract	0.3-1.5
Non-volatile ether extract	7.5-12

Source: Govindarajan (1985)

Table (3.2) :The approximate fatty acid composition of the fat of paprika powder (in mass per cent)

Lauric acid	0.9-2.5
Myristic acid	1.8-6.3
Palmitic acid	13.1-16.6
Hexadecenoic acid	1.3
Stearic acid	2.3-2.9
Oleic acid	9.4-12.1
Linoleic acid	53.0-63.5
Linolenic acid	6.6-7.7
Saturated acids	18.1-28.3
Unsaturated acids	70.3-84.6

Source: Govindarajan (1985)

Origin and Distribution

Since different varieties of bell peppers have been cultivated in America long before the arrival of the Europeans, their native countries cannot be unambiguously determined. South American origin is, however, established for all species of genus *Capsicum*, which emerged probably in the area bordering Southern Brazil and Bolivia. Then, the species moved to the North, being dispersed by birds. The three species *C. annuum, C. frutescens* and *C. chinense* evolved from a common ancestor located in the North of the Amazon basin (NW-Brazil, Columbia). Further evolution brought *C. annuum* and *C. frutescens* to Central America, where they were finally domesticated (in México and Panamá, respectively), whereas *C. chinense* moved to the West and was first put to cultivation in Perú; (although today it is not much cultivated in South America). Two other species were first cultivated in Western South America: *C. baccatum* in the

Peruvian lowlands and *C. pubescens* at higher elevations, in the Andes (Perú, Bolivia, Ecuador). As paprika plants tolerate nearly every climate, the fruits are produced all over the world. A fairly warm climate is, however, necessary for a strong aroma; therefore, in Europe, Hungarian paprika has best performance; the best comes from the Kalocsa region. In the Unites States, California and Texas are the main producers. Global paprika oleoresin production is about 7,000-8,000 tonne. Of this, China's contribution is in the range of 3,000-3,500 tonne while India's share is around 2,500 tonne. The rest is produced by Mexico, Peru and other countries.

Botanical Description

The genus *Capsicum* (pepper or paprika, 2n = 24) belongs to the family Solanaceae. There are about 25 known wild varieties, though most cultivated chillie peppers are variations of the *annuum* species, other cultivated varieties are *C. baccatum, C. chinense, C. frutescens*, and *C. pubescens*. The many hundreds of hybridized varieties, which come in all shapes, colours, sizes and degrees of pungency, make classification and nomenclature difficult. Most varieties are derived from the *annuum* species, though *C. frutescens* is also popular. The hottest chilies are *C. chinensis* varieties, popularly known as Habaneros. The fruits of paprika vary from roughly cherry shape about the size of apricot grown in Spain and Morocco to long conical anaheim shaped cultivated in Hungary USA and Canada. The dried ground product is available in sweet to mildly pungent forms with a wide range of colouring content capsaithin. Paprica and chilli fruit showed climacteric behaviour as long as they were attached to the plant, but when detached were non-climacteric (Mayuree Krajayklang *et al.*, 2000). Classification of capsicum on the basis of flower colour is useful for separating some Capsicum species into subgeneric categories. However, less than half of the commonly recognized species of *Capsicum* were included in these studies. In addition, some of the excluded species do not fit into this categorization, such as the yellow flowers of *C. ciliatum* and *C. scolnikianum* and the white flowers of *C. chacoense*, which seem to be more closely related to the purple-flowered group than to the white-flowered group (Walsh and Hoot, 2001; McLeod *et al.*, 1982).

Varieties

A large number of names for different cultivars of paprica is used in Latin America, especially México. Fresh and dried chillies are often referred to by different names. A list of some of the most common types of paprica cultivars is given in Table (3.3).

Table (3.3): A list of some of the most common types of paprica cultivars

Name	fresh pungency
Anaheim (California)	low
Chilaca	low
Jalapeño	medium
New Mexico	low
Poblano	low
Serrano	high

Cultural Practices

Raising the Nursery

Seedlings are raised in nursery bed. The seedlings are transplanted at the spacing of 60×30 cm and the recommended dose of N: P: K is 120:100:120 kg/ha. Presently consumers demand organic products because of more flavorful and safe to the environment and human health. In a study on the effects of conventional, integrated, and organic farming, grown in a controlled greenhouse of sweet pepper, it was found that organic farming provided peppers with the highest (a) intensities of red and yellow colors, (b) contents of minerals, and (c) total carotenoids. Integrated fruits presented intermediate values of the quality parameters under study, and conventional fruits were those with the lowest values of minerals, carotenoids, and color intensity. As an example, the concentrations of total carotenoids were 3231, 2493, and 1829 mg kg^{-1} for organic, integrated, and conventional sweet peppers, respectively (Pérez-López *et al*., 2007).

Use of plant growth regulators: Spraying of NAA @ 50 ppm showed increased in fruit weight, fresh fruit yield / plant, fresh yield / plot and dry yield / plant. It also enhanced the oleoresin, capasaicin and capsanthin content (Kannan *et al*., 2008).

Postharvest and Colour Extraction

Paprika extract is manufactured by solvent extraction of the dried capsicum pods. On harvesting, moisture content may be up to 90% which has to be reduced to at least 10%. The drying operation is carried out by sun-drying, in hot air-dryers or in drying chambers. Prime extract Paprika oleoresin is manufactured by solvent extraction of the dried Capsicum pods, followed by solvent removal. Typically, one kg of pods yields 90 to 120 g of extract. The pigment concentration in the extract depends mainly on two parameters, the composition of the fruit and the extraction technique employed. With respect to the fruit, the organic solvent will extract all of the lipophilic compounds, which are the pigments and the oil from the pepper pericarp. The oil is present in much higher quantity than the pigments. In addition, the pigments are located in cellular structures that are more difficult to access for the solvent. The oil is easily extracted at the early stage of the process and subsequently becomes richer in pigments. At the end of this process the solvent(s) is/are evaporated. New extraction methods have been investigated, e.g. fractionation of paprika extract by extraction with supercritical carbon dioxide. Higher extraction volumes, increasing extraction pressures, and similarly, the use of co-solvents such as 1% ethanol or acetone resulted in higher pigment yields. Pigments isolated at lower pressures consisted almost exclusively of β-carotene, while pigments obtained at higher pressures contained a greater proportion of red carotenoids (capsorubin, capsanthin, zeaxanthin, β-cryptoxanthin) and small amounts of β-carotene (Richard Cantrill, 2008).

Genetic Resources and Improvement

Forty indigenous and exotic genotypes collected from various sources were evaluated for three seasons. Variation was found in morphological and quantitative characters. Genotype Paprika King was found to be superior to other types in yield and colour value, followed by PBC 066, KT-pl-8 and KT-pl-20 (Anu *et al*., 2000). Capsaicin and ascorbic acid contents of seven Indian peppers vari-

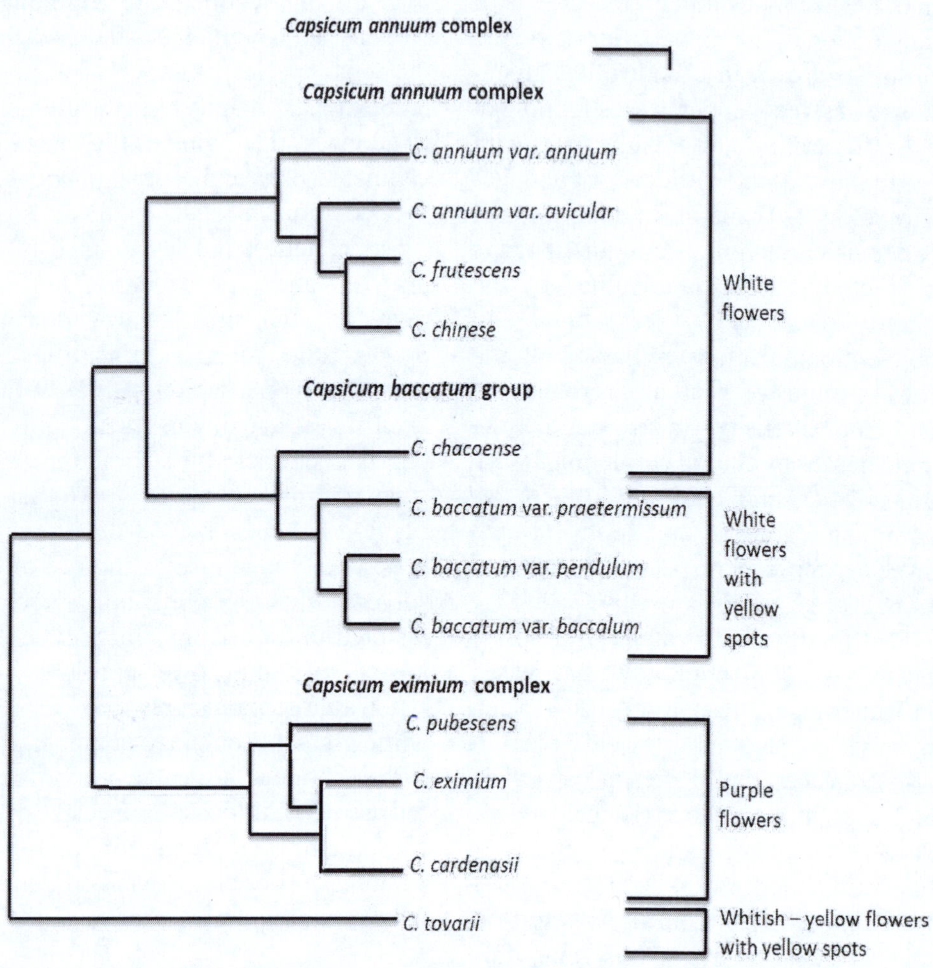

Fig. 3.1 Capsicum dendrogram constructed from standard genetic distance estimates based on isozyme data compared with a classification based on flower color (McLeod *et al.,* 1982)

eties/accessions from *Capsicum annuum* (CA 97, CCH, K1, KTPL19, Arka Abhir and Bayadagi Kaddi) and *C. frutescens* (CF1) species were determined using High Performance Liquid Chromatography (HPLC). Based on their pungency value, all the chilli accession/varieties (CA 97, CCH, K1, and CF 1) were classified as highly pungent peppers. The accession CF1 showed the highest concentration of capsaicin (445mg $100g^{-1}$ DW) with corresponding pungency value of 71,200 SHUs and Arka Abhir variety showed the lowest capsaicin concentration (29 mg $100g^{-1}$ DW) with 4,672 SHUs of pungency value. Similarly, Bayadaggi kaddi variety showed the highest ascorbic acid content (189 mg/ 100 FW) and the accession CA 97 showed the lowest ascorbic acid contents (55.3 mg/100 FW). The variability in capsaicin and ascorbic acid content presented in the pepper germplasm can be exploited for breeding cultivars with improved nutritional qualities. Moreover, CF1and Bayadagi kaddi can be used as a potential source for capsaicin and vitamin C, respectively (Tilahun *et al*., 2013).

Micropropagation

In the micropropagated plants, seeds were germinated on the 1/2 MS in a dark condition, and shoot tip explants were taken and cultured on the MS medium. Elongated shoots were cut into multiple single nodal cuttings and then were cultured on the fresh MS for shoot growth and rooting. Well-rooted plants were acclimatized in a greenhouse. In the seed propagated plant, seeds were germinated in rockwool cubes and then seedlings were transplanted in rockwool slabs after true leaves appeared. In the comparison of greenhouse performance of seed propagated plants (SPP) and micropropagated plants (MPP), no. of fruits from MPP was 49.8% higher than that of SPP, though its plant height and main stem length were smaller. Therefore, micropro-pagated plants can be expected to be used as alternative propagules and to increase grower's income by decreasing cost on propagules (Song *et al*., 2009).

Biotechnology

Anther derived spontaneous doubled haploid (DH-R1) plants were developed from an intercultivar F_1 hybrid (Holland) of 'blocky' type pepper cv. 'Mazurka' (*Capsicum annuum* L. var. annuum, 2n = 24). Regenerants were grown in greenhouses for seed production to R_2 generation (DH-R, prior to flow cytometric. Analysis which revealed D H-RI plants with haploid (n=12 (doubled haploid, D^H, (2n=24) tetraploid (2n=48); and aneuploid genomes. Pollen grains of anther cultures showed a high decrease in viability during the first two weeks of incubation time. The DNA samples of the 47 DH-individuals were bulked in nine groups and subjected to PCR analysis for estimating the level of homogeneity versus heterogeneity in comparison to the control of anther donor bulk of five plants of F_1 hybrid pepper cv. Mazurka (DH-Ro). In total 26 primers were applied, 22 of them revealed scorable PCR profiles with 54 scorable PCR bands. Of these, 19 (35.2%) bands generated by 7 primers were polymorphic (Gyulai *et al*., 2000).

References

Anu, A., Babu, K.N. and Peter, K.V. (2002). Evaluation of paprika genotypes in Kerala. Indian Journal of Plant Genetic Resources, 15(2):93-99.

Govindarajan, V.S. (1985). Capsicum production technology, Chemistry and Quality II CRC Crit. Reviews in Food Sci. and Nutrition. 22:109-176 .

Gyulai, G., Gemesne, J.A., Sagi, Zs, Venczel, G., Pinter, P., Kristof, Z., Torjek, O.,Heszky, L.Bottra, S., Kiss, J. and Zatyko, L. (2000). Double haploid development and PCR-analysis of F_1 hybrids derived DH-R2 paprica (*Capsicum annuum* L.) lines. J. Plant Physiol. 156:168-174.

Kannan, K., Jawaharlal, M., Prabhu, M. and Senthil Kumar, T. (2008). Effect of growth regulators on yield and quality of paprika cv. ktpi-19. Indian J. Agric. Res. 42(4):293-297.

Mayuree Krajayklang, Andreas Klieber and Peter R Dry (2000). Colour at harvest and post-harvest behaviour influence paprika and chilli spice quality. Postharvest Biology and Technology, 20(3):269-278.

McLeod, M. J., Guttman, S. I. and Eshbaugh, W. H. (1982). Early evolution of chili peppers (*Capsicum*). Econ. Bot. 36:361-368.

Pérez-López, A.J., López-Nicolas, J.M., Núñez-Delicado, E., Del Amor, F.M. and Carbonell-Barrachina, A.A. (2007). Effects of agricultural practices on color, carotenoids composition, and minerals contents of sweet peppers, cv. Almuden. J Agric Food Chem. 3:55(20):8158-64.

Song JuYeon, Sivanesan, I., An ChulGeon and Jeon Byoung Ryong, (2009). Micro-propagation of paprika (*Capsicum annuum*) and its subsequent performance in greenhouse cultivation. Korean Journal of Horticultural Science & Technology, 27(2):293-298.

Tilahun, S., Paramaguru, P. and Rajamani, K. (2013). Capsaicin and ascorbic acid variability in chilli and paprika cultivars as revealed by HPLC analysis. Journal of Plant Breeding and Genetics, 1(2):85-89.

Walsh, B.M. and Hoot, S.B. (2001). Phylogenetic relationships of *capsicum* (Solanaceae) using DNA sequences from two noncoding regions: the chloroplast atpb-rbcl Spacer region and nuclear waxy introns. Int. J. Plant Sci. 162(6):1409-1418.

●●●

4

SNOW PEA

Snow pea (*Pisum sativum* var. saccharatum) is a legume, more specifically a variety of pea eaten whole in its pod while still unripe. The name mangetout (French for eat all) can apply both to snow peas and to snap peas. The pods of snow peas are flat and thin with the bulge of the tiny seed barely visible at prime eating stage. The bright green pods should be turgid and crisp. They contain five to seven seeds and reach a length of 5-7 cm. The pods of sugar snaps are plump like regular English peas but are sweet and tender, thus eaten whole without being shelled, eaten raw or cooked. The snap pea is a type of edible-podded pea that is conspecific to field and garden peas (*Pisum sativum* L.). Edible-podded peas lack pod parchment or fiber, unlike field or garden peas that have fibrous pods. Within the edible-podded group, the most important trait that separates snap peas from snow peas is the thickened pod wall of the snap pea. Most contemporary snap pea cultivars have wrinkled seeds with green cotyledons, white flowers, and short internodes. The most recent introductions may also be stringless (lacking strings in the pod sutures). Presently popular as a vegetable, the snap pea has been cultivated at least since the 19th century, and probably even centuries earlier. In the 19th and early 20th centuries they were known as "butter" or "butter sugar" peas. Over the years, they enjoyed only limited popularity and may have disappeared from the vegetable seed trade in the 1960s and early 1970s prior to the cultivars released in 1979 and thereafter. The use of the term "snap pea" in America originated with the release of Gallatin Valley Seed Company commercial cultivars, the first of which was 'Sugar Snap'. Snap peas are sometimes referred to as "sugar snap" peas. The pea has been grown and consumed for at least 8,000 years, primarily as a pulse. Peas were the subject of one of the first systematic plant breeding efforts of Knight (Hedrick 1928), and provided the tool for Mendel (1866) to elucidate the science of genetics.

Table (4.1): Nutritional coposition of snow pea (per 63 g of edible portion)

Nutrients	Amounts
Macronutrients	
Water	56 g
Calories	26 Kcal
Protein	1.76 g
Carbohydrates	4.76 g
Fiber	1.6 g
Sugars	2.52 g
Total Fat	0.13 g
Saturated Fat	0.025 g
Monounsaturated Fat	0.013 g
Polyunsaturated Fat	0.056 g
Micronutrients	
Calcium	27 mg
Iron	1.31 mg
Magnesium	15 mg
Phosphorus	33 mg
Potassium	126 mg
Sodium	3 mg

Zinc	0.17 mg
Vitamin C	37.8 mg
Thiamin	0.095 mg
Riboflavin	0.050 mg
Niacin	0.378 mg
Pantothenic Acid	0.473 mg
Vitamin B_6	0.101 mg
Folate	26 µg
Vitamin A	685 IU
Vitamin E	0.25 mg
Vitamin K	15.8 µg
Phytonutrients	
β–Carotene	397 µg
Lutein and Zeaxanthin	466 mcg

Origin and Distribution

It is one of the earliest-known cultivated plants, with evidence of having been cultivated in a region that is now along the Thailand-Burma (Myanmar) border, 12,000 years ago. Snow peas and the similar sugar snap pea are part of a family with a long and exotic history. Yet while traces of primitive pea varieties dating back to 5000 BC have been found, the snow pea is a relative new comer, developed in China some thousands of years ago. It is still a popular addition to Chinese cooking today, and is widely available in Australia and many other countries.

Botanical Description

Duke (1981) reported that garden peas are treated as *P. sativum* ssp. *hortense* Asch. Graebn., field peas as *P. sativum* ssp. *arvense* (L.) Poir., and edible podded peas as *P. sativum* ssp. *macrocarpon*; early dwarf pea as *P. sativum* var humile. Later, Smart (1990), based on studies undertaken by Ben-Zeiev and Zohary (1973), and Polhill and van der Maesen *et al*. (1985) reported that pea comprises only two species, *viz; Pisum sativum* and *P. fulvum* Sibeth.Smith. It is a self pollinated annual herb, bushy or climbing, glabrous, usually glaucous; stems weak, round, and slender, 30-150 cm long; leaves alternate, pinnate with 1-3 pairs of leaflets and a terminal branched tendril leaflets ovate or elliptic, 1.5-6 cm long (Duke, 1981). The leaf type could be conventional, semi-leafless and leafless (Davies *et al*., 1985). Leaf size in most cases increases up to the first node bearing the first flower. Stipules are large, leaflike and up to 10 cm long. The inflorescence of pea is a raceme arising from the axil of the leaf. Corolla white, or pink, or purple; pods swollen or compressed, short-stalked, straight or curved, 4-15 cm long, 1.5-2.5 cm wide, 2-10 seeded, 2-valved, dehiscent on both sutures (Gritton, 1980; Duke, 1981). The node at which the first flower emerges is characteristic of a given variety; in temperate regions the number of nodes at which the first flower emerges is reported to vary from 4 in the earliest to about 25 in late maturing types under field conditions (Gritton, 1980). Flowers borne on the same peduncle produce pods that mature at different times, the youngest being at the tip. On a whole plant basis, flowering is sequential and upwards from node to node. Seeds are globose or angled, smooth or wrinkled, exalbuminous, whitish, gray, green, or brownish; 100 seeds can weigh from 10 to 36 g; germination cryptocotylar. Taxonomic position of the genus *Pisum* is given as under:

Taxonomic position of the genus *Pisum*

Kingdom	Plantae
Division	Magnoliophyta
Class	Magnoliopsida
Order	Fabales
Family	Fabaceae Lindl
Subfamily	Faboideae

Genus	*Pisum*
Species	*Pisum sativum*
Variety	*P. sativum* var. saccharatum
Trinomial name	*Pisum sativum* var. saccharatum

Morphological Description of the Plant

Annual, climbing, glabrous herb up to 2(-3) m tall (up to 1.3 m for sugar pea types); taproot well developed, up to 1.2 m long, with many lateral roots; stem terete, with no or few basal branches, internodes hollow. Leaves alternate, pinnate, with 1-3(-4) pairs of leaflets and ending in a usually branched tendril; stipules leaf-like, up to 8(-10) cm × 4 cm; petiole (2-)4-6 (-7.5) cm long; leaflets shortly stalked, ovate to elliptical, 1.5-8 cm × 0.5-4 cm, entire to toothed, sometimes converted into tendrils. Inflorescence an axillary, 1-3-flowered raceme. Flowers bisexual, papilionaceous; calyx with tube 4-8 mm long, lobes as long or longer than tube; corolla white to purple, standard 1-3 cm × 2.5-4.5 cm, wings a little shorter than standard, keel much shorter; stamens 10, 9 united and 1 free; ovary superior, 1-celled, style curved, longitudinally grooved. Fruit an oblong-ovate pod 3.5-15 cm × 1-2.5 cm, pendant, 2-11-seeded. Seeds globose, sometimes wrinkled, 5-8 mm in diameter, varying in colour from uniform yellow (sugar pea), green (crinkled garden pea) to purple or spotted or cream-white, sometimes with black hilum. Seedling with hypogeal germination; cotyledons remaining within testa; first 2 leaves simple (Messiaen *et al.*, 2006).

Climate and Soil

It is a cool season crop and temperatures above 30°C will cause poor pollination, early maturity and lower yields. Very low temperature prevailing for longer period may also cause curled pods. The minimum soil temperature for growth is 10°C. The plants are not affected by frost, but this may damage flowers and pods. Peas thrive on a wide range of soil types and ideal pH range is between 6.0 and 7.0.

Varieties

Cultivars can be either the bush type which are usually grown without a trellis or the tall type. Vine types may grow to 2 m high. There are several commercial varieties of snow peas. Dwarf Oregon is a very popular variety of snow pea and it grows to 1.0-1.8 m high.

Some of the commonly available snow pea cultivars are decribed as under:

Snowbird : This variety is ready in 60 days, short plants only reach 45-50 cm tall, pods average 7.5 cm long, prolific producer, pods set on plant in groups of 2-3, moderately sweet.

Gray Sugar : This variety is ready in 65 days, traditional snow pea, tender pods average 7.5-8.0 cm long, plants reach 45-60 cm tall, moderately sweet.

Sugar Daddy : This variety of peas features vines that reach 60-70 cm tall, ready in 75 days, good disease resistance, pods average 7.5-8.5 cm long, sweet and tender.

Oregon Sugar Pods : This variety is ready in 70 days, prolific producer, plants reach 60-75 cm tall, pods average 8-10 cm long, disease resistant, very sweet and tender.

Mammoth Melting Sugar : This heirloom variety features very tall vines that reach 120-150 cm tall, ready in 70 days, huge pods average 10-12 cm long, pods are thick, holds up very well to longer cooking methods, sweet flavor.

Oregon Sugar Pod 2 : This variety is ready in 70 days, plants reach 60-70 cm tall pods average 7.5-8.0 cm, prolific producer, good disease resistance, usually 2 pods per group, very sweet and tender.

Avalanche : This snow pea variety produces huge dark green pods, sweet, 7.5-8.0 cm and tender, ready in 60 days, disease resistant, prolific producer, plants reach 100-120 cm tall.

Seed Rate and Spacing

Peas are sown directly into the field. About 80-100 kg of seed is needed for a hectare. Seeds are sown 2 to 3 cm deep. Spacing within the rows should be 7 to 10 cm. The distance between rows of bush varieties should be 60 to 70 cm. For trellis production of tall varieties, rows should be 1 to 1.5 m apart.

Trellising

Snow peas are vigorous plants and trellising is necessary to manage the growth, to keep the pods off the ground and to facilitate harvesting. Trellising is constructed by using 50 to 100×25 to 50 mm posts every 6 to 10 m and a strainer post at each end. Two tie wires or strings are then run at vertical intervals of 20 to 40 cm to support the growing vines. Growers may have three to seven tiers of pairs of wires or strings, which are pulled tight on each side of the plant so they are only about 5 cm apart. The tops of the vines may be pinched out if they are too high.

Manures and Fertilizers

Apply the following rates of magnesium and trace elements to the soil before planting:

- 50 kg/ha magnesium sulphate to supply magnesium;
- 20 kg/ha manganese sulphate to supply manganese;
- 18 kg/ha borax to supply boron;
- 18 kg/ha iron sulphate to supply iron;
- 18 kg/ha copper sulphate to supply copper;
- 18 kg/ha zinc sulphate to supply zinc; and
- 2 kg/ha sodium molybdate to supply molybdenum.

The use of compost at up to 10-15 tonnes FYM per hectare to other crops in the rotation will be beneficial. It will supply organic manure, add nutrients and help to retain moisture in the soil. Levels of double superphosphate can be reduced if a soil test shows that phosphorus levels are high. Double superphosphate is preferred to superphosphate, as the latter contains higher levels of cadmium, which is a toxic heavy metal. Peas are a legume crop, and so are capable of tapping their own nitrogen, but to do this the seeds need to be inoculated with a suitable *Rhizobium bacterium.* Snow pea being a legume vegetable needs low doze of nitrogen fertilizer. In general, at the time of sowing 20 kg N, 40 kg P_2O_5 and 30 kg K_2O per ha should be given. When plants are 30 days old , apply 20 kg N/ha as top dressing.

Irrigation

Watering through sprinklers are mainly used for irrigation. In general, 1st irrigation should be given when crops are 50 per cent in flowering stage. Thereafter, next irrigation should be given after 1st picking.

Harvesting

Peas are self-pollinated and snow peas are harvested approximately 10 days after flowering. Harvesting commences at 8 to 12 weeks after planting and may continue for 8 to 10 weeks. Average pod weight is 7 g. Pickers can harvest approximately 11 kg per hour. This may take longer in wet weather in winter, as the old flowers may stick to the pod. Yields of snow peas range from 2 to 7 t/ha. Snow peas are harvested every two days when the pods are flat, without any development of seeds and are 7.5 to 10 cm long and 2.5 cm wide. Old pods should be picked and discarded. Sugar snap peas are harvested when the pods are up to 7.5 cm long and the seeds are almost at full size. These may be harvested for four to six weeks. A good picking rate is about 20 kg per hour per person.

Genetic Resources and Improvement

The world collection of cultivars and mutant forms of *Pisum sativum* is housed at the Nordic Gene Bank, Alnarp, Sweden (about 2700 accessions). Emphasis in the collection is on lines with multiple disease resistance, wild and primitive types, lines carrying structural mutations, breeding lines and cultivars of special interest. Large *Pisum sativum* collections are held in Australia (Australian Temperate Field Crops Collection, Horsham, Victoria, 6300 accessions), the Russian Federation (N.I. Vavilov All-Russian Scientific Research Institute of Plant Industry, St. Petersburg, 6200 accessions), Italy (CNR - Istituto di Genetica Vegetale, Bari, 4100 accessions), the United States (Western Regional Plant Introduction Station, Pullman, 3500 accessions; Horticultural Sciences Department, NY State Agricultural Experiment Station, Geneva, 2500 accessions), China (Institute of Crop Germplasm Resources (CAAS), Beijing, 3400 accessions), and the United Kingdom (John Innes Centre, Department of Applied Genetics, Norwich, 2700 accessions). The largest collection of *Pisum sativum* germplasm in Africa is located at the Institute of Biodiversity Conservation, Addis Ababa, Ethiopia, with over 1600 accessions. The 'edible pod' character (absence of 'parchment' in the pod walls) is induced by two recessive genes. A mutation inducing thickening of this wall of up to 3 mm was recently introduced in American cultivars, giving rise to the 'sugar snap pea'. The 'sugar snap' character will be interesting if it appears attractive to consumers. It might also be interesting to introduce more new characters into sugar pea, e.g. true dwarfs which could be grown without support, or climbing semi-leafless types in order to increase yields by higher plant densities and to make fruit picking easier. Well-known cultivars of sugar pea in Africa are 'Sugar Snap', 'Carouby de Maussane', 'Oregon Sugar Pod', 'Shield' and 'Sugar Queen'. Some cultivars of garden pea are 'Alderman', 'Télévision' and 'Green Feast'. Many growers use their own seed originating from old introductions (Messiaen *et al.*, 2006). De *et al.* (2005) evaluated 33 edible-pod pea lines selected from single plants within 11 snow pea landraces and three elite cultivars for their horticultural value in three field trials. Lines, such as MB-0298, MB-0324, MB-0325, MB-0326, MB-0332, and MB-0334 were appropriate for vegetable production as edible pod snow pea varieties and for use in breeding programs. The lines MB-0298, MB-0321, MB-0322, and MB-0324 showed stable earliness and MB-0330 and MB-0332 stable pod quality across the three environments evaluated.

References

Athar N., Taylor G., McLaughlin J, and Skinner, J. (2004). FOOD files 2004. New Zealand Institute for Crop and Food Research Limited and New Zealand Ministry of Health.

Ben-Zeíev, N. and Zohary, D. (1973). Species relationships in the Genus *Pisum*. Israel J. Botany 22: 73-91.

Davies, D.R., Berry, G.J., Heath, M.C. and Dawkins, T.C.K. (1985). Pea (*Pisum sativum* L.). p. 147-198. In: R.J. Summerfield and EH Roberts, (eds.), Williams Collins Sons and Co. Ltd, London, UK.

De Ron Antonio, M., Magallanes Jorge J., Marinez, Oscar, Rodino Paula, santalla Marta (2005). Identifying superior snow pea breeding lines. Hort science, 40(5):1216-1220.

Duke, J.A. (1981). Hand Book of Legumes of World Economic Importance. Plenum Press, New York, pp. 199-265.

Gritton, E.T. (1980). Field Pea. Hybridization of Crop Plants. p. 347-356. In: W.R. Fehr and H.H. Hadley (eds.), American Society of Agronomy, Inc., and Crop Science Society of America, Inc., Wisconsin, USA.

Hedrick, U.P. (1928). Vegetable of New York, Vol.I.Part.I.J.B.Lyon Co. Albony, New York.

Messiaen, C.M., Seif, A.A., Jarso, M. and Keneni, G. (2006). *Pisum sativum* L. In: Brink, M. & Belay, G. (eds). PROTA 1: Cereals and pulses/Céréales et légumes secs., Wageningen, Netherlands.

NHMRC (2006). Nutrient Reference Values for Australia and New Zealand. 6 October 2006.

Polhill, R.M. and van der Maesen, L.J.G. (1985). Taxonomy of Grain Legumes. p. 3-36. In: Summerfield and Roberts (eds.), Grain Legume Crops. Collins, London, UK.

Smart, J. (1990). Grain Legumes: Evolution and genetic resources. Cambridige University Press, Cambridge, UK. 200.

•••

5

GHERKIN

Gherkin is popularly called pickling cu cumber. It has less than 12 cm long fruit with a length and diameter ratio of 2.8-3.2 usually with white spines, however, some primitive type cultivars possess black spines on pronounced warts, green outer skin often slightly striped. It has very much potential for processing. It is also consumed as fresh salad. Fresh fruits of gherkin are rich in vitamin B, C and minerals such as calcium, phosphorus, iron and potassium. The consumption of fresh fruits of gherkin as salad prevents constipation, jaundice and indigestion. Mainly two compounds 2, 6 non adienal and 2, 6 nonadienol are responsible for flavour. The pleasant aroma of fruit is mainly due to 2, 6-dieadienal. The nutritive value of the fruit is given in table (5.1).

Table (5.1). Nutritional composition of fresh fruits of gherkin (per 100 g of edible portion)

Constituents	Amount
Water	95.1 per cent
Food energy	15 Kcal
Protein	0.9 g
Fat	0.1 g
Carbohydrate	3.4 g
Fibre	0.6 g
Ash	0.5 g
Calcium	25 mg
Phosphorus	27 mg
Iron	1.1 mg
Sodium	6.0 mg
Potassium	160 mg
Vitamin 'A'	250 I.U.
Thiamine	0.03 mg
Riboflavin	0.04 mg
Niacin	0.2 mg
Ascorbic acid	4.0 mg

Origin and Distribution

The gherkin originates from south-western Africa. Domestication of this species included selecting for sweeter fruits with shorter spines. The wild form is known as *Cucumis anguria* var. longaculeatus and in southern Africa occurs in Namibia, Botswana, Swaziland and South Africa (Limpopo, Mpumalanga and KwaZulu-Natal) (Welman, 2003). It is the only wild *Cucumis* species with prickly stems and leaf stalks. The gherkin is thought to have been brought to the West Indies in the slave trade days (probably from Angola), where it became popular as a vegetable. The West Indian gherkin is also called burr cucumber.

International Trade

In recent past, gherkin has gained much popularity in India and has emerged as a potential export oriented commodity, earning foreign exchange of Rs. 150 crores from 50,000 tonnes every year. Gherkin was introduced during 1990 in Tamil Nadu and presently its cultivation has spread in Karnataka, Tamil Nadu, Andhra Pradesh, Maharashtra and West Bengal. Presently 35 exporters are exporting gherkin to European countries, USA, Russia, Australia and Asian countries. Although India is a traditional producer of cucumber, its export potential was discovered during the late1980s, and since then their exports have been increasing progressively. When the cost of production of gherkins in the Europe became too high, their production shifted to Turkey. Later, Turkish farmers found tomato-growing to be more profitable and shifted to tomato production (Anonymous, 2007). Farmers in India seized upon this opportunity to produce gherkins. The production of gherkins in India is concentrated in the three southern states, viz. Karnataka, Tamil Nadu and Andhra Pradesh.

Karnataka accounts for almost 60 per cent of the total gherkin production and Tamil Nadu and Andhra Pradesh account for 20 per cent each (Anonymous, 2007). It is not palatable with Indian taste, but is a major dietary constituent to many European countries and the USA. Hence, almost the entire volume of gherkin produced in India is exported, with little or no domestic demand, except for some five star hotels (Acharya, 2006). In the emerging trade scenario, cost and quality of a commodity determine the flow and dynamics of its trade in the world market. India with favourable agro-climatic conditions and surplus labour has the potential to produce high-quality gherkin round the year and has the capacity to export it to the international market. Gherkin satisfies all the criteria described by Vyas (1994) for the export-orientation, viz. a genuine and growing surplus after meeting domestic requirements, favourable ratio of export to domestic price, and growing international demand. The Govt. of India has rightly identified the export potential of this commodity and has established Agri- Export Zones (AEZs) for the crop in the states of Karnataka and Andhra Pradesh (Kumar *et al.*, 2008). Export scenario of cucumber and gherkin is given in table (5.2)

Table(5.2) : Export of cucumber and gherkin

Year	Export (tonnes)	Value in 000 $
1994	570	518
1996	4057	422
1998	1720	235
2000	2747	369
2002	9748	350
2004	1933	269
2006	327	314
2008	1894	478
2010	194	265

Source: Vanitha *et al.*(2013)

Botanical Description

Gherkin belongs to the genus *Cucumis* and specxies *anguria*. Its chromosome number X=n= is 7. Taxonomic position of the genus is given as under:

Taxonomic position of the genus *Cucumis*

Kingdom	Plantae
Subkingdom	Tracheobionta
Superdivision	Spermatophyta
Division	Magnoliophyta
Class	Magnoliopsida
Subclass	Dilleniidae
Order	Violales
Family	Cucurbitaceae
Genus	*Cucumis* L
Species	*Cucumis anguria* L.

Annual monoecious herb with trailing or scandent stems, having solitary, simple, setose tendrils 3-6 cm long; stems grooved, with bristle-like hairs. Leaves alternate, simple; stipules absent; petiole (2-)6-13 cm long, hispid to setose; blade broadly ovate in outline, 3-12 cm × 2-12 cm, shallowly to deeply palmately 3-5(-7)-lobed, with punctate to hispidulous hairs on both surfaces. Flowers unisexual, regular, 5-merous; sepals narrowly triangular, 1-3 mm long; petals united at base, 4-8 mm long, yellow; male flowers in 2-10-flowered fascicles, with pedicel 0.5-3 cm long, stamens 3; female flowers solitary, with pedicel 2-10 cm long, ovary inferior, ellipsoid, 7-9 mm long, softly spiny, stigma 3-lobed. Fruit an ellipsoid to sub-globose berry 3-4.5 cm × 2-3.5 cm, on a stalk 2.5-21 cm long, beset with soft, thin spines with transparent tips, green, ripening yellow, many-seeded. Seeds ellipsoid, 5-6 mm long, compressed with rounded margins, smooth (Wilkins, 2004).

Varieties

Cucumber and gherkin plants are similar. They are vine plants with large leaves and long petioles. The vine contains separate male and female flowers, with the flowers produced being males. A tiny fruit develops at the base of the female flower and these produce the edible fruit. Gherkin fruit is generally round more warted than cucumber and generally 1-3 inch long. Some of the promising hybrids are described as under.

Calypso: It is a gynoecious hybrid. This was a promising hybrid developed during early 1990 in U.S. As there was no promising varieties/hybrids of gherkin, Calypso was adopted at very large scale.

Ajax: This hybrid was introduced by Nunhems Seeds Pvt. in India during 1997. Its yield is very high with good fruits having excellent processing quality and less wastage.

Other important varieties/ hybrids are Venolo Picking, NCVH-32, NCVH-3 and NCVH-41(Nath *et al.*, 2007)

Paulista:

It is a new gherkin type originally derived from the crossing of *Cucumis anguria* var. *anguria* x *C.* var. longaculeatus, with distinct characteristics of fruits and leaves. Paulista gherkins are spineless and have larger fruit than common gherkins.

Climate and Soil

Gherkin is a warm season crop. For germination of seed 22-35°C soil temperature is essential. If the soil temperature is below 16°C seeds do not germinate properly. For the vegetative growth of plant 30-35°C temperature is considered ideal. Gherkin is very much sensitive to frost and cold; and temperature below 4.4°C may prove injurious to young seedlings of plants. Low soil moisture followed by higher temperature exceeding 22°C favours seed formation in the fruit, resulting in poor quality. Well-drained red sandy loam soil with pH in the range of 6.5 to 7.5 is best suited for cultivation of gherkin. Soil should be ploughed to a depth of 30 to 45 cm till a fine tilth is obtained. Then, ridges and furrows are formed across the slope of the land at a distance of 1 to 1.2 metre. It is advisable to avoid plots that had cucurbitaceous crops like pumpkin, cucumber, watermelon, muskmelon or gourd in the previous season or such crops in the neighbourhood to prevent recurrence of pests and diseases.

Manures and Fertilizers

Well decomposed farm yard manure @ 15-20 t /ha should be given at the time of field preparation. Basal dose of fertilisers NPK @ 150:50:80 kg/ha (half dose of N and full dose of P & K of the recommended) and neem cake are applied in the furrow at 7.5 cm deep and covered with soil. A complete package for application of fertilizer and other plant protection chemicals has

Table (5.3) : Application of fertilizer in gherkin

Period from sowing	Manures, fertilizer and micro nutrient	Rate kg/ha	Methods of application
1st day	DAP	250 kg	Basal
	Neem cake	150 kg	Basal
15th day	NPK Mixture (17:17:17)	150 kg	Top dressing
27th day	NPK Mixture (178:17:17)	150 kg	Top dressing
	Potassium	250 kg	Top dressing
45th day	NPK Mixture (178:17:17)	150 kg	Top dressing
	Ammonium Sulphate	150 kg	Top dressing
	Muriate of Potash	150 kg	Top dressing
50th day	Cucumix+ Rider (micro nutrients)	2g /litre	Foliar spray
60th day	Calcium Ammonium Nitrate	250 Kg	Top dressing
	Muriate of Potash	150 Kg	Top dressing
65th day	GA/ Potassium Nitrate	2g /litre	Foliar spray

Source: Pandey(2003)

been suggested by Pandey (2003) (Table 5.3).

Seed Rate and Sowing

For a hectare planting 1-1.25 kg seeds are required. Seeds are treated prior to sowing with captan or thiram @ 2g/kg seeds. Seeds are sown in well prepared field in rows prepared at the spacing of 1-1.30 m and plants are spaced at 30-35 cm in rows. Seeds germinate within a week after sowing. Resowing is done to fill in the gaps within 3-days after germination to maintain optimum plant population. Remove the plants if found very close (less than 30 cm) to each other. Soon after the fourth leaf stage, leaf miners start infesting the cotyledon leaves. Hence these leaves must be clipped, and destroyed by burning.

Irrigation

Gherkin is a comparatively shallow rooted crop and requires irrigation frequently. However, moisture on the leaves from rain, dew or irrigation encourages diseases.

Pollination and Fruit set

In gherkin monoecious, gynoecious, andromonoecious or gynomonoecious flowers appear. There are hybrids expressing exclusive female flowers for them it is essential to put at least 10-15 per cent monoecious plants for pollination and fruit set. For enhanced fruit set, 3-5 colonies of honey bee should be kept. However, parthenocarpic strains do not require any pollination aids. Avoid spraying chemicals during morning hours which affects honey bee activity. Spray should be given during late evening .

Staking

The crop requires staking, being a creeper, and if not staked, the fruits will be formed on the ground. Staking is done between 16 to 20 days after sowing. Staking is done using wooden stakes held serially by Galvonised Iron (Gl) wire and the vines are trained to grow along the jute thread tied on to the Gl wires connecting the top and bottom Gl wires. Usually stakes are of 2.4 metre length and 5 cm diameter to support the weight of the crop till the end. They are grouted 30 cm below the ground level at a distance of 3.5 metres. Stakes at the end of the row are strengthened with good gauge of Gl wire (10 to 12 gauge for border and 16 to 18 gauge along the rows) to hold the crop firmly. Staking, besides supporting the plant to stand straight, facilitates ventilation, cultural operations and quick harvest of fruits. This ensures sunlight reaching every fruit, thereby avoiding bleaching. It has been observed that unstaked vine is more prone to pest and disease infestation.

The trellised system could be an excellent choice for the Paulista gherkin support. Most fruit production are concentrated in lateral shoots as a result of the strong suppression of apical dominance. As the trellised net had vertical and horizontal strings, secondary and tertiary shoots are well distributed along the net. The trellised net avoids fruit contact with the soil, improving its quality and making harvesting easier, without pruning, and consequently, reducing labor costs (Valéria Aparecida Modolo and Cyro Paulino da Costa, 2004).

Harvesting and Yield

Harvesting at right stage is utmost essential in gherkin. Approximately 35-50 fruits are harvested from a single vine. For processing, fruits are harvested as per the requirements of produce. In general, gherkin is grown in farmers' fields on contract basis. Green fruits of gherkins are collected from the fields every day. There is a well-established buyback arrangement at a price fixed by the company. The average yield of gherkin varies from 200-250 quintals per hectare.

Grade

Gherkins are sold on the basis of counts (number of fruits per kg). The lower the count, the lower the price and *vice versa.* The popular counts in commercial transactions vary as follows: 5-10, 10-20, 20-30, 30-40, 40-60, 60-80, 80-100, 80-120, 120-160, 150-300, 160-300, and 300+. Some buyers may combine the above counts and ask for assorted counts such as 80-120, 150-300 etc. Therefore, the processing and trading / exporting of gherkins are wholly customer - focussed.

Hydro-cooling

Post harvest cooling of gherkin fruit can be separated into two processes (i) the removal of field heat to the desired product storage temperature and (ii) the maintenance of optimum temperature and relative humidity through continued removal of product respiratory heat. The first process is known as pre-cooling. Fruit such as gherkin that cannot tolerate short-term turgidity loss without detrimental physiological or biochemical changes are often pre-cooled. Precooling of gherkin fruit is accomplished almost exclusively by submersion of fruit in cool water in a single controlled handling step. Cool air can also be used. However, since water has a much higher heat transfer coefficient than air, hydro-cooling is a more rapid, thorough and efficient methods of eliminating field heat. For hydro-cooling to be effective, it must satisfy the following criteria (i) contact between water and the fruit surface must be uniform, (ii) water temperature must be reduced to near chilling temperature and (iii) duration of the hydro-cooling treatment must be enough to uniformly cool the exterior and interior of the fruit. This requires that the re-circulated hydro-cooling water must be re-cooled after each cooling cycle. Since, chilling injury can occur when gherkin fruits are held at or below 7°C, hydro cooling temperature is kept slightly above 7°C (~11°C is customary). Chilling injury may result in fruit pitting, water soaked lesions, shriveling and decay.

Salt brining

Salt brine storage offers a method for preserving large amounts of gherkin fruits during and after the harvest season. The salted product can be held at relatively low costs and processed further during the year to meet consumer demands. Three methods are characteristic at the fermentation of gherkin in a salt solution (i) salt stock, (ii) genuine dill, and (iii) overnight processing. Salt stock fermentation is done by submersion of washed and graded fruits of gherkin in tanks of 5-8 per cent concentration strength. Process is completed till all fermentable sugars have been converted to acids and other end products. When the fermented product is needed by the processor, it is then desalted by leaching in water to an acceptable organoleptic level (2-2.5 per cent). Generally, dill pickles are produced by fermentation in a 4-5 per cent salt solution that contains dill weed, garlic, and other spices. The product is completely fermented in about 3-6 weeks, depending on fruit size, fermentation solution (1-1.5% lactic acid and 3-3.5 per cent salt) and fermentation temperature. In contrast to salt stock brining procedures, genuine dill pickles do not require desalting and are often salt with the filtered fermentation liquor that is produced during the process. The product should be used within a year of preparation. The overnight process involves fermentation of fruit in a 2-4 per cent salt solution containing dill weed, garlic and other flavorings until the acidity reaches a desirable level (0.75-1% tritable acidity as ascetic acid). This usually requires about one week but is dependent upon fruit size and curing conditions. This type of product should be stored in refrigerators as they have short shelf life.

Major Pests

Gherkin Fruit Borer (*Diaphania indica* (Saunders) Lepidoptera: Pyralidae)

This is the most serious pest of gherkins in southern India, which occurs throughout the year. Its larvae feed on leaves as well as on buds. The early instar larvae are restricted to tender leaves where they scrape the green matter and feed, the damaged portion develops into lace-like patches with main veins intact. The larvae also feed on early stages of fruits by tunnelling into them. Fruits that are in contact with leaves are damaged most. Affected fruits show a bored hole usually at the base.

Though the pest occurs throughout the year on one or the other cucurbit crop, its incidence on gherkins is highest during rainy season. In addition to gherkins, it also feeds on ash gourd, watermelon, musk-melon, squash, cucumber, pumpkin, bitter gourd, bottle gourd, sponge gourd, ribbed gourd, *Coccinia*, snake gourd, chow chow and other cucurbitaceous plants.

Natural enemies

Larval parasitoids, *Apanteles taragamae* and *Goniozus sensorius* are common on this pest.

Management

1. Destroy infected fruits
2. Follow deep ploughing during sum mer season
3. Spray 20 ml Malathion 50EC+200g sugar dissolved in 20 l water on selected plants to trap the insects

Melon fly or Fruit Fly [*Bactrocera cucurbitae* (Coquillett) Diptera: Tephritidae].

This is the second most serious pest of gherkins in southern India. It occurs from March to May and July to October. It is a polyphagous insect affecting more than 125 host plants. Both adults and larvae are the damaging stages. Males are attracted to pheromone lures. Melon fly damages gherkin in three ways: i) oviposition injury by the female on fruits and vegetative parts ii) larval feeding on fruit pulp and iii) decomposition of fruit tissue by invading saprophytic micro-organisms. Oviposition injury- female fly puncture the fruit surface with their pointed ovipositor and deposit eggs inside the tissue. A gummy exudate covers the punctured hole of larval feeding injury .The spot at which the eggs are laid sinks and forms a shallow depression which is covered by greyish brown exudatee. After hatching the maggots that emerge feeds on the fruit tissues and the fruits become crooked, malformed. Following the feeding damage by the maggots, micro-organisms invade, leading to rotting of the tissues, which later turn brown and give out a foul smell. Cucurbits are the primary host plants of melon fly. Other hosts include ash gourd, watermelon, muskmelon, squash, cucumber, pumpkin, bitter gourd, bottle gourd, sponge gourd, ribbed gourd, *Coccinia*, snake gourd, chow chow and other cucurbitaceous plants. It also breeds on some of the fruits such as banana, tomato, bell pepper, guava, egg plant, papaya, mango, sweet orange etc.

Natural enemies

Opius fletcheri (Silvestri), a pupal parasite occasionally parasitizes this insect but does not exercise control.

Management

1. Avoid continuous cultivation of cucumbers at the same place since this may lead to fruit fly outbreaks.
2. Destroy all infested fruit.
3. Spray with a pyrethrum solution in the evenings after the bees are mostly back in their hives

Cucurbit Fly [*Dacus (Didacus) ciliatus* Loew Diptera: Tephritidae].

This is an occasional pest of gherkin and when it occurs become serious as melon fly. It is also known as Lesser pumpkin fly or Ethiopian fly. There is no pheromone lure for this pest. Eggs, larvae and pupae are similar to those of melon fly. The adults are smaller than melon fly, measuring 8-10 mm. It is predominantly an orange coloured species with facial spots. The band along the front margin of the wings extends up to the tip and forms a small apical spot, and a basal oblique spot. Abdomen especially in female has two black spots. Mainly cucurbitaceous plants are listed under melon fly. It also attacks tomato, okra, cotton, bell pepper and beans.

Management

When detected, all infected fruit should be destroyed. Insecticidal protection is possible by using a full cover spray or a bait spray. Malathion is the usual choice of insecticide for fruit fly control and this is usually combined with protein hydrolysate to form a bait spray.

Serpentine Leaf Miner [*Liriomyza trifolii* (Burgess) Diptera: Agromyzidae].

It is a highly polyphagous insect; the damage is more severe in summer months. It is resistant to several insecticides, hence, it is necessary to use only recommended chemicals for its eradication. The eggs are minute (0.5 mm long, 0.2 mm broad), laid within leaf lamina by the female with the aid of the ovipositor. The young maggots are colourless, whereas full-grown maggots (2 mm long) are orange-yellow in colour. The pupae are small barrel shaped golden yellow to begin with and turn brown later. Adults are very small, black with prominent yellow patches on head and thorax. The leaves show narrow whitish serpentine mines. Heavy infestation results in large number of mines that coalesce and form brown blotch, leading to premature leaf fall. It is a polyphagous pest affecting more than 80 hosts across several families. Tomato, beans, cucurbits and castor are most preferred hosts.

Natural enemies

Some of the important parasitoids are the species of *Chrysocharis, Chrysonotomyia* and *Hemitarsenus*.

Management

1) Castor can be grown as a trap crop around gherkin fields to distract the fly away from gherkin.
2) Remove cotyledons and affected lower leaves and burn.

Melon Thrips [*Thrips palmi* Karny Thysanoptera: Thripidae].

It is a polyphagous, widely distributed species attacking most of the ornamental and cucurbit plants. Their life cycle consists of egg, nymph, pre-pupa, pupa and adults. Adult thrips are small, pale yellow and body is elongated with fringed wings. The nymphs are pale yellow. Both the stages are found among the buds and on lower surface of leaves. This species may be mistaken for another similarly coloured species, *Thrips flavus* from which it differs in having the ocellar setae within the ocellar triangle. Both nymphs and adults feed mostly on plant tissues by using their rasping and sucking mouthparts to extract plant fluids. Feeding injury appears as coarse stippling on the leaf surface. Large population of thrips causes severe injury, which results in silvery appearance of leaf surface. Thrips also feed on flowers causing marginal necrosis. Thrips pose a serious threat to gherkin crop as they transmit virus diseases. They are capable of transmitting tospo viruses, which are fatal to the gherkin crop. Once a plant is infected, it cannot be cured. Only adult thrips are capable of transmitting the virus disease, on the other hand only the first instar nymphs can acquire the virus during the feeding. Control options for tospo viruses are limited to suppression of vector population in the early stage of the crop and destruction of infected plant material to reduce innoculum.

Aphids [*Aphis gossypii* Glover, Hemiptera: Aphididae].

It is a widely distributed species found throughout the year. Its population peaks during long dry spells. High humidity coupled with cloudy/rainless cool months is congenial for faster build up of aphid colonies. Adults and nymphs are soft bodied and pear shaped. Nymphs, winged and wingless adults are commonly found in colonies. Both nymphs and adults suck sap from lower surface of leaves, shoots, flowers and fruits. Affected leaves curl downwards, get wrinkled, deformed and turn yellow. Severe infestation during the seedling stage leads to death of the plant. The aphids also excrete honeydew on which sooty moulds grow affecting photosynthesis. Aphids act as vectors of Cucumber Mosaic Virus (CMV), and Zucchini Yellow Mosaic Virus (ZYMV), which are severe during summer months. Effective management of aphids in the initial stage of the gherkin crop is very important to reduce the severity and further spread of virus diseases.

Whitefly [*Bemisia tabaci* (Gennadius) Hemiptera: Aleyrodidae].

It is a polyphagous pest found infesting most of the field and vegetable crops. The eggs are pear shaped, light yellow with a short stalk that is inserted into leaves. The nymphs are oval, scale-like, greenish white and are found on lower surface of the leaves. The adults are tiny, moth-like, covered with white waxy bloom on entire body including wings. The nymphs suck sap from young and succulent leaves. Affected leaves become yellowish, wrinkle and curl downwards. They excrete honeydew, which encourages growth of sooty mould on the leaves.

Management

Spray Dimethoate 1.5 ml/ per litre

Root Grubs [*Holotrichia serrata* (Fabricius) Coleoptera: Scarabaeidae].

It is a minor pest of gherkin occurring during Kharif season. These pests originate from non decomposed farm yard manure. The eggs are spherical, pearly white in colour and are laid in the soil. The larvae are C-shaped and pass through three instars. They are slightly yellowish creamy white, measuring 11 to 47 mm in length and 3 to 12 mm in width. They are found at the base of plants close to the root zone. They pupate in the soil at a depth of 30-40 cm in an eartthen shell. The pupae are creamy white to begin with and later turn brown. All the appendages are free. The adults are large, 22-25 mm long, reddish brown with dark brown elytra (forewing). The adults emerge from the soil during the month of April- May i.e., immediately after first rains, feed on foliage of trees (neem, *Acacia, Cassia* etc.) and lay eggs in the soil. Larvae are the damaging stage and are known to damage most of the field crops including cucurbits. They feed on the roots that leads to the death of the plants in early stages of the crop. Often plants in a row of 1.0 to1.5 m show wilting symptoms.

Management

If the field has a previous history of root grub infestation, after every rain during April-May freshly cut neem twigs may be placed in the field between 7 am & 9 pm to attract the beetles, which may be collected and destroyed.

Red spider mite [*Tetranychus urticae* Koch Acari: Tetranychid].

It is a widely distributed and highly polyphagous pest. The mite populations are very high during summer, especially during long dry spells. The eggs are globular and whitish. First stage larvae are whitish with three pairs of legs whereas, later instar nymphs are greenish with four pairs of legs. The adult female mites are very tiny, 1.0-1.5 mm long, with two black patches on either side of the body, these have four pairs of legs. These are found in large numbers on the lower surface of leaves. These produce thin silken thread, which form a thick web when the infestation is high. The mites feed on ventral surface of leaves under fine silken web, which forms a protective cover for mite colonies. Continuous feeding results in development of yellow spots on the leaves, severely affected leaves gradually dry. During high incidence, even gherkin fruits are infested with mite colonies. In severe stage, the entire leaf dries quickly. Severely infected vines bear a few small sized fruits.

Management

1. Conserve natural enemies. Predatory mites and anthocorid bugs are important in natural control of mites.
2. Avoid use of broad-spectrum pesticides. They may kill natural enemies and may lead to mite outbreaks.
3. Follow good growing conditions for plants. Adequate irrigation is particularly important.

Major Diseases

Downy Mildew

Attack by the fungus causes yellow, water soaked lesions or spots or angular patches on upper surface of leaf and on the corresponding lower surface, greyish or purplish downy growth is seen. Later grayish or purplish colour tums to black because of sporulation. In severe stage, the entire leaf dries quickly. Severely infected vines bear a few small sized fruits. The infection by the pathogen is conducive during cool wet nights and warm humid days with a relative humidity of 95-100%, free moisture or film of water on leaves for, at least, 2 hours and temperature between 18-28°C. The disease occurs during rainy season and it is severe during winter. The pathogen spreads by wind, rain splashes and contact of infected leaf with healthy ones.

Management

1. Avoid flooding, water logging in the field and mono-cropping or relay cropping with other cucurbits.
2. Plucking and destroying the diseased leaves periodically control the spread of the disease.
3. Use biocontrol agent like *Trichoderma* @ 3.5g/l of water

Powdery Mildew

The disease is prevalent during warm rain-free months with low relative humidity. The mycelium lives on the host surface and produces abundant spores. White powdery growth is seen, especially on the upper surface of the leaves and stem. Leaves show yellow lesions, which later turn brown and dry out. In early infection, growth is checked and plants remain stunted, leaves curl, wither and drop-off. Flowers remain unopened. Field sanitation helps in minimising the available innoculum in the field. Excess application of nitrogen and high density cropping favours the spread of the fungus.

Management

1. Use resistant varieties, if available.
2. Leave wide spacing between plants.
3. Avoid overlap cucumber plantings.
4. Apply copper fungicides at 0.1 per cent

***Fusarium* Wilt**

Fusarium wilt causes rotting of seeds in the soil, damping off of seedlings and wilting of the vines. *Fusarium* infects all stages of crop starting from cotyledon stage. After germination the cotyledon leaves become yellow, the hypocotyls are girdled by watery soft rot and ultimately shrink and decay. In infected older plant, leaves droop at mid day for a couple of days, which is similar to water deficiency and within 3-4 days; the vines shrivel permanently and die. Vascular bundles in the collar region become yellow or brown. At later stage, growth of white mycelium may be observed on the infected parts. Soil temperature of 20-30°C is conducive for disease development.

Management

1. Removal of old plant debris before sowing of gherkin crop
2. Seed treatmnt with hot water @ 50°C temperature for 20-25 minutes.

Tomato Spotted Wilt Virus (Tospo virus)

Young leaves of infected plants exhibit chlorosis, yellowing and upward curling of the margins. Growing tips are affected with systemic necrosis leading to tip dieback and stunted growth. Flower buds dry and drop, leading to severe yield loss. Fruits develop light green rings with raised centres and become mottled and spineless. Vines remain stunted and fail to put forth side shoots. The virus is transmitted by thrips. Virulence lasts for 20-40 days.

Management

1. Removing and burning of infected plants at frequent intervals will check further spread of the disease.

Cucumber Mosaic Virus (CMV)

Cucumber Mosaic Virus is transmitted by aphids (*Aphis gossypii*), in nonpersistent manner and also by humans, mechanical means and other sucking vectors. Initially leaves become malformed, mottled, distorted and wrinkled. Leaf edges curl downward. In later stages, plants remain stunted with shorter internodes. In severe infection, plants do not produce side shoots thus reducing number of flowers and fruits produced. Older leaves become chlorotic and then necrotic areas develop along the margins. The infected fruits have pale green or white areas intermingled with dark green, bumpy areas. Virus infected plants produce malformed, smooth grey-white fruits with irregular green areas at a later stage. Such fruits taste bitter and make soggy pickles.

Management

1. Removal and burning of weed hosts and infected plants near gherkin fields prevents further spread of the disease.

Zucchini Yellow Mosaic Virus (ZYMV)

The ZYMV is transmitted by aphids. Infected plants produce mosaic, yellowing, severe malformation, blisters and extreme reduction in the size of leaf lamina. Plant growth is drastically reduced. Fruits develop knob-like areas, which cause prominent deformation.

Management

Uprooting and destroying the early infected plants prevents spreading of infection.

Root-knot nematode

Infestation by root-knot nematodes is severe during summer season and is high in red sandy soils. The nematodes spread through irrigation water. In affected plants, older leaves turn yellow and plants remain stunted. Roots develop characteristic knots or galls. In severe cases, plants wilt

Management

Suspected plants are to be pulled out and examined for the presence of root-knots or galls. To prevent the attack, the crop is to kept with crops like marigold and sweet potato helps in reducing the severity of the nematode infestation.

Genetic Resources and Improvement

Cucumis anguria is not in danger of extinction in its native habitat. The National Plant Germplasm System of the USA Department of Agriculture maintains numerous accessions of cultivated types of *Cucumis anguria* at its regional plant introduction station in Ames, Iowa. Another collection is maintained at the Centro Agronómico Tropical de Investigacion y Enseñanza (CATIE), Turrialba, Costa Rica (Wilkins, 2004). Various Western seed companies offer seed of West Indian gherkin, including **'African Heirloom'**, and **'West Indian Burr Gherkin'**. *Cucumis anguria* proved to be totally immune to cucumber green mottle mosaic virus (CGMV). Resistance also occurred to root-knot nematodes and powdery mildew. In a study in South Africa, where fungal diseases and fruit parasitisation by Trypetid larvae is usually severe in Cucurbitaceae (*Cucumis anguria*) showed a high resistance to both fungi and Trypetids. Research efforts to transfer resistances into cucumber and melon have been undertaken. Repeated attempts to hybridize different *Cucumis* species have not been entirely successful; some species have never been successfully crossed to produce a fertile F_1 generation, whereas other species have been crossed to a limited extent. The possibility of using *Cucumis anguria* as a rootstock has been suggested, where scions of desirable crop species are grafted to it. In populations of some cucurbit species that normally produce bitter or toxic fruits, individuals may occasionally arise spontaneously which produce non-bitter, edible fruit. These variants are genetically stable when removed from the bitter gene pool. In *Cucumis anguria* a single gene distinguishes the bitter from the non-bitter type, the gene producing bitterness being dominant. Multiple factors appear to be involved in controlling bitterness, including various physiological conditions.

Refereneces

Acharya, K. (2006). A long look at a small vegetable, Deccan Herald, Spectrum supplement, 21 February, (http://www.deccanherald.com/)

Anonymous (2007). Fall in EU sugar output likely; Ministry flags scope for export of gherkin, Hindu Business line, 4th June. (http://www.thehindubusinessline.com).

Kumar, N. R., Rai, A.B. and Mathura Rai (2008). Export of cucumber and gherkin from India: performance destinations, competitiveness and determinants. Agricultural Economics Research Review, 21:130-138.

Nath, P., Srivastava, V.K., Dutta, O.P. and Swamy, K.R.M. (2007). Vegetable crops-improvement and production. Dr. Prem Nath Agriculture Science Foundation, Bangalore.

Pandey, S. (2003). *Gherkin ka Gunvattayukt Utpadan*, In: production and marketing of export oriented ghekin. Compendium of Meeting Organized at Pilibhit, U.P. November, 7, 2003, pp. 24-27.

Valéria Aparecida Modolo and Cyro Paulino da Costa (2004). Production of paulista gherkin using trellis net support. Sci. Agric. (Piracicaba, Braz.), 61(1):43-46. Jan./Fev. 2004.

Vanitha, S.M.,Chaurasia,S.N.S., Singh,P.M. and Prakash S. Naik. (2013). Vegetable Statistics.Tech. Bull. No. 51, IIVR, Varanasi, pp. 250.

Vyas, V. S. (1994). Agriculture price policy: Need for reformulation In: Economic liberalization and Indian Agriculture, Ed: G. S. Bhalla, Institute for Studies in Industrial Development, New Delhi.

Welman, W.G. (2003). Cucurbitaceae. In: Germishuizen, G. & Meyer, N.L. (eds), Plants of southern Africa: an annotated checklist. Strelitzia 14:413-417. National Botanical Institute, Pretoria.

Wilkins Ellert, M.H. (2004). *Cucumis anguria* L. In: Grubben, G. J. H. & Denton, O.A. (Eds.). PROTA 2: Vegetables/Légumes. PROTA, Wageningen, Netherlands.

6

BABY CORN

Baby corn is a form of maize where young, unfertilized, tender cobs are utilized either as fresh or canned. Presently this type of maize or baby corn is utilized for salad, soup, pickles, baby foods and other several Chinese dishes. It has emerged one of the most preferred items in multi-star hotels. Its cultivation has gained much popularity in Thailand where it is being cultivated in around 21,049 hectare area with the annual production of 129647 tonnes. Today, Thailand and China are the world leaders in baby corn production. The growth of baby corn exports from Thailand has been multifold from 67 tonnes worth U.S.$38,059 in 1974 their exports had risen to 3676 tonnes worth of U.S.$ 33 million in 1992. As the production of baby corn increased, country established several processing industries. In India, its cultivation is now picking up in a serious way in Meghalaya, western Uttar Pradesh, Haryana, Maharashtra, Karnataka and Andhra Pradesh. Dehusked cobs are very nutritious and its nutritional value is comparable with other important vegetables. Further, crop is almost free from pesticide because edible cob is completely covered with several layers of husk and does not require frequent spray of harmful chemicals to avoid the infestation of pests etc. Nutritive value of the baby corn is given in table (6.1).

Table (6.1): Nutritive value of baby corn
(per 100 g of edible portion)

Constituents	Amount
Moisture	89.1 per cent
Fat	0.20 g
Protein	1.90 g
Carbohydrates	8.20 mg
Calcium	28.0 mg
Phosphorus	86.0 mg
Vitamin A	64.0 I.U.
Thiamine	0.05 mg
Riboflavin	0.08 mg
Ascorbic acid	11.0 mg
Niacin	0.03 mg

Production and Export of Baby Corn

It has great potential both for internal consumption as well as for export round the year as 3-4 crops of baby corn can be taken in one year. Thailand is the major baby corn producing and exporting country in the world. Recently, India has emerged as one of the potential baby corn producing countries because of the low cost of production as compared to many other countries. Major baby corn producing, exporting and importing countries are mentioned in table (6.2).

Table (6.2): Major baby corn producing, exporting and importing countries

Particulars	Countries
Producing countries	Thailand, Taiwan, Guatemala, South Africa, Zambia, Zimbabwe, Peoples Republic of China, Nepal, Japan, Sri Lanka and India
Exporting countries	Thailand, Taiwan, Guatemala, S. Africa, Zambia, Zimbabwe and Peoples Republic of China
Importing countries	U.S.A., Japan, Hong Kong, Singapore, Australia, Malaysia, Canada, Saudi Arabia, New Zealand, European countries and India

Botanical Description

Baby corn is a member of family Poaceae (Gramineae). Its scientific name is *Zea mays* and chromosome number (2n) is = 20.

Varieties

Although any kind of corn variety can be grown for baby corn production in kitchen garden. For commercial purpose varieties should be selected which are early in maturity and having potential to bear at least 3 cobs per plant without loosing quality, size and shape. Further, varieties with yellow immature kernel with uniform row arrangements is preferred as it has been prescribed standard for the international market.

For Indian climatic conditions an open pollinated variety VL-42 has been developed from VPKAS, Almora. MEH-114 is a potential hybrid while COBC-1 is a potential synthetic variety.

HM 4

This single cross hybrid has an attractive creamish to light yellow colour with desirable size of 6.0 to 11.0 cm in length and 1.0 to 1.5 cm in diameter with regular row arrangement. It possesses all the desirable traits of an ideal baby corn hybrid. Its cultivation is gaining momentum in nearby areas of cities. It is highly nutritive and sweet in taste. It has most desirable medium height plant type, lodging resistant, prolific, responsive to high dose of fertilizers and plant remains green after picking of baby corn. Three to four pickings are common. Green plants also provides quality fodder to livestock.

Climate and Soil

Baby corn as like common maize is a warm season crop. It can be sown round the year in southern India. Baby corn can be grown on a wide variety of soil types but grows best on well drained soils that have good water holding capacity. Sandy soils are often preferred for early crop. Soil pH for growing baby corn should be between 5.8 and 6.5.

Planting Season

In northern India, it can be sown from February to November. Bay corn can be raised through transplanting in furrows in December-January in northern India. For this purpose nursery should be raised in November. Generally, August to November planting yield best quality baby corn. Seed germination is poor below 10°C. However, in the prevailing climatic conditions of north India, its 2-3 crops can easily be taken.

Manures and Fertilizers

For obtaining high yield of baby corn, well rotten FYM @ 10-15 t /ha should be mixed in the soil at least one month before sowing. Besides this, N:P:K fertilizer should be given @ 120:60:60 kg/ha. At the time of sowing one third dose of nitrogen and full dose of phosphorus and potassium should be applied while rest 2/3rd amount should be given in the form of top dressing at an interval of 20 and 35 days from the date of sowing. Saha *et al.* (2007) reported that application of vermicompost contributed more to concentration of iron, zinc and manganese in baby corn, whereas, total P content was more due to application of poultry manure. Nitrogen, irrespective of sources, has been highly correlated with baby corn yield.

Among the various treatment combinations the higher dose of chemical fertilizer (150% of RF) + 100% of inorganic source was found the highest yielding treatment for baby corn at Ambikapur. In another trial at Arbhavi the 150% organic fertilization alone followed by 125% organic and 62.5% organic + 62.5% inorganic + bio-fertilizer were found the highest yield treatments for baby corn yield. At Chhindwara, again the higher dose of chemical fertilizer (150% of RF) + 100% of inorganic source was found the highest yielding treatment. At Udaipur, the population density of 166 thousand/ha was found the best population at all the level of fertilizer application, and highest baby corn yield was obtained at recommended dose of N and P. In NxG trial on baby corn at Udaipur the entry Madhuri with 180 kg/ha N was found the highest yielder of baby corn. At Hyderabad, the 150% dose of recommended fertilizer (50% organic and 50% inorganic) was identified the best treatment for baby corn yield (Anonymous, 2012).

Seed Rate and Spacing

Optimum seed rate is 22-25 kg/ha depending upon the test weight of the hybrid. Seeds should be treated with fungicides and insecticides before sowing to protect it from seed and soil borne diseases and some insect-pests. Bavistin + Captan in 1:1 ratio @ 2 g/kg seed for *Turcicum* leaf blight (TLB,) Banded leaf and sheath blight (BLSB), Maydis leaf blight(MLB), etc. Apron 35 SD @ 4 g/kg seed for Brown stripe downy mildew (BSDM). Captan 2.5 g/kg for *Pythium* stalk rot. Fipronil @ 4 ml/ kg for termite and shoot fly. Seeds should be sown at a distance of 40-60 cm x 10-15 cm, depending upon variety. Thavaprakash *et al.* (2008) obtained higher yield of bay corn as well as fodder yields follwing a spacing of 60×19 cm.

Irrigation

Proper moisture should be maintained in the field. In general, kharif crop does not require irrigation. First irrigation should be applied very carefully. Water should not overflow on the ridges. The irrigation should be applied in furrows upto 2/ 3rd height of the ridges. Irrigation should be given as and when required by the crop depending upon the rains and moisture holding capacity of the soil. Young seedlings, knee high stage, silking and picking are the most critical stages for water stress for crops and irrigation should be ensured at these stages. Light and frequent irrigations are desirable for crop. During winter (mid December to mid February) soil should be kept wet to avoid frost injury.

Weed control : Spraying Simazine /Atrazine at the rate of 2.5 kg/ha (for sandy loam) to 3 kg/ha (for black soil) dissolved in 750 litres of water on the soil on the day of sowing or the next day after irrigation has been found most effective. One to two hoeings are recommended for aeration and uprooting of the remaining weeds. While doing hoeing, the person should move backward to avoid compaction of soil and to facilitate better aeration

Plant Protection : Baby corn is a 60 days crops and thus chances of being infested by pests and diseases is less but any attack by pest and disease would reduce the plant's ability to grow and hence reduce yield. Thus, preventive measures are always recommended.

Cross-pollination: If field corn or specialty '*Su*' baby corn varieties are grown next to any sweet corn varieties grown for mature sweet corn production, the varieties will cross-pollinate and the sweet corn will lose its sugariness. Detasseling the field corn and baby corn varieties will prevent undesirable cross-pollination and will not affect baby corn production (Mansour and Hemphill, 1999).

Detasseling

Crop of baby corn requires detasseling before flowering. Detasseling leads to increased number with better development of cob by preventing fertilization and making harvesting faster. It is done by removing the tassel of the plant as soon as it emerges from the flag leaf. It should be practiced row-wise. While detasseling, leaf should not be removed which will otherwise affect net photosynthesis and ultimately reduce average baby corn yield. It has been observed that the removal of 1 to 3 leaves along with tassel reduces 5-15% yield of baby corn. The removed tassel should not be thrown in the field as it is nutrient rich and should be fed to the cattle.

Intercropping

Intercropping with baby corn has been found very remunerative, if it is cultivated with intercrop. A number of crops *viz.*, potato, green pea, French bean , palak, cabbage, cauliflower, sugar beet, celery, etc. have been successfully tried in the winter season. There is no adverse affect of intercrops on baby corn and *vice-versa*, rather, some of the intercrops help in improving soil fertility and protect the baby corn crop from cold injury. Intercrops protect the baby corn from northern cold wind because baby corn is planted on southern side and intercrops in northern side of the ridge. In general, short duration varieties of intercrops are preferred for intercropping with baby corn. Recommended dose of fertilizers of intercrops should be applied in addition to the recommended dose of fertilizers of baby corn. In kharif season, cowpea for green pods and fodder purposes, urd, mung, etc. can be intercropped with baby corn. Numbers of intercrops are option for the farmers but for commercial purpose, pea and potato can be taken on large scale during winter season.

Harvesting and Yield

Harvesting is most important practice in production baby corn. Cob should be harvested every day in the early morning from the date of silk emergence for fresh market and 2 or 3 days after silk emergence for processing purpose. Generally, 6-7 pickings are required to harvest complete crop in 45-60 days. Yield of baby corn varies from 60-80 quintals per hectare. Young cob is highly perishable owing to its high rate of respiration. Hence, cob should be stored at 5-7°C temperature with 90 per cent relative humidity. Low temperature storage also favours to maintain the sugar and conversion of sugar into starch. It has been estimated that if the freshly harvested cob is stored at 0°C it maintains 80 per cent sugar upto 4 days while there is 80 per cent loss of sugar if cobs are stored at 30°C for only 24 hours.

Improvement

Genotype-location interaction and phenotypic stability of cob yield and maturity parameters of eight baby corn genotypes including hybrids and composite varieties were assessed. The genotype Khoibhutta was found stable for days to tasselling, days to harvesting and yield of by product and suited to unfavourable environment. On the other hand, the BBC1 was stable across environments for days to harvesting and cob yield. Genotype NS pop corn was stable for days to tasselling and suited to unfavourable environment and stable for yield of by-product and cob yield across the environments. The genotype BHM5 showed stable performance only for cob yield across the environments (Nahar *et al.*, 2010).

References

Anonymous (2012). Directorate of Maize Research, Pusa Campus, New Delhi - 110012, India) Website:- http://www.maizeindia.org

Mansour, N. S. and Hemphill, D. (1999). Commercial Vegetable Production Guides: Sweet Corn. Oregon State University. http://osu.orst.edu/Dept/NWREC/corn-pr.html.

Nahar, K., Ahmed, S., Akanda, M.A.L., Mondal, M. A. A and Islam, M. A. (2010). Genotype- environment interaction for cob yield and maturity in baby corn (*Zea mays* L.). Bangladesh J. Agril. Res. 35(3):489-496.

Saha, S., Appireddy, G.K., Kundu, S. and Gupta, H.S. (2007). Comparative efficiency of three organic manures at varying rates of its application to baby corn. Archives of Agronomy and Soil Science, 53(55):507-517.

Thavaprakash, N., Velayudham, K. and Muthukumar, V.B. (2008). Response of crop geometry, intercropping systems and important practices on yield and fodder quality of baby corn. Asian J. Scientific Res. 1: 153-159.

7

SWEET CORN

Sweet corn (*Zea mays* convar. *saccharata* var. *rugosa* is a variety of maize with a high sugar content. Sweet corn is the result of a naturally occurring recessive mutation in the genes which control conversion of sugar to starch inside the endosperm of the corn kernel. Unlike field corn varieties, which are harvested when the kernels are dry and mature (dent stage), sweet corn is picked when immature (milk stage) and prepared and eaten as a vegetable. Since the process of maturation involves converting sugar to starch, sweet corn stores poorly and must be eaten fresh, canned, or frozen, before the kernels become tough and starchy. Nutritional composition of the sweet corn is given in the table (7.1).

Table (7.1). Nutritional composition of the sweet corn (per 100 g of edible portion)

Components	Value
Water	76 g
Energy	360 kJ
Protein	3.0 g
Fat	1.0g
Carbohydrates	19.0 g
Fiber	3.0 g
Sugars	3.0 g
Iron	0.5 g
Manganese	0.2 g
Calcium	2.0 mg
Magnesium	37.0 mg
Phosphorus	89.0mg
Potassium	270.0 mg
Zinc	0.5 mg
Panthothenic acid	0.7 mg
Vit B_6	0.1 mg
Folate	42 µg
Thiamin	0.2 mg
Riboflavin	0.1 mg
Niacin	1.8 mg

Origin and Distribution

Sweet corn occurs as a spontaneous mutation in field corn and was grown by several Native American tribes. The Iroquois gave the first recorded sweet corn to European settlers in 1779 (Karl, 2010). It soon became a popular vegetable in southern and central regions of the United States Open pollinated varieties of white sweet corn started to become widely available in the United States in the 19th century. Two of the most enduring varieties, still available today, are Country Gentleman (a Shoepeg corn with small, white kernels in irregular rows) and Stowell's Evergreen. Sweet corn production in the 20th century was influenced by the following key developments: (i) hybridization allowed for more uniform maturity, improved quality and disease resistance, (ii) identification of the separate gene mutations responsible for sweetness in corn and the ability to breed varicties based on these characteristics:

i) *su* (normal sugary)
ii) *se* (sugary enhanced, originally called Everlasting Heritage)
iii) *sh2* (shrunken-2)

Sweet corn is a plant of family Poaceae (Gramineae). Taxonomic position of the genus *Zea* isgiven as under

Table (7.2). Taxonomic position of the genus Zea

Kingdom	Plantae
Order	Poales
Family	Poaceae
Subfamily	Panicoideae
Tribe	Andropogoneae
Genus	*Zea*
Species	*Z. mays*
Binomial name	*Zea mays* L.

Varieties

There are currently hundreds of varieties, with more constantly being developed.

Win Orange Sweet Corn: It is recommended for cultivation in J & K, Uttarakhand, North Eastern Region and Himachal Pradesh under rainy season. It is a Medium, shriveled and sugary dent, tolerance to MLB, TLB, ESR.

Priya Sweet Corn : Recommended for cultivation in Peninsular, and Andhra Pradesh. Medium maturing, yellow dent seeded, tolerance to diseases like *Turcicum* leaf blight, Charcoal rot, Late wilt and other stalk rots

Description of Plant

Maize stems resemble bamboo canes and the internodes can reach 44.5 cm. It has a distinct growth form; the lower leaves being like broad flags, generally 50-100 cm long and 5-10 cm wide ;the stems are erect, conventionally 2-3 m in height, with many nodes, casting off flag-leaves at every node. Under these leaves and close to the stem grow the ears. They grow about 3 mm a day. The ears are female inflorescences, tightly covered over by several layers of leaves. The silks are elongated stigmas that look like tufts of thread, at first green, and later red or yellow. The apex of the stem ends in the tassel, an inflorescence of male flowers. When the tassel is mature and conditions are suitably warm and dry, anthers on the tassel dehisce and release pollen. Maize pollen is anemophilous (dispersed by wind) and because of its large settling velocity most pollen falls within a few meters of the tassel. Each silk may become pollinated to produce one kernel of maize. Young ears can be consumed raw, with the cob and silk, but as the plant matures, the cob becomes tougher and the silk dries to inedibility. The kernel of maize has a pericarp of the fruit fused with the seed coat, entire kernel is often referred to as the seed. The cob is close to a multiple fruit in structure, except that the individual fruits (the kernels) never fuse into a single mass. The grains are adhere in regular rows round a white pithy substance, which forms the ear. An ear contains from 200 to 400 kernels, and is from 10-25 cm in length. Seed colour varies *viz.*, blackish, bluish-gray, purple, green, red, white and yellow.

Immature maize shoots accumulate a powerful antibiotic substance, DIMBOA (2,4-dihydroxy-7-methoxy-1,4-benzoxazin-3-one). DIMBOA is a member of a group of hydroxamic acids (also known as benzoxazinoids) that serve as a natural defense against a wide range of pests including insects, pathogenic fungi and bacteria. DIMBOA is also found in related grasses, particularly wheat. A maize mutant (bx) lacking DIMBOA is highly susceptible to be attacked by aphids and fungi. DIMBOA is also responsible for the relative resistance of immature maize to the European corn borer (family Crambidae). As maize matures, DIMBOA levels and resistance to the corn borer decline. Maize is a facultative long-night plant and flowers in a certain number of growing degree days > 10°C in the environment to which it is adapted (Brown, 2009). The magnitude of the influence that long nights have on the number of days that must pass before maize flowers is genetically prescribed and regulated by the phytochrome system.

Genetics

Many forms of maize are used for food, sometimes classified as various subspecies related to the amount of starch each had:

- Flour corn - *Zea mays* var. amylac*ea*
- Popcorn - *Zea mays* var. everta
- Dent corn - *Zea mays* var. indentata
- Flint corn - *Zea mays* var. indurata
- Sweet corn - *Zea mays* var. saccharata and *Zea mays* var. rugosa
- Waxy corn - *Zea mays* var. ceratina
- Amylomaize - *Zea mays*
- Pod corn - *Zea mays* var. tunicata Larrañaga ex A. St. Hil.
- Striped maize - *Zea mays* var. japonica

Maize has 10 chromosomes (n=10). The combined length of the chromosomes is 1500 cM. Some of the maize chromosomes have what are known as 'chromosomal knobs': highly repetitive heterochromatic domains that stain darkly. Individual knobs are polymorphic among strains of both maize and teosinte. Barbara McClintock used these knob markers to validate her transposon theory of "jumping genes", for which she won the 1983 Nobel Prize in Physiology or Medicine. Maize is still an important model organism for genetics and developmental biology today (Brown, 2009). The Maize Genetics Cooperation Stock Center, funded by the USDA Agricultural Research Service and located in the Department of Crop Sciences at the University of Illinois at Urbana-Champaign, is a stock center of maize mutants. The total collection has nearly 80,000 samples. The bulk of the collection consists of several hundred named genes, plus additional gene combinations and other heritable variants. There are about 1000 chromosomal aberrations (e.g., translocations and inversions) and stocks with abnormal chromosome numbers (e.g., tetraploids).

In 2005, the U.S. National Science Foundation (NSF), Department of Agriculture (USDA) and the Department of Energy (DOE) formed a consortium to sequence the B73 maize genome. The resulting DNA sequence data was deposited into GenBank, a public repository for genome-sequence data. Sequences and genome annotations have also been made available throughout the project's lifetime at the project's official site, MaizeSequence.org. Primary sequencing of the maize genome was completed in 2008. On November 20, 2009, the consortium published results of its sequencing effort in Science. The genome, 85% of which is composed of transposons, was found to contain 32,540 genes. Much of the maize genome has been duplicated and reshuffled by helitrons a group of rolling circle transposons (Feschotte and Pritham, 2009).

There are three known genetic mutations responsible for the various types of sweet corn. Early varieties, such as those used by Native Americans, were the result of the mutant *su* (sugary) allele. They contain about 5-10% sugar by weight. Super sweet corn are varieties of sweet corn which produce higher than normal levels of sugar (Anonymous , 2003). Super sweet corn was developed by University of Illinois at Urbana-Champaign professor John Laughnan. He was investigating two specific genes in sweet corn, one of which, the *sh2* gene, caused the corn to shrivel when dry. After further investigation Laughnan discovered that the endosperm of *sh2* sweet corn kernels store less amounts of starch and from 4 to 10 times more sugar than normal *su* sweet corn. The popularly of super sweet corn rose due to its long shelf life and large sugar content when compared to conventional sweet corn. This has allowed the long-distance shipping of sweet corn and has enabled manufacturers to can sweet corn without adding extra sugar or salt. The third gene mutation to be discovered is the *se* or 'sugary enhanced' allele, responsible for so-called 'Everlasting Heritage' varieties, such as Kandy Korn. Varieties with the *se* alleles have a longer storage life and contain 12-20% sugar compared to *su* varictics. All of the alleles responsible for sweet corn are recessive, so it must be isolated from any field corn varieties that release pollen at the same time; the endosperm develops from genes from both parents, and heterozygous kernels will be tough and starchy. The *se* and *su* alleles do not need to be isolated from each other. However, super sweet varieties containing the *sh2* allele must be grown in isolation from other varieties to avoid cross-pollination and resulting starchiness, either in space minimum quarantine distances from 100 to 400 feet or 30 to 120 m) or in time (i.e., the super sweet corn does not pollinate at the same time as other corn in nearby fields). Modern breeding methods have also introduced varieties incorporating multiple gene types:

- *sy* (for synergistic) adds the *sh2* gene to some kernels (usually 25%) on the same cob as a *se* base (either homozygous or heterozygous)

- **augmented** *sh2* adds the *se* and *su* gene to a *sh2* parent.

Climate and Soil

Sweet corn is a warm season crop. Its seed germinates best at the temperature range from 20°C - 25°C. The weather should not be too cold; otherwise, it will be caused slow germination, poor growth, bearing few and small ears and even died when damaged by the frost. It can be grown in different kinds of soil conditions, but best at deep fertile and good drainage loam or sandy loam soil. Crop rotation with rice-sweet potato, sugarcane, soy bean and vegetables has been found useful to avoid the infestation of disease, pests and nematodes.

Seed Rate

Fresh seed should be used each year as seed quality (vigour) gets reduced substantially within a year, especially in the case of super sweets. Recommended plant population for optimum yield is 45000-60000 plants per hectare with row spacing of 75-100 cm and intra row spacing of 15-30 cm. This will require 10-11 kg of *su* or *se* corn seed per ha. Super sweet corn seed contains a large quantity of sugar, which causes the seed to be crinkled and smaller. Therefore, super sweet sweet corn varieties have a higher seed count (500 seed /kg) than other sweet corn types (325 seed /kg) and will require less seed (5 -6 kg/ha). Seed treatment with Imidachloprid 70WS @5g/kg seed avoids the infestation of insect pests up to 30 DAS and fungicide treatment helps to prevent the attack of damping-off fungi.

Method of Sowing: Two seeds per hill are dibbled manually or mechanically one third from the top on the side of the ridge. Sowing on ridges conserve moisture and give protection from water logging in the initial stages to which sweet corn is highly susceptible. Sowing depth should be 3-4 cm for all cultivars except for super sweets, which are planted 2.5 cm deep. Plants are thinned to one plant per hill 10-12 days after emergence. Normal (*su*) and sugary enhanced (*se*) sweet corn should not be planted earlier than 7 to 10 days before the average date of the last severe frost. At least 3 rows of each variety should be sown at each planting to assure good pollination. Seeds of *sh-2* varieties are less vigorous than other sweet corn types, which can lead to reduced and uneven crop stand.

Manures and Fertilizers

Application of FYM or well rotten cow dung @ 20-25 t/ha before a month of sowing should be given. If the report of soil test is available in that case fertilizer should be applied as per the recommendation. In general, N, P and K @ 120:60:80kg/ ha is applied. Full dose of phosphorus and potassium, while one third dose of nitrogen should be applied at the time of sowing. Rest of the nitrogen should be applied at knee high stage and at the time of tasseling.

Weed Management

Crop rotation is important for limiting the build up of troublesome weeds. Shallow cultivation should be used in concert with chemicals for weed control. The crop should remain weed free during the early stages of plant growth; otherwise, yields might be substantially reduced. Several herbicides are available for weed control. Efficient weed control is achieved for 30-35 days through a spray of the herbicide Atrazine @ 1 kg ai/ha 1-2 days after the initial irrigation. If double cropping is practiced, it must be borne in mind that the crop is sensitive to herbicide carryover, particularly atrazine or atrazine containing products. Therefore, selection of herbicides should be made accordingly.

Water Management: Depending on the soil type, the number of irrigations varies from 4-5 irrigations in heavy soils and 7-8 irrigations in light soils. A grower should be prepared to irrigate at least 2.5-4.0 cm a week in order to produce high quality sweet corn. The most critical time period to have adequate moisture is during tasseling and silking.

Inter-cropping: Sweet corn being a highly remunerative crop can successfully be cultivated in peri urban agriculture. However, instead of cultivating sweet corn as sole crop it may be intercropped with other highly remunerative crops like marigold, tuberose, gladiolus, spices, vegetable pea etc. This provides an additional income to the farmers from unit area.

Harvesting of Cobs

In sweet corn about 45 days after sowing (DAS), tassel emerges and 2-3 days later the silks emergence takes place. Sweet corn is an abundant pollen producer and seed set is usually very high. As an ear approaches maturity, sugar changes to starch, the hull becomes tougher, and the kernels pass through stages called pre-milk, milk, early dough, and dough. At field temperatures of 16°C, an ear may remain in prime condition for as long as 5 days, while at 30°C it might remain in prime condition for only 1-2 days. Corn will be ready for harvest approximately 18-22 days after completion of pollination (indicated by drying of silk). As the field nears maturity, a few ears should be examined daily to determine the time for the first picking. Corn is ready for harvest when the ear is full size for the variety, has a tight husk, and has somewhat dried silks. The kernels are fully developed and exude a milky liquid when punctured. Delaying the harvest will progressively reduce the sugar content in the kernels. The ear husks are still green at this stage and the kernels remain lustrous at milky stage. Keep harvested ears cool and in shaded areas. Ears can be eaten raw, as a dessert or ears can even be steam boiled, salted and eaten, which is more relishing. Under good management, two ears are borne per plant especially in the rabi (winter) season. Staggered planting of sweet corn at intervals of 7-10 days will enable farmers to harvest sweet corn ears at intervals for continuous supply to the market and thereby get continuous and higher returns.

Post-harvest Handling

Sweet corn has a high respiration rate, it produces heat which can cause ears in bulk trailer loads to heat up considerably during delays between picking and precooling. The longer the delay, the greater the heating, conversion of sugar to starch, and subsequent quality loss. Sweet corn must be moved quickly from the field to packing sheds, where it should be rapidly sorted, packed, and cooled. Sweet corn is generally packed in wire bound wooden crates, which can hold from 4 to 6 dozen ears, depending on the size of the crate or ears. The ears are graded according to the raw ear size and packed in polythene bags and sold in the market. Under room temperatures sweet corn will lose 50% or more of its sugars in 24 hours. Sweet corn should be precooled to as close to 0°C as possible, although it is rarely cooled below 4°C in commercial practice.

Several methods of cooling sweet corn are available:

Hydro-cooling

It consists of precooling by either showering the corn or immersing it in cold water. It is the most popular method of precooling. Immersing the corn in cold water is much more efficient than showering the corn.

Packaging

Packaging of the sweet corn with crushed ice is quite popular among the traders. In this method, 7-10 kg of crushed ice is distributed throughout the container (box or crate) during the packaging process. Typically about 500 g of ice for 2 kg of sweet corn is sufficient.

Cold Storage

To maintain best quality, sweet corn is placed in cold storage immediately after precooling. Storage can be in a refrigerated truck or in a room. Temperature is maintained as close to 0°C as possible without freezing the corn, and relative humidity of the air in the cold room at 95 per cent or higher to keep the corn fresh. The corn is to be taken out as soon as possible since more than a few days in cold storage will decrease quality.

Sweet corn should be kept cold in transit otherwise it losses its original quality and sweetness. Developing the cool chain system in transpotation is most desired.

Corn Colour

Sweet corn comes in three colors: yellow, white and bicolour (yellow and white). Cross-pollination of yellow kernel varieties with white kernel varieties will result in production of bicolour corn. Also, if a bicolour is cross pollinated with a yellow variety, kernel colour will be predominantly yellow. Although there are geographical preferences for certain kernel colors, there is no relationship between colour and sweetness of sweet corn.

Yields

In general, sweet corn yields about 66000 ears per ha. High density planting gives higher number of harvested ears.

Pollination Management

Sweet corn is wind pollinated and must be planted in blocks. Sweet corn of any type downwind from popcorn or field corn develops starchy kernels. Ears of supersweet *sh-2* corn pollinated by a normal (*su*) sweet corn cultivar or a cultivar with the *se* gene will develop hard, ugly and starchy dent kernels. To avoid this cross pollination, different kernel types must be separated by at least 250 m or staggered planted so that they flower at least 14 days apart.

Precautions with Modified Endosperm Sweet Corn to Avoid Xenia Isolation of 'sweet corn' cultivar plantings of different genetic types is necessary to prevent cross-pollination. Xenia is the immediate effect of foreign pollen on a variety; on sweet corn (*su*), it will produce a starchy kernel. Isolation can be obtained by planting at a different time, planting cultivars of different maturities, or providing barriers and border rows. All of these methods will reduce the isolation distances necessary. On a practical basis, commercial growers should provide at least 50 m separation, plant upwind of normal field corn, and use four or more border rows.

Major Pests

Important pest of sweet corn are listed as under:

- Corn earworm (*Helicoverpa zea*)
- Fall armyworm (*Spodoptera frugiperda*)
- Common armyworm (*Pseudaletia unipuncta*)
- Stalk borer (*Papaipema nebris*)
- Corn leaf aphid (*Rhopalosiphum maidis*)
- European corn borer (*Ostrinia nubilalis*) (ECB)
- Corn silkfly (*Euxesta stigmatis*)
- Lesser cornstalk borer (*Elasmopalpus lignosellus*)
- Corn delphacid (*Peregrinus maidis*)
- Western corn rootworm (*Diabrotica virgifera virgifera* LeConte)
- Southwestern corn borer (*Diatraea grandiosella*)
- Maize weevil (*Sitophilus zeamais*)

Major Diseases

Downey mildew (*Peronosolerospora sacchari*):

Seed should invariably be treated with metalaxyl @ 2.5g/kg seed and need based foliar sprays of systemic fungicide such as metalaxyl @ 2-2.5g/l is recommended at first appearance of disease symptoms.

***Helminthosporium* leaf spot** (*H.maydis* and *H. turcicum*): Spray with 80 % Maneb or 65 % Zineb.

Rust (*Puccinis sorghi*): This rust attacks in the maize growing areas with subtropical temperate and high land environment of Jammu & Kashmir, Himachal Pradesh, Sikkim, Meghalaya, West Bengal, Punjab (Rabi), Haryana. It appears at the time of tasseling. The circular to elongate, golden brown to cinnamon brown pustules are visible over both leaf surfaces changing to brownish black at plant maturity.

Control Measures

i) Adopt disease resistant varieties.

ii) Spray of Mancozeb@ 2.0g /litre of water at first appearance of pustules.

- Prefer early maturing varieties.

Bacterial stripe and leaf bright (*Pseudomonas andropogonis* and *P. avenae*):

Control : Avoid to plant in hot and humid season.

Improvement

In sweet corn, the sugary gene prevents the normal conversion of sugar into starch during endosperm development, and the kernel accumulates a water-soluble polysaccharide called 'phytoglycogen.'As a result, the dry, sugary kernels are wrinkled and glassy. The higher content of water-soluble polysaccharide adds a texture quality factor in addition to sweetness. Today, the standard sugary corns are being modified with other endosperm genes and gene combinations that control sweetness to develop new cultivars. As a result, growers must consider genetic type while making selections for planting. The genetic type is not readily identifiable by cultivar name alone. At least 13 endosperm mutants, in combination with sugary, have been studied for improving sweet corn. Except for sugary, the genes used in breeding act differently to produce the taste and texture deemed desirable for sweet corn. Major modifier genes of kernel sweetness are shrunken-2 (*sh2*) and sugary enhancer (*se*). In partial modifications, the sugary (*su*) kernels are modified by the segregation of major modifier genes such that about 25 per cent of the kernels are double-mutant endosperm types possessing the enhanced benefits of the modifier. The addition of the sugary enhancer (*se*) gene along with one of the major modifier genes (e.g., *sh2*) will further modify some of the sugary kernels to about 44 per cent doublemutant endosperm types (Kumar *et al.*, 2012).

References

Anonymous (2003). Maize. The Oxford English Dictionary, online Edition, 1989.

Brown, David (2009). Scientists have highopes for corn genome. Washington Pos http://www.washingtonpost.com/wp-dyn/content/article/2009/11//AR2009/11/19/AR2009111903190html.

Feschotte, C. and Pritham, E. (2009). A cornucopia of helitrons shapes the maize genome. Proceedings of the National Academy of Science, 106(470: 19747-19748.

Karl, J. (2010). Tallest Maize Data.Maize. The Oxford English Dictionary, online edition. December 2007. Accessed December 16, 2007.

Kumar, B., Karjagi, C.G. , Jat, S.L., Parihar, C.M., Yathish, K.R., Vishal Singh, V., Hooda, K.S. Dass, A.K., Ganapati Mukri, J.C. Sekhar, J.C, Kumar, R. and Sai Kumar, R. (2012). Maize biology: An Introduction, Directorate of Maize Research, (ICAR), Technical Bulletin, 2012/2, pp. 32.

8

BROCCOLI

Broccoli is grown for its large, edible, young inflorescence (heads). Like cauliflower, broccoli is consumed as cooked vegetable as well as soup, salad and as mixed vegetable. Other use include raw broccoli for dipping or adding to a salad along with macaroni and cheese. Broccoli has become popular as a quick-frozen vegetable, particularly in the United States and Europe due to its high nutritive value. It is also processed in dried mixtures vegetables soup. In broccoli, a kind of head consisting of green buds and thick fleshy flower bud stalks are formed. The terminal head is rather loose and green in colour, and the flower stalks are longer than those of cauliflower. The internodes are long and the plant produces axillary flowering shoots in addition to the terminal inflorescence. The colour of the flower buds, forming the edible portion varies from white to green, depending on cultivar. The inflorescence is made of fully differentiated flower-buds, although, it is considered to be a good quality fruit if these reach the open stage. The sprouts in the leaf axil develop especially after removal of the terminal head.

In broccoli, both the terminal head and sprouts with bud clusters are consumed as food. Nutritional composition of broccoli resembles that of cauliflowers. Glucosinolates (thioglucosides) are sulphur containing compounds present in the broccoli. The glucosinolates are anions and occur in plants mostly as potassium salts. Breakdown products of glucosinolates possess important sensory properties such as odour and flavour and they may induce physiological changes in humans, including carcinogenesis inhibition. The glucosinolate content of purple headed broccoli has been found in the range of 72-212 mg/100g. The National Research Council Committee on Diet, Nutrition and Cancer has recommended increased consumption of vegetables especially broccoli to decrease the incidence of cancer. The American Cancer Society suggested inclusion of *Brassica* vegetables in the diet to reduce the risk of developing cancer. Study indicates that tumorogenesis is inhibited by carcinogen metabolizing system induce by compounds in *Brassica* plants. Sulphur containing phytochemicals of two different kinds are present in broccoli and other *Brassica* vegetables: glucosinolate and S. metholcysteine sulfoxide. Numerous findings indicate that the hydrolytic products of some of the glucosinolates have anticarcinogenic activity. *Brassica* vegetables contain high concentrations of carotenoids, which are believed to be chemopreventive and associated with a decreased risk for various human cancer in epidemiological studies.

Person consuming greater quantity of broccoli had the highest concentration of plasma carotenoid. Hence, it is suggested that plasma carotenes may be useful biomarkers of carotenoid-rich food intake. Study indicates that increased consumption of broccoli, particularly among women, appeared to improve survival in lung cancer. Broccoli also plays an important role in reducing the serum cholesterol. Broccoli was reported to contain about 3.5 g per kg of D. glucoric acid. Purified diets containing calcium D. glucorate or potassium hydrogen D. glucarate lowers the serum levels of cholesterol. Nutritional composition of the broccoli is given in table (8.1).

Table (8.1): Nutritive value of broccoli (per 100 g of edible portion).

Constituents	Amounts
Moisture	89 per cent
Protein	3.6 g
Fat	0.30 g
Carbohydrates	5.9 g
Energy	32 Kcal
Calcium	103 mg
Phosphorus	78 mg
Iron	1.1 mg
Sodium	15 mg
Potassium	382 mg
Vitamin-A	2500 I.U.
Thiamine	0.10 mg
Riboflavin	0.23 mg
Niacin	0.9 mg
Ascorbic acid	113 mg

Glucosinolates

Glucosinolates are a class of about 100 naturally occurring thioglucosides that are characteristic of *Brassica* and related families. The presence of some glucosinolates in agricultural crop plants, such as oilseed rape (*Brassica napus*) and *Brassica* vegetables, is undesirable because of the toxicological effects of their breakdown products. Such breakdown products include nitriles, isothiocyanates, thiocyanates, epithionitriles and vinyl oxazolidinethiones. Some glucosinolates, especially those in broccoli, have anticarcinogenic properties and are being studied for their potential therapeutic use. The enzymes catalysing the hydrolysis of glucosinolates are known as myrosinases.

Origin and Distribution

The name broccoli refers to the young shoots that develop on same species of the genus *Brassica* (brocco is Italian for shoot). The sprouting broccoli (*B. oleracea* var. italica) is believed to be the ancestor of the present quick growing cauliflowers. Broccoli with multiple green, purple or even white flower-heads (sprouting broccoli) became popular in northern Europe in the 18th century. Broccoli with one main green head (Calabrase) was introduced in the United States by Italian immigrants during the early 20th Century. From thc United States it has spread to northern Europe, Japan and other regions in the last century. According to FAO (2008), world estimate of broccoli production is around 20.0 million tones. China is largest broccoli producer (8.0 million tones, whereas India is second largest broccoli producing country in the world with annual estimate of 5.0 million tones. Other important broccoli producing countries in the world are United State (1.0 million ton), Spain and Italy (400, 000tonnes) and France and Mexico (300, 000 tonnes).

Botanical Description

Broccoli (*Brassica oleracea* L. var. italica Plenck) is a member of the Brassicaceae or cole crop family. Its chromosome No. (2n) is = 18.

Taxonomic position of broccoli is given as under:

Kingdom	Plantae
Subkingdom	Tracheobionta
Superdivision	Spermatophyta
Division	Magnoliophyta
Class	Magnoliopsida
Subclass	Dilleniidae
Order	Capparales
Family	Brassicaceae
Genus	*Brassica* L.
Species	*Brassica oleracea* L.
Variety	*Brassica oleracea* L. var. botrytis L. – broccoli

Morphological Description of Plant: Erect, glabrous, annual or biennial herb up to 80 cm tall at the mature vegetative stage, up to 150 cm when flowering, with unbranched stem thickening upwards; root system strongly branched. Leaves alternate but closely arranged and more or less erect, forming a rosette surrounding the young inflorescence especially in cauliflower, usually simple; stipules absent; leaves almost sessile, but often shortly petiolate in broccoli; blade ovate to oblong, up to 80 cm × 40 cm, undulate or irregularly incised to almost entire, coated with a layer of wax, blue-green with whitish veins. Inflorescence a terminal paniculate raceme up to 70 cm long, when young (curd or head) composed of more or less densely arranged branching partial inflorescences and fleshy peduncles, in cauliflower up to 30 cm in diameter, very solid and globular to rather loose and flat, white to green, in broccoli less densely arranged with longer peduncles, green to purple. Flowers bisexual, regular, 4-merous; pedicel up to 2 cm long, ascending; sepals oblong, c. 1 cm long, erect; petals obovate, 1.5–2.5 cm long, clawed, pale to bright yellow or whitish; stamens 6; ovary superior, cylindrical, 2-celled, stigma globose. Fruit a linear silique 5–10 cm × c. 5 mm, with a tapering beak 5–15 mm long, dehiscent, up to 30-seeded. Seeds globose, 2–4 mm in diameter, finely reticulate, brown. Seedling with epigeal germination, with a taproot and lateral roots; hypocotyl 3–5 cm long, epicotyl absent; cotyledons with petiole 1–2 cm long, blade cordate, 1–1.5 cm long, cuneate at base, notched at apex. In broccoli the single headed 'calabrese' type with a compact dome-shaped head with small beads (flower buds) is grown worldwide, under conditions comparable to the summer types of temperate cauliflower. The buds are preferably dark green (Tjeertes, 2004)

Varieties

1. Pusa Broccoli (KTS-1)

This variety was developed by recurrent selection from an exotic material. It is a medium-tall variety (65-70 cm), dark green waxy foliage with slightly wavy margins, heads are solid, main head size and weight about 6.0-15.4 cm and 350-450 gm respectively, matures in 90-105 days after transplanting. Average yield potential is 125q/ha.

2. Palam Samridhi

This variety was developed from an exotic material by mass selection. Himachal Pradesh Agricultural University, Palampur. It bears green terminal heads each weighing approximately 300-400gm, which get ready for market in about 85-90 days after transplanting. Its average yield is 150 to 200 q/ha. It is rich in vitamins and minerals (Ca, Fe and P). Nutritional composition of this variety is given in table (8.2).

Table (8.2): Nutritional composition of broccoli var. Palam Samridhi (per 100 g of edible portion).

Particulars	Quantity
Ascorbic acid	101.51 mg
B carotene	3.50 ppm
Protein	3.63%
Calcium	54.48 mg
Iron	0.92 mg
Phosphorus	61.20 mg
Sulphur	19.70 mg
Potassium	123.47 mg
Magnesium	18.04 mg

Source: Kalia (2002).

3.Palam Kanchan

It is a heading broccoli with long, broad, bluish green upright leaves having prominent white midrib and veins. The head is large in size, compact, attractive yellowish green in colour. It is rich in vitamin A. The heads attain marketable size in about 140-145 days after transplanting. Average yield potential is 250-275q/ha. Nutritional composition Palam Kanchan variety of broccoli is given in table (8.3).

Table (8.3): Nutritional composition of broccoli var. Palam Kanchan(per 100 g of edible portion)

Particulars	Quantity
Protein	3.18%
Dry matter	9.07%
Ascorbic acid	62.83 mg
Calcium	54.48 mg
Iron	1.40 mg
Phosphorus	52.27mg
Potassium	326.18mg
Copper	3.65mg
Manganese	0.32mg
Zinc	0.34 mg
Chlorophyll 'a'	0.11
Chlorophyll 'b'	0.10
Total chlorophyll	0.21

Source : Kalia and Singh (2004).

4. Palam Haritika

It is a sprouting broccoli with dark green upright leaves having purple reddish tinge. The heads attain marketable stage in about 145-150 days. This variety has yield potential of 230-250q/ha.

5. Palam Vichitra

It is a heading broccoli, medium-sized, open dark green leaves. Purple tinge is prominent on stem at seedling stage and on leaf margins at full growing stage. The head is purple and compact. The variety is rich in nutrients especially vitamins and minerals. It takes about 115-120 days to first harvest with average yield of 200-225q/ha.

6. Calabrese: is one of the most popular cultivars in the United States. The cultivar De Cicco is light green in colour and produces abundant lateral shoots, it is the popular winter variety grown in Texas. Waltham-29 is another popular and widely adapted variety. It is late maturing with uniform maturity. Other varieties which are popular in India are described as under:

7. Lucky: It is also another important variety matures in approximately 60-65 days after transplanting. Head weight is 600-800 g and gives compact curd with fine heads.

8. Fiesta: It matures in 65-70 days after transplanting. Head weight ranges from 800 g to 1.0 kg. It is very compact and has fine heads and very high yielding.

Cultural Practices:

Broccoli is a cool season crop. For good quality heads, average day temperature of 15-20°C is ideal. Broccoli comes fairly well during winter season in plains of northern India. In Himachal Pradesh, its seeds are sown depending upon the attitude of the zones. In lower hills (zone-I), it is sown in September-October; in mid hills (zone-II) August-September and in high hills (Zone-III and IV), sowing is completed between March and April. Broccoli can be grown on a wide range of well drained soils. Since broccoli has a shallow root system, a constant supply of moisture and nutrient is required for succulent growth. A well-drained, medium heavy soil with a high organic content is considered ideal for broccoli growing. Optimum pH for broccoli is 5.5 to 6.5. For raising the crop of broccoli, nursery is prepared like cauliflower. One gram seeds of broccoli contain approximately 375 seeds. About 450-500 g seeds are required for one hectare planting. Irrigation is provided for good seed germination. The seeds germinate best at 23°C accordingly suitable time for seed sowing in particular climate may be adjusted. Transplanting to the field should be done when seedlings are 4-5 weeks old and has 7-9 true leaves. Soil preparation includes deep digging, mixing with FYM @ 20 tonnes per hectare at least one month before transplanting. Rate of NPK fertilizer depends upon soil fertility, however, 150 kg nitrogen, 80 kg P_2O_5 and 60 kg K_2O should be applied. Broccoli is very much nitrogen responsive crop. In field trial conducted at IIVR, Varanasi, highest yield was obtained with the application N @ 200 kg per hectare (Table 8.4).

Table (8.4): Response of nitrogen fertilization on vegetative growth and yield of broccoli

N level (kg/ha)		Plant height (cm)
50		55.66
100		59.33
150		68.00
200		58.33
SE +-		2.73
CD at 5%		6.295
Head size (cm^2)	**Head weight (g)**	**Yield (q/ha)**
252.33	356.66	47.72
309.00	396.67	62.50
367.33	466.67	85.73
209.5	273.33	88.25
25.58	29.24	10.50
58.987	67.247	24.21

Healthy saplings of broccoli are transplanted in well prepared field at the spacing of 60x60 cm or 60x45 cm. The size of head can be controlled by planting distance i.e. for larger heads production a wider spacing is given and *vice-versa* for suitable heads. Irrigation is often given after transplanting to establish the plants in the field. Moisture stress during early stage of growth can reduce vegetative growth and quality of the heads. Adequate moisture is especially necessary during head initiation to produce large quality heads.

Application of Micronutrients

Application of different micro elements is required in broccoli fields as crop is very sensitive to physiological disorders.

Boron

Deficiency may cause hollow stem, stem discolouration, cracking, leaf rolling, deformed buds as well as browning of broccoli heads. Application of boron may be made on the basis of soil analysis and if the soil test indicates low levels of boron apply 2.5 – 3.0 kg of boron/ha before planting. Boron should never be banded, however it can be foliar applied.

Molybdenum

Molybdenum deficiency causes whiptail in broccoli. Whiptail results in a deformed growing point causing no head to develop, as well as leaf blades consisting mostly of midribs. Molybdenum may be supplied as a as a foliar spray to transplants before field setting, in the transplant water or as a foliar spray. Apply 30-45 g of Sodium Molybdate per 100 L of transplant water or 280 g of Sodium molybdate as a spray in 1100 L of water per hectare. (Sodium molybdate is approximately 40 % molybdenum).

Manganese

Deficiencies may occur on sandy, over limed soils. Manganese deficiency causes yellowing between veins of young leaves. Leaves gradually turn pale-green with darker green next to the veins, petioles and stems. Foliar sprays of manganese sulfate may be necessary to correct a deficiency.

Weed Management

For realizing the good yield potential, field should be kept free from seasonal weeds by removing the weeds and hoeing. Herbicides have been effective in managing the weeds in broccoli. Good control of weeds was obtained with Tok E 25 (nitrofen) at 3 lb/ vergee (4/9 acre) and Brasoran (aziprotryne) at 1 lb 10 oz and 3 lb 4 oz/ vergee without damage to the leaves of the broccoli plants. At recommended rates, Semeron (desmetryne) caused foliage injury and Ramrod (propachlor) checked growth. Treflan (trifluralin) failed to control groundsel (*Senecio vulgaris*) and shepherd's purse (*Capsella bursa-pastoris*), possibly because it was incorporated too deeply (Anonymous, 1970). Trifluralin in combination with chlorthal, irrespective of the time and method of application gave excellent weed control and high broccoli yields (Chitapong *et al.,*1983). In broccoli crops, by 8 weeks after application of oxyfluorfen resulted in 93-100% control of redroot pigweed (*Amaranthus retroflexus*), common lambsquarters (*Chenopodium album*), common purslane (*Portulaca oleracea*), large crabgrass (*Digitaria sanguinalis*) and goosegrass (*Eleusine indica*). Earthing is required to prevent the collapse of plants from wind after irrigation. In broccoli, Devrinol at 2 lb a.i./acre and Lasso at 2-3 lb a.i./acre provided fair to excellent control of *Cyperus* spp. and *Eclipta alba*. *Mollugo verticillata* was controlled by all treatments except low rates of Prowl or Amex. All herbicides controlled *Brachiaria* spp. and *Echinochloa crusgalli*, *Portulaca oleracea* and M. verticillata. Amaranthus hybridus was controlled by all herbicides except Devrinol. Herbicides did not affect broccoli vigour or yield (Porter, 1983).

Harvesting and Yield

Broccoli takes considerable time to mature, generally harvesting starts from 60-65 days after transplanting. Broccoli is harvested by hand while the clusters are still compact and before the individual flowers begin to mature and turn yellow. In broccoli unlike cauliflower, lateral shoots develop in the axil of the leaves into marketable heads 3-5 cm in diameter, producing a continuous harvest for several weeks. The average yield of broccoli varies 150-200 quintals per hectare.

Post -Harvest Management

Broccoli is highly perishable and should be cooled immediately following harvest. The approximate storage life of broccoli under ideal conditions is approximately 10-14 days. In addition to icing, hydro cooling and forced-air cooling can also be used, but good temperature management must be maintained following cooling. If held at 0^0C and near 100% relative humidity, broccoli can be stored for 3 to 4 weeks. Exposure to ethylene (from apples, other ethylene producing fruit or engine exhaust) will accelerate the yellowing of flower buds and reduce storage life and should be avoided. Controlled atmosphere (CA) storage atmospheres of 1-2% oxygen with 5-10% carbon dioxide at temperatures of 0-5^0C will benefit broccoli and can double storage life, especially when held above optimum temperatures. Crushed ice or slurry ice is usually added to packed cartons to keep produce fresh during shipping, especially when adequate refrigeration is not available. Storage of broccoli heads in low-density polyethylenes (LDPE) without ethylene absorber resulted in the overall longest shelf-life. Broccoli stored in polyvinyl chloride (PVC) deteriorated faster than broccoli packaged in the other materials. The influence of packaging material was greater at the higher temperature (Jacobsson *et al.*, 2004). Lucera *et al.* (2011) observed that , micro-perforated film that had the lowest OTR value showed the best performance in prolonging of the product shelf-life if compared to either control samples and fresh-cut broccoli packaged in the non-perforated film. The results highlighted that an approximately 50% shelf-life increase of fresh-cut florets broccoli compared to whole broccoli, and of about 30% respect to the unpackaged fresh-cut produce.

Seed Production

Since broccoli is insect pollinated, the crop has to be isolated from most of the other common *Brassicas* that may be in flower. Pollinating insects do get around (they can travel and move pollen around over many hundreds, perhaps thousands of feet) and if other members of the broccoli species are in flower at the same time, they will be crossed. Cabbage, cauliflower, kohlrabi, some kinds of kale, Brussels sprouts are some members of the *Brassica oleracea* species. Once broccoli starts to flower, it bolts quickly and attracts many kinds of insect pollinators. We get an equal number of bumble bee and honey bees working the flowers as well as flies and cabbage butterflies. Rarely, broccoli can self-pollinate but mostly it doesn't. Sometimes pollination is hit or miss and seed production can be erratic. Pollen will generally have to be transferred among several plants. This self-incompatibility may help to facilitate chance hybrids that show greater vigor.

Seed Yield

Average of 800 -1 200 kg/ha, depending on variety.

Major Diseases

1. Black Rot

It is caused by *Xanthomonas campestris.* Symptoms appear as irregularly shaped dull yellow areas along leaf margins which expand to leaf midrib and create a characteristic "V-shaped" lesion; lesions may coalesce along the leaf margin to give plant a scorched appearance. Pathogen is spread *via* infected seed or by splashing water and insect movement; disease emergence favored by warm and humid conditions.

Management

i) Use disease free seed or treat with hot water to remove fungus prior to planting.

ii) Remove and destroy crop debris after harvest or plow deeply into soil.

2. Damping-off

It is caused by *Rhizoctonia solani.* Symptoms appear as death of seedlings after germination; brown-red or black rot girdling stem; seedling may remain upright but stem is constricted and twisted (wirestem)

Management

i) Plant pathogen-free seed or transplants that have been produced in sterilized soil.

ii) Apply fungicide to seed to kill off any fungi.

iii) Shallow planting of seeds in nursery beds.

3. Black leg

It is caused by *Phoma lingam* fungus. Symptoms appear like damping-off of seedlings; round or irregularly shaped gray necrotic lesions on leaves with dark margins; lesions may be covered in pink masses in favorable weather conditions. Warm, wet conditions and higher temperatures result in the development of more visible symptoms

Management

i) Use disease free seed or treat with hot water to remove fungus prior to planting.

ii) Remove and destroy crop debris after harvest.

***4. Flea beetle* (Phyllotreta cruciferae)**

It causes small holes or pits in leaves that give the foliage a characteristic "shothole" appearance; young plants and seedlings are particularly susceptible; plant growth may be reduced; if damage is severe the plant may be killed; the pest responsible for the damage is a small (1.5–3.0 mm) dark colored beetle which jumps when disturbed; the beetles are often shiny in appearance.

Management

i) Apply neem based formulation.

ii) Application of insecticides containing carbaryl, and permethrin can provide adequate control of beetles for up to a week.

5. Diamond Back moth

Young larvae feed between upper and lower leaf surface and may be visible when they emerge from small holes on the underside of the leaf; older larvae leave large, irregularly shaped shotholes on leaf undersides, may leave the upper surface intact; larvae may drop from the plant on silk threads if the leaf is disturbed; larvae are small (1 cm/ 0.3 in) and tapered at both ends; larvae have to prolegs at the rear end that are arranged in a distinctive V-shape

Management

i) Larvae can be controlled organically by applications of *Bacillus thurengiensis* or Entrust; application of appropriate chemical insecticide is only necessary if larvae are damaging the growing tips of the plants.

Physiological Disorders

Broccoli crops show various non-parasitic disorders which cause tissues to die off. In some cases, these deviations have been shown to depend mainly on heritable characters; whereas in other cases external factors had a least marked effect.

1. Blindness

Plants do not form heads, but produce many shoots at ground level. This may be caused by insects or damage to the growing point early in the plants life.

2. Leafy Heads

Small leaves develop and protrude through the head during high temperatures, drastic fluctuations in day and night temperatures or improper nitrogen balance.

3. Buttoning

Nitrogen deficiency at the early growth stage causes 'buttoning', stunted plants with reduced leaf and head development.

4. Lack of heads

During periods of extremely warm weather (days over 30 ^{0}C and nights of 25 ^{0}C) broccoli can remain vegetative (does not head) since they do not receive enough cold for head formation. This can cause a problem in scheduling the marketing of even volumes of crop.

5. Hollow stem

Symptoms are internal only. This condition starts with gaps that develop in the tissues, and gradually they enlarge to create a hollow stem, sometimes from the bottom of the stalk into the head. Ordinarily, there is no discolouration of the surface of these openings at harvest, but both discolouration and tissue breakdown may develop soon after harvest. Avoid excessive nitrogen after head initiation. Dense plantings will maintain even growth rates and decrease the occurrence of hollow stem.

Genetic Resources and Improvement

Presently F_1 hybrids are occupying most of the areas hence, primitive land races and OP varieties need to be conserved. Germplasm collections of broccoli are available in several genebanks, particularly in Europe, the United States, India, China and Japan. In Europe a *Brassica* gene bank has been established in co-operation with private companies. A central electronic catalogue of the collections is available at the Centre for Genetic Resources (CGN), Wageningen, Netherlands. While genetic erosion reduces the existing variability, interspecific crosses within the *Brassicaceae* family widens the gene pool available to breeders (Tjeertes, 2004).

Biotechnology

To distinguish new broccoli cultivar 'HAI LV' from existing varieties and ensure the quality of commercial seeds, genetic purity assessment of 300 individuals of 'HAI LV' were made using Simple Sequence

Repeat (SSR) markers and a grow-out test (GOT). The DNA samples of female and male parents were analysed with 150 pairs of SSR primers and eight pairs presented polymorphism. Two pairs of primers, which produced suitable male and female-specific markers after PCR, were chosen for purity testing of the F_1 -hybrid seeds. The studies demonstrated that SSR markers and GOT could both identify sibs. The use of SSR markers is more convenient and faster than GOT, but the identified SSR markers could not differentiate other loci mutants or plants with epigenetic changes from the hybrids (Yu *et al.* 2013).

References

Chitapong, P., Ilnicki, R.D. and Kupatt, C. (1983). Effects of time and method of application of several herbicides for weed control in broccoli. Proceedings of the 37th annual meeting of the Northeastern Weed Science Society, 1983, 180-186.

Eaton, W., Coffey, D. L., Mullins, C. A. and Rhodes, G. N. Jr (1990). Weed control with oxyfluorfen in transplanted broccoli. Tennessee Farm and Home Science. No. 154, 17-21.

Jacobsson, A. Nielsen, T. and Sjöholm, I. (2004). Effects of type of packaging material on shelf-life of fresh broccoli by means of changes in weight, colour and texture. European Food Research and Technology, 218(2): 157-163.

Kalia, P. and Singh, Y. (2004). Palam Kanchan: a new nutritious broccoli. Indian Horticulture, 49(2):36-37.

Kalia, P. (2002). Palam Samridhi: a new variety of Sprouting broccoli. Indian Horticulture, 47(1):24-25.

Lucera, A., Costa, C., Mastromatteo, M., Conte, A. and Del Nobile, M. A. (2011). Fresh-cut broccoli florets shelf-life as affected by packaging film mass transport properties. J. Food Eng. 102(2):122-129.

Porter, W. C. (1983). Weed control in broccoli and cauliflower. Louisiana-Agriculture, 27(2):14-16.

Tjeertes, P. (2004). Brassica oleracea L. (cauliflower and broccoli) Protabase. Grubben, G.J.H. & Denton, O.A. (Eds.). PROTA (Plant Resources of Tropical Africa, Wageningen, Netherlands.

Yu, H.F., Wang, J.S., Zhao, Z.Q., Sheng, X.G. and Gu, H.H. (2013). DNA fingerprinting and genetic purity testing of a new broccoli hybrid using SSR markers. Seed Science and Technology, 41(3):464-468(5).

•••

9

ROMANESCO CAULIFLOWER

Romanesco cauliflower is a kind of cauli flower, having attractive ornamental curd shape like corals. Heads are very soft and sweet in taste. Its culinary preparation is similar to cauliflower and broccoli. Romanesco cauliflower is not commercially grown as cauliflower and broccoli. It is much popular in Belgium, where a number of its cultivars have been evolved. The Romanesco type is a traditional crop in the area around Rome and Naples, dating back at least to the 16th century. But its, area has been spread since the development of hybrid varieties like Veronica and the greater availability of open-pollinated varieties such as Minaret and Natalino. Nutritional composition of romanesco cauliflower is very similar to common cauliflower of *B. oleracea* L. var. botrytis group. Heads of romanesco contains fairly high amount of minerals and vitamin C. Nutritive value of romanesco cauliflower is given in table (9.1).

Table (9.1). Nutritive value of romanesco cauliflower (per 100 g of edible portion)

Constituents	Amount
Protein	2.7 g
Fat	0.2 g
Carbohydrates	5.2 g
Dietary Fibre	3.0 g
Energy	25 kcal
Calcium	0 mg
Phosphorus	56 mg
Sodium	0 mg
Potassium	350 mg

The volatile oil composition of three different romanesco cauliflower seeds was analysed. The major compounds were cyanides such as 4-(methylthio)butyl cyanide (61.3, 66.3 and 79.6%, respectively), 3-(methylthio) propyl cyanide (21.7, 21.6 and 10.7%), and isothiocyanates such as 4-(methylthio) butyl isothiocyanate (5.3, 4.0 and 6.7%) for the three oils. This study indicates presence of 12 glucosynolate in Romanesco seeds (Laurent Valette *et al.*, 2005).

Botanical Description

Romanesco cauliflower (*Brassica oleracea* L. var. botrytis) possesses chromosome number (2n) = 18. Genetic analyses have confirmed that Romanesco is just a mutation of cauliflower in which the branch apical meristems, the growing stems, develop in a way that produces angular, pyramidal curds in a regular mathematical pattern. There are also green varieties of cauliflower with the conventional round shape and a whitish yellow Italian type, di Jesi, that has spiraling florets. Romanesco is highly polymorphic and includes varieties which exhibit a headed phenotype (a large perinflorescence): the curd of cauliflower and romanesco (var. botrytis), and the spear of broccoli (var. italica). This headed phenotype results from highly interactive patterns of activity at the primary meristems. Differences in the morphology of curds and spear are accounted for by three quantitative variables: the rate of production of branch primordia and the flanks of the apical meristems (RPP); the number of branch primordia produce before the first formed begin producing their own branch primordial (the interaction interval, ITI); and the duration of the pre-inflorescence stage (before production of flower perimordia). Relatively stable interaction parameters (RPP and ITI) during curd development lead to the production of semi-spherical curds with a smooth surface in cauliflower and broccoli whereas in romanesco RPP and ITI increase throughout curd development, inducing pyramidal curd with an angular surface. A relatively long pre-inflorescence stage in cauliflower and romanesco results in the curd surface being composed largely of branch primordia, whereas in broccoli this stage is short and the spear surface is made of flower buds.

Varieties

Intensive breeding within the broccoli/cauliflower complex has eroded the boundary between the two crops. Anamolies in the listing of some of cultivars have also arisen in the EEC Common Catalogue for vegetables, particularly those that belong to the forms of Purple Cauliflower and Romanesco. Comparative ontogeny and morphological studies have been carried out to resolve the classification of these forms. Multivariate analysis are presented which support the view that cultivars grouped within the form purple cauliflower should be placed in the same botanical variety as broccoli/calabrase (var. botrytis). In Belgium, Amfora, Minaret, Navona and Shannon are commonly grown varieties of romanesco cauliflower. In India, Shannon cultivar matures in 90-100 days after transplanting and it has a very attractive shape.

Cultural practices

Cultural practices of romanesco is very similar to common cauliflower. For raising the crops, seedlings are grown by sowing the seeds in the raised nursery beds in month of October-November in plains of Northern India. Approximately 450 g seeds are required for one hectare planting. Seedlings become ready for transplanting in 3-4 weeks after .

Romanesco cauliflower needs cool climate for its head formation. Hence, it can be grown as a winter vegetable in the plains of north India. A good fertile soil with rich in organic matter is considered ideal for raising the crop. For realizing the good yield potential, apply FYM @10-12 t/ha and it should be mixed in the soil at least one month earlier from the date of planting. For a hectare field, 150 kg N, 80 kg P_2O_5 and 60 kg K_2O is recommended. Half of the nitrogen, full amount of phosphorus and potash should be given at the time of planting. The rest amount of nitrogen is applied in two split doses at the interval of 30 days after transplanting. In well prepared field, seedlings are transplanted at the spacing 60x60 cm or 60x45 cm depending upon the fertility status and preference of head size. Generally wider spacing is adopted for bigger size head and *vice versa.* Judicious application of water is required to maintain the optimum moisture in the field.

Harvesting and Yield

It matures in 90-100 days after transplanting. Harvesting should be done at the proper stage when heads are very soft. Care should be taken to avoid the delay in harvesting otherwise heads become loose and discolour. The average yield varies from 150-200 quintals per hectare.

Plant Protection

Bacterial Blight

First report of bacterial blight of romanesco cauliflower caused by *Pseudomonas syringae* pv. alisalensis was reported by Koike and Kammeijer (2006). Initial symptoms consisted of small (1 to 2 mm in diameter), angular, water-soaked flecks. These flecks developed into tan to gray, angular lesions measuring as much as 5 mm in diameter. Lesions were usually surrounded by chlorotic borders. Coalescing lesions caused the leaf to turn papery in texture and have a blighted appearance.

MANAGEMENT

1. Plant clean and disease-free transplants.
2. Rotation away from fields where the disease has recently occurred may reduce inoculum levels in soil or infected debris.
3. A change from sprinkler to furrow or drip irrigation may limit its spread. Foliar application of *Pseudomons solanacearum* @3.5

g/litre of water.

Genetic Resources and Improvement

Romanesco cauliflower attracts attention for its branching, spiraling pattern of growth. Scientists have traced the origin of Romanesco and related cauliflowers to selective breeding by 15th century Italian farmers. The vegetable's striking shape highlights the incredible amount of variation that can occur within a single plant species; studying such natural variation can provide clues as to how traits develop in plants. A head of cauliflower or broccoli is actually a collection of immature flower buds. Studying Romanesco and its relatives has provided clues to how flowers are formed. Researchers identified a gene that allowed them to induce *Arabidopsis* a weed not closely related to cauliflower, to produce cauliflower-like structures instead of flowers. They dubbed the gene responsible cauliflower.

References

Koike, S.T. and Kammeijer, K. (2006). First Report of bacterial blight of romanesco cauliflower (*Brassica oleracea* var. botrytis) caused by *Pseudomonas syringae* pv. alisalensis in California. University of California Cooperative Extension, Salinas 93901; C.T. Bull, USDA Agricultural Research Station, Salinas 93905; and D. O'Brien, Doug O'Brien Agricultural Consulting, Santa Cruz, CA 95062.

Laurent Valette, Xavier Fernandez, Sophie Poulain, Louisette Lizzani-Cuvelier and André-Michel Loiseau (2005). Chemical composition of the volatile extracts from *Brassica oleracea* L. var. botrytis 'Romanesco' cauliflower seeds. Flavour and Fragrance Journal, 21(1):107-110.

10

RED CABBAGE

Red cabbage has attractive dark red colour head, which is very compact. It is used for salad decoration and cooked vegetables like common cabbage. It is not very common in India and other South-East Asian countries. It is cultivated in Himachal Pradesh on limited scale. It is one of the vegetables with economic importance mainly in Europe and America. It is very heavy yielder.

Red cabbage is a rich source of anthocyanin pigments. Both mono and diacylated cyanidin derivaties are found. Some of the anthocynanins are characterized as - 3-O-(6-O-E-Ferulyl-β-D-glucopyranosyl)-5-O-(β-D-glucopyranosyl) cyanidin, 3-O-(2-O- -D-glucopyranosyl)-6-O-(4-O-β-D-E-P-coumaryl glucopyranosyl)-β-D-glucopyranosyl-5-O-(β-D-glucopyranosyl) cyanidin, 3-O-(6-O-E-sinapyl-2-O-β-D-glucopyranosyl)- -glucopyranosyl-5-O-(β-D-glucopyranosyl) cyanidin, 3-O-(6-O-E-ferulyl-2-O-(2-O-E-Sinapyl-β-D-glucopyranosyl)-5-O-(β-D-glucopyranosyl) cyanidin, 3-)-(6-O-E-P-coumaryl-2-O-(2-O-E-sinapyl-β-D-glucopyranosyl)-5-O-(β-D-glucopyranosyl) cyanidin, and 3-O- (6-O-E- sinapyl- 2-O- (2-O-E - Sinapyl-β-D-glucopyranosyl)-5-O(-D-glucopyranosyl) cyanidin.

Anthocyanin is an antioxidant that can help protect brain cells, thus can help prevent Alzheimer's disease. Red cabbage has more phytonutrients than the green cabbage. The vitamin C content of red cabbage is 6-8 times higher than that of the green cabbage. Recently, there has been an increased interest in red cabbage pigments as a possible source of natural colorants. The best carrier for red cabbage anthocyanin pigment was dextrin, followed by cellulose, soluble starch and glucose. The colour of anthocyanin pigment was stable at pH1.0-4.0. Further, pigment degradation was not observed at 40-80°C. (Rizk *et al.*, 2009) .The pigments of red cabbage preparation have been reported to impart a colour superior to that obtained with the synthetic colour Red No. 40 and maintain satisfactory colour stability in beverages. Presently, a red cabbage colorant is marketed under the name San Red RC, recommended for use in beverages, chewing gums, candies, sherbets, dressings, yogurts and other fermented products. Flavour in red cabbage is mainly due to the presence of 2-propenyl isothiocyanate and 3-butenyl isothiocynate. Nutritional composition of the red cabbage is given in table (10.1).

Table (10.1). Nutritive value of red cabbage (per 100 g of edible portion)

Constituents	Amount
Water	90 per cent
Protein	2.0 g
Fat	0.2 g
Carbohydrates	6.9 g
Energy	31 k cal
Calcium	42 mg
Phosphorus	35 mg
Iron	0.8 mg
Sodium	26.0 mg
Potassium	268.0 mg
Vitamin 'A'	40 I.U.
Thiamine	0.09 mg
Riboflavin	0.06 mg
Niacin	0.4 mg
Ascorbic acid	61 mg

Botanical Description

Red cabbage belongs to the family Brassicaceae and has been grouped under the genus *Brassica oleracea* sub group rubra. Its chromosome number (2n) is = 18.

Cultural Practices

Red cabbage can be planted in winter season in northern plains of India. In Himachal Pradesh, its 3 crops are raised by October-November sowing in lower hills, August-September sowing in mid hills and April-May sowing in high hills. In the plains of north India , October-November are best months for raising the nursery. Approximately 450 g seeds are required to raise the seedlings for one hectare planting. There are approximately 375 seeds per gram. Nursery is raised by dibling the seeds in well prepared nursery beds. Seedlings attain transplanting stage within 3-4 weeks. Red cabbage has very high tonnage yielding ability and needs heavy soil, rich in organic matter with ample provision of drainage. Apply FYM@ 20 tonnes per hectare at the time of land preparation. Further, at the time of final land preparation, apply 100 kg N, 100 kg P_2O_5 and 100 kg K_2O per hectare and make light ridges and furrows. Then 30 days after transplanting, apply 100 kg N per hectare for dressing. If the soil is acidic, apply micronutrient for boron and molybdenum. Ready seedlings are transplanted in well prepared field at the spacing of 45×45 cm or 60×30 cm. The size of the head depends on variety and can be controlled by planting distance i.e. for large heads wider spacing is given and *vice versa.*

Systematic irrigation is a crucial factor in the cultivation cycle. During the growing season, 8 to 12 irrigations are given to the crop with a water application rate of 350-400 l per 10 m^2 (350-400 m^3 ha) per irrigation prior to the formation of heads and 400-450 l per 10 m^2 (400-450 m^3/ha) after the formation. Combining drip irrigation with application of water-soluble fertilizers through the drip irrigation system (fertigation) is a highly efficient technique resulting in a more uniform moisture and fertilizer distribution in the root zone, more efficient water use, less soil compaction, and no soil crust formed. Readily soluble mineral fertilizers are given with each water application. Light irrigation should be given according to the needs of the crop depending on soil type. A dry spell followed by sudden heavy irrigation may result head cracking.

Varieties

Red Ruby Ball

It is an early-maturing cabbage with purple-red leaves. The head is a very tight ball type weighing just over 1 kg.

Cardinal Red

It has a large round head (average 1.5 to 2 kg). This variety makes better size when growing into warm season.

Harvesting and Yield

Harvest the crop when heads are hard before they start loosing colour and flavour. Cut the heads along with outer leaves for marketing. It takes about 12-15 weeks from sowing to harvest. The average yield ranges from 600-700 quintals per hectare. Red cabbage is almost free from the attack of DBM and aphids. Verkerk and Dekker (2004) suggested that higher retention of GSs and controllable amounts of active myrosinase can offer increasing health-promoting properties of microwave-prepared.

Major Diseases

1. Black rot

It is caused by *Xanthomonas campestris.* Infection occurs on the leaves through marginal waterpores or wounds. The bacteria move down the leaf veins into the stem and then invade other leaves. The movement of the bacteria causes the leaf to turn yellow, then brown and finally dry out. Movement is usually in a V-shape. Black rot is encouraged by warm, moist weather and rapidly growing soft tissue. It is carried both in and on the seed, and in crop debris. It can survive from year to year in the soil on leaves from diseased crops. Older plants carry the infection and it is transferred to young plants. Insects, water droplets, drainage water and windblown dust help spread this disease.

Control

1) Treat seed with hot water.
2) Sterilize the seedbed.

3) Avoid using seedbeds where *Brassica* vegetables have been grown during previous year.
4) Remove infected plants, and rotate crops on a 4 year pattern.

2. *Rhizoctonia* disease

It causes damping-off in young seedlings, while older seedlings become stunted and the soft tissue at ground level dies, leaving the symptom known as 'wire stem'. Older plants are prone to stem rot and root rot. Leaves usually take on a purplish red colour. The disease is favoured by cool, wet conditions and is spread by wind-carried spores. Contaminated soil also act as a source of infection.

Control:

1) Seedbed sterilization.
2) Wire stem can be checked in the seedbed by drenching the base of the plant with a registered chemical.
3) Apply Indofil M.45 @3.0g/ litre of water.

3. Club root

It is caused by *Plasmodiophora brassicae*. It is the most important disease of crucifer crops. Swellings develop on the tap root, secondary roots and even parts of the under ground portions of the stem. Roots are often spindle-shaped with thick centres and tapered ends. Diseased roots often decay before the end of harvest. Plants are usually stunted and wilt on hotter days. Plant collapse occurs with advanced decay and enlargement of roots. High soil moisture, acid soil and temperatures between 18.5°C and 25.5°C favour the disease. The fungus survives for long periods in the soil and on diseased crop trash. Club root is spread by infected seedlings, windblown soil and contaminated farm machinery.

Control:

1) Use disease-free seedlings and rotate crops so that crucifers are not grown for 3-4 years.
2) Lightly infested soil can be treated with lime.
3) Apply *Trichoderma* @3.5 g/ litre of water.

4. Turnip mosaic virus

It is caused by turnip mosaic virus transmitted on seed and by green peach aphids (*Myzus persicae*). There is a yellow ring spotting of the younger leaves, which later become mottled with light and dark green rings and blotches. These symptoms are most prominent in temperatures over 18°C. In lower temperatures the virus shows a definite black ring spotting of the outer leaves. The disease is spread by the green peach aphid feeding on infected plants and weeds then transmitting the disease to healthy plants. Aphids acquire the turnip mosaic virus after 10 seconds of feeding on infected plants, and transmit it after 5 seconds of feeding on healthy plants.

Control:

1) Plant disease-free seedlings.
2) Produce seedlings away from infected plants
3) Avoid planting near diseased crops or residue, and remove all cruciferous weeds as these carry the virus.
4) Regular spraying will help control green peach aphid populations and reduce the spread of the virus.

5. *Sclerotinia* rot

It is caused by *Sclerotinia sclerotiorum*. A soft, rapidly spreading, light brown, watery rot develops. Under humid conditions this rotted area becomes covered with a white growth of mycelium in which irregular-shaped bodies similar to seeds develop. These are the sclerotia and are the means by which the disease survives in the soil for several years. Cool, moist conditions favour the development of the disease. *Sclerotinia* attacks almost all vegetables and a wide range of other plants.

Control:

1) Removal and destruction of diseased plants prevents the sclerotia from developing.

***In vitro* propagation**

By manipulation of various growth

regulators and physical conditions, plants were regenerated from excised roots, stem segments, cotyledons, leaves, and callus cultures of red cabbage (*Brassica oleracea* var. capitata) grown under *in vitro* conditions. Shoot buds were induced on isolated root segments (1 cm long) cultured on Murashige and Skoog's medium and the frequency of bud formation was greatly enhanced by the addition of kinetin (0.5 part 10^{-6}). Callus obtained from the seeds, cotyledons, and hypocotyl segments cultured on a medium fortified with 2,4-D (1 part 10^{-6}), kinetin (0.1 part 10^{-6}), and coconut milk (10%, v/v) was repeatedly subcultured. The callus was slow growing, and on transference to a kinetin (2 parts 10^{-6}) and IAA (2 parts 10^{-6}) medium underwent morphogenesis to give rise to plants (Bajaj and Nietsch, 1975).

References

Bajaj, Y. P. S. and Nietsch, P. (1975). *In vitro* propagation of red cabbage (*Brassica oleracea* L. var. capitata). Journal of Experimental Botany, 26(6): 883-890.

Rizk, E.M., Azouz, A. and Lobna, A.M.H. (2009). Evaluation of red cabbage anthocyanin pigments and its potential uses as antioxidant and natural food colorants. Arab Universities Journal of Agricultural Sciences, 17(2):361-372.

Verkerk, R. and Dekker, M. (2004). Glucosinolates and myrosinase activity in red cabbage (*Brassica oleracea* L. var. capitata f. rubra DC.) after various microwave treatments. J. Agric. Food Chem., 52(24):7318-7323.

11

SAVOY CABBAGE

Savoy cabbage unlike the smooth leafy green type common cabbage has beautifully crinkled savoy leaves. Leaves are leathery and dark in colour. It is not much popular like common cabbage and cultivation is restricted to paltry areas. Wrinkled leaves are used for decorating the salad and as cooked vegetable. Savoy cabbage is rich in nutritional constituents and contains fairly high amount of vitamin A, C and iron. Anna PodseÛdek *et al.* (2006) estimated total phenolic and vitamin C contents which varied from 21 to 171 mg and from 18 to 129 mg/ 100 g, respectively. Levels of carotenoids ranged from 0.009 to 1.16 mg while alpha-tocopherol levels varied from 0.008 to 0.82 mg per 100 g . Nutritive value of savoy cabbage is given in table (11.1)

Table(11.1). Nutritive value of savoy cabbage (per 100 g of edible portion)

Constituents	Amount
Water	92 per cent
Protein	2.4 g
Fat	0.2 g
Carbohydrates	4.6 g
Energy	24 K cal
Calcium	67 mg
Phosphorus	54 mg
Iron	0.9 mg
Sodium	22 mg
Potassium	269 mg
Vitamin 'A'	200 I.U.
Thiamine	0.05 mg
Riboflavin	0.08 mg
Niacin	0.3 mg
Ascorbic acid	55 mg

Origin and Distribution

Savoy cabbage is mainly grown in Western Europe, but also in small areas in North and East Europe, in Western Mediterranean areas, and in North America. In U.K., savoy cabbage is considered one of the hardiest cole crops, where it may mature from October to January and may stand upto March or even April.

Botanical Description

Savoy cabbage is a plant of family Brassicaceae and it has been grouped under the genus *Brassica*, species *oleracea* and sub group *Sabauda*. Its chromosome number (2n) is =18

Varieties

Chieftain savoy is a common savoy cultivar. In our country, Melissa variety is marketed by Bejo Sheetal Seeds Pvt. Ltd., which has wrinkled leaves with dark green outer leaves and a dense interior structure.

Climate and Soil

Savoy cabbage can be grown in soils ranging from a pH of 7 (neutral ranges from 6.6 to 7.5) to 8.5 (alkaline ranges from 8.1 to 8.5). It is adapted to clay loam, loam, loamy sand, sandy clay, sandy clay loam, sandy loam and silt loam soils and prefers high fertility. Cultural practices of savoy cabbage is very similar to cabbage. For raising the crop seedlings are raised in month of October-November in the plains of northern India.

Nursery Raising and Planting

Approximately 300-400 g seeds are required for one hectare planting. Seeds are sown in raised bed measuring 1 m width and of required length. Cover the bed with straw/ mulch and apply frequent water for proper germination of seeds. Within 3-4 weeks, plants become ready for transplanting. Seedlings are transplanted in rows of 60 cm apart and 45 cm within the row. Proper moisture should be maintained in the field by providing the irrigation at definite interval. Field should be free from seasonal weeds and one hoeing following by earthing is needed.

Manures and Fertilizers

Savoy cabbage is high yielder crop and needs well fertile heavy soil for better growth and yield. Ample drainage facility is required. Mix well rotten FYM@10-12t/ha prior to one month of planting at the time of field preparation. Among the chemical fertilizers, 150 kg N, 80 kg P_2O_5 and 60 kg K_2O/ha should be given for realizing the optimum yield. One third of nitrogen should be applied one day before transplanting and remaining amount of nitrogen should be given at an interval of 30 days from the date of planting and another after 30 days of first application. Yields of heads (fresh matter) were 29-31 t ha^{-1} without fertilizer and 56-59 t ha^{-1} with highest fertilization. The average weights of heads were 469 to 947 g, dry matter content of heads 101 (no fertilizer) to 69 g kg^{-1} on highest fertilized plots. Total DM production was from 4.6 t ha^{-1} on control plots to about 7 t ha^{-1} on highest fertilization with removal of 100-206 kg N ha^{-1}, 43-73 kg P_2O_5 ha^{-1} and 161-261 kg K_2O ha^{-1}. According to yield increase, total above ground mass and head weight, the optimum nitrogen fertilizer dose for Savoy cabbage was between 200 and 300 kg ha^{-1}. The yield response to fertilization was higher on soil with lower N min. P and K contents before fertilization. Increased fertilization resulted in lower DM content but higher DM production. Nutrient removal per ton of head yield was the highest for potassium (4.2-5.6 kg K_2O) followed by nitrogen (3.3-3.5 kg N) and the lowest for phosphorus (1.2-1.5 kg P_2O_5). Crop residues of savoy cabbage contained 42-83 kg N ha^{-1}, 13 - 24 kg P_2O_5 ha^{-1} and 51 - 96 kg K_2O ha^{-1} (Loncaric *et al.*, 2003).

Harvesting and Yield

Crop becomes ready in 90-100 days after transplanting. Weight of individual head ranges from 3 to 3.5 kg and approximately 62,500 plants are accommodated in one hectare area, giving the yield 400-500 quintals per hectare.

Genetic Resources and Improvement

Among the different cole crops (*Brassica oleracea* L.) have been derived from the wild forms of the species by artificial selection and hybridisation are 'savoy' (var. bullata DC. et var. sabauda L.) and 'cabbage'. They are distinguished from each other in that 'the leaves of savoy are puckered or rugose and the head is generally less compact' (Higgins *et al.*,1986). Cabbage is considered to comprise two cultivar groups namely 'white cabbage' (f. *alba* DC.) and 'red cabbage' (f. *rubra* L.). White cabbage has 'green outer leaves and white inner leaves' whereas red cabbage has 'outer leaves which are red or purple and red inner leaves.' While applying the numerical methods to re-examine the crop identifications of 17 'intermediate' cultivars between white cabbage and savoy cabage, Higgins *et al.* (1986) reported that 16 cultivars were intermediate between savoy and white cabbage. On the basis of the sum of the means of the diagnostic characters and the results of a principal components analysis, one was identified as savoy and 15 as white cabbage. Twelve of the 15 had previously been identified as savoy. Ten cultivars were intermediate between savoy or white cabbage and red cabbage but none was identified as red cabbage. Most markets prefer a dark green savoy cabbage. A firm and well covered head is mainly preferred for autumn and winter cabbage.

Biotechnology

Agrobacterium rhizogenes A4M70 GUS-mediated transformation of savoy cabbage and two local lines of cabbage was obtained using hypocotyl and cotyledon explants. The percentage of explants which formed roots was very high in all genotypes: 92.3 per cent in savoy Gg-1, 64.4 per cent in cabbage P22I5, and 87.2 per cent in P34I5. Spontaneous shoot regeneration of excised root cultures grown on the hormone-free medium occurred in all three genotypes. In cabbage lines P22I5 and P34I5 shoot regeneration was higher (9.3 and 2.6 per cent respectively) than in savoy cabbage Gg-1 (1.3 per cent). Transgenic nature of hairy root-derived plants was evaluated by GUS histological test and PCR analysis. All the tested cabbage shoots were GUS positive whilst in a savoy cabbage GUS expression was registered only in 55 per cent of tested clones. PCR analysis demonstrated the presence of the GUS gene in regenerated shoot clones and in T_1 progeny (Sretenovi Raji *et al.*, 2006).

References

Anna PodseÛdek, Dorota Sosnowska and MaBgorzata Redzynia, Barbara (2006). Antioxidant capacity and content of *Brassica oleracea* dietary antioxidants. International Journal of Food Science & Technology, 41(Spl. 1):49-58.

Higgins, J., Sparks, T.H., Evans, J.L. and Law, J.R. (1986). Crop identifications of some *Brassica oleracea* cultivars. Acta Hort. (ISHS) 182:285-292.

Loncaric, Z., Teklic, T., Paradjikovic, N. and Jug, I. (2003). Influence of fertilization on early savoy cabbage yield. Acta Hort. (ISHS) 627:145-152.

Sretenovi Raji, T., Ninkovi, S., Milju-uki, J., Vinterhalter, B. and Vinterhalter, D. (2006). *Agrobacterium rhizogenes* mediated transformation of *Brassica oleracea* var. sabauda and *B. oleracea* var. capitata. Biologia Plantarum, 50(4): 525-529.

12

CHINESE CABBAGE

Leaves of Chinese cabbage during the vegetative stage are arranged in an enlarged rosette, forming a short conical more or less compact head, with ill defined nodes and internodes and alternate heading and non-heading leaves. Chinese cabbage is also used like common cabbage in form of cooked, soup or stir-fried. Raw succulent white midribs of the leaves are also sliced or coarsely shredded for salad, or cut in strips for platters. Leaves are also dried and kept for week before being used as vegetables. Leaves of Chinese cabbage are very soft, crisp and contain fairly high amount of vitamins A and C. Seeds of Chinese cabbage contain 35-40% oil and lose their viability during storage. Hence the seeds can be stored at low relative humidity and low temperature. Nutritional composition of the Chinese cabbage is given in table (12.1).

Table (12.1). Nutritive value of Chinese cabbage (per 100 g of edible portion)

Constituents	Amount
Moisture	95.0 g
Protein	1.2 g
Fat	0.1 g
Carbohydrates	3.0 g
Energy	14 kcal
Calcium	43 mg
Phosphorus	40 mg
Iron	0.6 mg
Sodium	23 mg
Potassium	253 mg
Vitamin 'A'	150 I.U.
Thiamine	0.05 mg
Riboflavin	0.04 mg
Niacin	0.6 mg
Ascorbic acid	25 mg

Origin and Distribution

Chinese cabbage is a native of China; it probably evolved from the natural crossing of Pak Choi (non-headed Chinese cabbage), which was cultivated in southern China for more than 1600 years, and turnip, which was grown in northern China. Much of its variety differentiation took place in China during the past 600 years. Its derivatives were introduced into Korea in the13th century, the countries of south-eastern Asia in the 15th century and Japan in the 19th century. Chinese cabbage is one of the most important vegetables in South East-Asia. It is widely cultivated in China, Japan, Korea and Thailand. In India, it is not popular and cultivation is restricted to temperate regions.

Botanical Description

Chinese cabbage is member of family Brassicaceae and its scientific name is *Brassicarapa* L. ssp. *pekinensis* (Lour.) Hanelt; having the chromosome no. (2n) = 20.

Morphological Description of Plant

Chinese cabbage is a biennial herb, cultivated as an annual, 200-500 mm tall during the vegetative stage and reaching up to 1.5 m in the reproductive state. The taproot and lateral roots are prominent in older plants, forming an extensive, fibrous, finely branched system. During-vegetative stages, the leaves are arranged in an enlarged rosette. This forms a short conical more or less compact head, with ill-defined nodes and internodes and alternate heading and non-heading leaves; the leaves are 200-900×150-350 mm in size. Leaves vary in shape with different growth stage; the dark green outer leaves are narrowly ovate with long laminae and winged petioles. Inner heading leaves are broad, sub-circular, whitish-green. The flowering stem carries lanceolate leaves, much smaller than heading leaves, with broad compressed petioles and blades, clasping the stem (Dixon,1970).

Climate and Soil

Chinese cabbage responds to inductive temperatures from germination; apparently there is no juvenile stage, with flower initials forming when 'vernalization' is completed. Minimum plant raising temperatures of 18°C are required in order to reduce bolting in the field. The growing point height is determined at or before transplanting. A linear model incorporating temperature and solar radiation describes subsequent extension of thegrowing point (Dixon,1970). Chinese cabbage can be grown almost in all types of soils. But heavy soil, rich in organic matter with provision of ample drainage is considered best.

Varieties

Generally two types of cultivars are raised. In first type, heads are formed like common cabbage whereas in other type leaves are fleshy and centre is open. In head forming group of varieties are Chihili, Michihili and Wang Bok and non-heading type includes Pakchoi.

Hybrid Optiko

It is marked by BejoSheetal Seed Pvt. Ltd. It is an early hybrid, barrel shaped, short and compact with vigorous growing habit. Light green colour and good in taste. It is resistant to *Fusarium* wilt.

Cultural Practices

Chinese cabbage grows best and forms heads in climate with temperature in the range of 12-22°C. Hence, it can successfully be grown during winter season in the plains of northern part of the country. Cultural practices of Chinese cabbage are very much similar to cabbage. For raising the crop, seeds are sown in raised bed and then covered with a thin layer of soil, compost or rice hulls. Seedlings at the 2-3 true leaves stage are thinned to one seedling per hole. Approximately 400 g seeds are required for planting of one hectare field. About 3-4 weeks after sowing, seedlings with 5-8 leaves are transplanted to the field, usually in the late afternoon at 45 cm between rows and 30 cm within row. October-November are ideal months for raising the seedlings and transplanting. For realizing the good yield at least 10-12 tonnes well rotten FYM should be incorporated in the soil at least one month before transplanting. Besides, 120 kg N, 60 kg P_2O_5 and 50 kg K_2O should be given at the time of planting. While applying the nitrogen doses, care should be taken to apply 1/3rd dose at the time of transplanting and remaining 2/3rd dose in two split applications at the interval of 30 days from the date of planting. In Chinese cabbage most of the feeder roots grow to a depth of 30 cm, irrigation to maintain the top 30 cm at a soil moisture level between 65-85 per cent of field capacity is of great importance.

Major Diseases

Black rot (*Xanthomonascampestris*): It is caused by *Xanthomonascampestris* and is the most severe bacterial disease of *Brassica* vegetables worldwide. When there is plentiful rainfall or heavy dews and temperatures are between 16-21^{0}C, the disease spread rapidly and cause high crop losses. In young seedlings, cotyledon margins turn black and the cotyledons shrivel and fall off. In plants with more fully developed leaves the internal tissue near the infection site turns black or brown. In the field, large yellow sections at the leaf margins are noticeable, especially in plants that remain wet for extended periods. The bacteria are spread throughout the plant *via* the xylem (water carrying) vessels to the stem, roots and leaves. In plants raised from infected seed, bacteria move from the cotyledons to the young leaves as the seedlings emerge and begin to grow. When an early infection occurs in the field, the bacteria move from the infection site throughout the plant.

Control Measure

i. Crops should be grown in soils with good drainage and good sanitation practices such as removing infected plants, ploughing debris into soil after harvest and the use of clean equipment should be employed.
ii. Grow resistant varieties like Spectrum and Green Rocket.

***Cercosporella* leaf spot:** It is caused by the fungus. Symptoms appear as circular spots with grey to brown or almost papery white centres with slightly darkened margins. If the infection is severe, the affected leaves may turn yellow and drop to the ground. The species of *Cercospora* which infects Chinese cabbage is called *Cercospora brassicicola* and also infects mustard, rape, turnip and to a small extent, cabbage. It causes circular to angular spots on leaves which are pale green to grey or white, usually with a brown border. These fungi are carried in a small percentage of seeds and can overwinter in cruciferous weeds. Spores are carried by wind or rain and the greatest infection occurs in conditions with abundant moisture and temperatures between 13 and 18^{0}C.

Control Measure

i. Remove cruciferous weeds from around crops.
ii. Maintain good drainage in the field.

Important Pests

Diamond back moth (*Plutella xylostella*): It is a most destructive pest of the Chinese cabbage (Talekarand Shelton, 1993; McKinlay, 1992) and feeds only on cruciferous crops and weeds (Hely *et al.,* 1982; Talekar and Shelton, 1993). The caterpillar phase of *Plutella* attacks and skeletonises leaves of cruciferous crops from the seedling stage to crop maturity and can result in widespread losses. Adult moths are about 9 mm long and greyish brown, with three light brown to white, triangular marks on the trailing edge of each forewing which are visible as diamond shapes when the wings are folded at rest.

Control Measure

Plutella xylostella is attacked by a

range of parasites although only a few are effective in the field (McKinlay, 1992). Ichneumon wasp parasitoids of the genera *Diadegma*, *Cotesia* and *Diadromas* are among the most prominent and effective biological controls.

Cabbage Aphid (*Brevicoryne brassicae*) and Green Peach Aphid (*Myzuspersicae*):

They are serious pests during the February-March. The wingless form (aptera) of the cabbage aphid is between 1.6 and 2.6 mm long and is grey-green in colour, with characteristic black, transverse, dorsal bars on its abdomen and thorax. The winged form is 1.6 to 2.8 mm long, with a dark coloured head and thorax and black, transverse, dorsal bars only on its abdomen. The wingless form of the green peach aphid is smaller than the cabbage aphid, between 1.25 and 2.5 mm long and has a barrel shaped body which can vary in colour from a uniform pale green to pink. The winged form has a similar size and colour to the wingless form, except it has a black patch on the centre of its abdomen. Infestations are usually seen on the leaf as small, bleached, necrotic spots at the feeding sites. The leaves turn yellowish and curl in, thus protecting the aphid colonies. In addition to direct crop damage, aphids transmit plant viruses. *B. brassicae* and *M. persicaeare* vectors of the non-persistent cruciferous viruses, Cauliflower Mosaic Virus (CaMV) and Turnip Mosaic Virus (TuMV).

Control Measures

Aphids are attacked by several natural predators including the common spotted ladybird, *Harmonia conformis* and wasp parasitoids such as *Diaeretusrapae*.

Harvesting and Yield

Heads are harvested when they are well compact. Head is cut at the base, keeping the entire head intact. Outer leaves are trimmed to 2-3 non-heading leaves for protection during transport. Chinese cabbage is high yielder and average yield varies from 450-500 quintals per hectare.

Seed Production

Chinese cabbage normally requires cold period, from a few days to several weeks, below 5 to 12 ^{0}C for flower induction and development. The seed-to-seed or head-to- seed method of temperate Chinese cabbage seed production, using natural cold periods, is not feasible in the tropics, even at altitudes between 800 and 1200 m. Chinese cabbage must be either vernalized at the seedling stage or their mature plants uprooted for vernalization in a cold storage, in order to induce flowering. Tropical Chinese cabbage on the other hand can produce seed without requiring very low temperature. The vernalization requirement of these tropical types can he fulfilled at rather high temperatures, i.e. near or below 20^{0}C for a period of time, supplemented with artificially extended photoperiod, canbe effective (Opena *et al.*, 1988).

Physiological Disorder

Tipburn : It is a field disorder which causes browning of the margins of young inner leaves and in its mostsevere form results in necrotic breakdown of leaf tissue and subsequent secondary invasion by bacteria. The primary cause is a low supply of calcium to the young, inner leaves. particularly during rapid growth and head formation. Climatic factors which include water stress or hightranspiration rates increase tipburn while large diurnal fluctuations in relative humidity reduces tipburn.

Control Measure

Maintain proper moisture in the field and foliar applications of calcium may be given

Genetic Improvement

Breeding of Chinese cabbage has received much attention from AVRDC (Taiwan) and from Asian seed companies. In the tropics, breeding is limited by difficult vernalization at ambient temperatures. If flowering can be induced in the field, plants are bagged for controlled pollination. If flowering has to be induced artificially, plants are dug out and vernalized in cold rooms. Both genetic and cytoplasmatic male sterility occur naturally in *Brassica* species. Cytoplasmatic male sterility is now widely used to produce hybrids, systems based on genetic male sterility being impractical and costly. Anther microspore culture is used for the fast creation of completely homozygous lines. The breeding efforts resulted in early producing F_1 hybrid cultivars for tropical lowland conditions, that combine medium firm heads with heat tolerance and resistance to major diseases likesoft rot, anthracnose, downy mildew, leaf spot, viruses (Toxopeus and Baas, 2004). Now days Red color heading type Chinese cabbage is receiving greater attention by seed companies due to the increased market demand. Most of the commercial cultivars of Chinese cabbage are F_1 hybrids, which require a sufficient degree of homozygous parental lines. Conventional breeding in Chinese cabbage is time-consuming and labor-intensive. Haploid breeding is more efficient than conventional plant breeding for the generation of diploid homozygous lines. Generally, to obtain a pure homozygous line from a hybrid, it needs at least six or seven generations by self fertilization, while it only needs a single or several generations by haploid breeding (Lee *et al.*, 2014).

Biotechnology

Genetic modification of Chinese cabbage was made by LEA protein gene which holds considerable potentiality for crop improvement toward environment-stress tolerance. Transgenic Chinese cabbage (*Brassica campestris* ssp. pekinensis) expressing a *B. napus* late embryogenesis abundant protein gene was generated. Infection by *Agrobacterium* strain LBA4404 containing the binary vector pIG121-LEA, which carried LEA protein gene linking to CaMV promoter and terminator sequences, and the neomycin phosphotransferase II (*NP* TII) gene, was applied. Transgenic Chinese cabbage plants expressed enhanced growth ability under salt and drought-stress conditions. The increased tolerance was reflected by delayed development of damage symptoms caused by stress as well as recovery upon the removal of stress condition. These results suggest that the genetic modification of Chinese cabbage by LEA protein gene holds considerable potentiality for crop improvement toward environment-stress tolerance (Park *et al.*, 2005). While studying the morphological analysis in heading type of Chinese cabbage, Wang *et al.* (2014) observed that the significant earliness of heading in the transgenic plants over expressing BrpSPL9-2 gene was produced because the juvenile phase was absent and the early adult phase shortened, whereas the significant delay of folding in the transgenic plants overexpressing Brp-MIR156a was due to prolongation of the juvenile and early adult phases. Thus, miR156 and BrpSPL9 genes are potentially important for genetic improvement of earliness of Chinese cabbage and other crops.

References

Dixon, Geoffrey R. (1970). Vegetable Brassicas And Related Crucifers CABI Head Office, CABI, North American Office, Nosworthy Way Wallingford Oxford Shire OX10, DEUK.

Hely, P.C., Pasfield, G. and Gellatley, J.G. (1982). Insect pests of fruit and vegetables in N.S.W., Inkata Press, pp. 312.

Lee, M.H., Lim, C.J , Lee, I.H. Song, J.H. (2014). High-purity seed production of doubled haploid chinese cabbage [*Brassica rapa* L. ssp. *pekinensis* (Lour.)] through microspore culture. Plant Breed, Biotech, 2014(2):167-175.

McKay, A. and Phillips, D. (1990). Chinese cabbage. Department of Agriculture, Western Australia, Bulletin No. 4197, pp. 13.

McKinlay, R.G. (1992). Vegetable crop pests. CRC Press, Inc., pp. 406.

Opena, R.T. , Kuo C.G. and Voon, J.Y. (1988). Breeding and seed production of Chinese cabbage in the Tropics and Subtropics.Technical Bulletin No. 17. Asian Vegetable Research and Development Center, Tropical Vegetable Information Service.

Park, B.J., Zaochang Liu, Z., Kanno, A. and Toshiaki Kameya, T. (2005). Genetic improvement of Chinese cabbage for salt and drought tolerance by constitutive expression of a *B. napus* LEA gene. Plant Science , 169 (3): 553-558.

Talekar, N.S. and Shelton, A.M. (1993). Biology, ecology, and the management of the diamond backmoth. Annu. Rev. Entomol., 38: 275-301.

Toxopeus, H. and Baas, J. (2004). *Brassica rapa* L. In: Grubben, G.J.H. & Denton, O.A. (Editors). PROTA 2: Vegetables/ Légumes. PROTA, Wageningen, Netherlands.

Wang Y., Wu, F., Bai, J. and He, Y. (2014). BrpSPL9 (*Brassica rapa* ssp. pekinensis SPL9) controls the earliness of heading time in Chinese cabbage. Plant Biotechnol J., 12(3):312-21.

13

BRUSSELS SPROUTS

Brussels sprouts is a cool season crop, belonging to the cabbage family, and closely related to cauliflower, broccoli, kale, collards, etc. Like cauliflower, it thrives best in a cool humid climate. The edible portion of this crop is the bud or small cabbage-like head which grows in the axils of each leaf. Occasionally the tops are used as greens. Brussels sprouts, as with broccoli and other brassicas, contains sulforaphane, a chemical possessing anti-cancer properties. Although boiling reduces the level of the anti-cancer compounds, steaming, microwaving, and stir frying does not result in significant loss. Brussels sprouts and other *Brassicas* are also a source of indole-3-carbinol, (-)5-vinyloxazolidine -2-thione (goitrin), a well-known goitrogen, has been identified as a bitter principle in cooked Brussels sprouts (Roger and Griffiths, 1981). Nutritive value of Brussels sprouts is given in table (13.1).

Table (13.1): Nutritive value of Brussels sprouts (per 100 g of edible portion)

Constituents	Amount
Water	85.2 per cent
Food energy	45 K cal
Protein	4.9 g
Fat	0.4 g
Carbohydrates	8.3 g
Fibre	1.6 g
Ash	1.2 g
Calcium	36 mg
Phosphorus	80 mg
Iron	1.5 mg
Sodium	14 mg
Potassium	390 mg
Vitamin 'A'	350 I.U.
Thiamine	0.10 mg
Riboflavin	0.10 mg
Niacin	0.9 mg
Ascorbic acid	102 mg

Source : Gopalan *et al.* (2004)

Origin and Distribution

While the origin of Brussels sprouts is unknown, its first mention can be traced to the late 16th century. It is native to Belgium, specifically to a region near its capital, Brussels, derived the name. Brussels sprouts remained a local crop in this area until its use spread across Europe during world war I. Brussels sprouts are now cultivated throughout Europe and the United States. In the U.S., almost all Brussels sprouts are grown in California. In India , its cultivation is restricted to Himachal Pradesh, J&K and in some parts of Utterakhand.

Botanical Description

Brussels sprouts has been classified as *Brassica oleracea* L. var. gemmifera DC. or convar. gemmifera (DC.) Gladis. It can best be considered as a cultivar-group and as such has been called Brussels Sprout Group.Taxonomic position of the genus *Brassica* is given as under:

Taxonomic position of the genus *Brassica*

Kingdom	Plantae
Subkingdom	Tracheobionta
Superdivision	Spermatophyta
Division	Magnoliophyta
Class	Magnoliopsida
Subclass	Dilleniidae
Order	Capparales
Family	Brassicaceae
Genus	*Brassica* L.
Species	*Brassica oleracea* L.
Variety	*Brassica oleracea* L. var. gemmifera DC.

Morphological Description of the Plant

Erect, glabrous, biennial herb up to 120 cm tall, with unbranched stem, developing lateral buds (sprouts) in leaf axils; root system strong and branched. Leaves alternate, simple or lower ones with some small side lobes at base; stipules absent; all leaves with distinct petiole, but upper ones with short petiole giving the top of the plant a rosette-like appearance; blade more or less circular, undulate or irregularly incised, blue-green. Inflorescence a terminal paniculate raceme. Flowers bisexual, regular, tetramerous; pedicel up to 2 cm long, ascending; sepals oblong, c. 1 cm long, erect; petals obovate, 1.5-2.0 cm long, clawed, pale to bright yellow or whitish; stamens 6; ovary superior, cylindrical, 2-celled, stigma globose. Fruit a linear silique 5-10 cm × c. 5 mm, with a tapering beak 5-15 mm long, dehiscent, up to 30-seeded. Seeds globose, 1.5-2.0 mm in diameter, finely reticulate, dark brown. Seedling with epigeal germination, with a taproot and lateral roots; hypocotyl 3-5 cm long, epicotyl absent; cotyledons with petiole 1-2 cm long, blade cordate, 1.0-1.5 cm long, cuneate at base, notched at apex (Tjeertes, 2004). Its Chromosome number (2n) is =18.

Varieties

Varieties of Brussels sprouts are mainly grouped in tall and dwarf groups. Among the tall cultivars Evesham, Bedfordshire, Cambridge No.1, 3 and 5, Irish Elegance and Sherradian are important cultivars. Dwarf cultivars have short stem, mostly less than 50 cm long. In this group Improved Long Island, Catskill, Early Morn Dwarf Improved are important varieties.

Climate and Soil

Brussels sprouts are a cool season crop. They require a long, cool growing season and can withstand considerable frost. In the plains of India, where there is prolong winter, it can be raised successfully. Brussels sprouts grow in temperature ranges of 7-24°C, with highest yields at 15-18°C In Himachal Pradesh, in lower hills, it is sown in October and in mid hills in the months of August-September, while in high hills sowing is mainly done in month of February and March-April. Fields are ready for harvest 90 to 180 days after planting. This crop may be grown successfully on a wide variety of soils; however, it performs best on a medium to heavy soil that is high in organic matter and fairly high in nitrogen. The soil pH should be between 6.0 and 6.8.

Nursery and Transplanting

For raising the crop, nursery is prepared in raised bed. Approximately 450 g seeds are required for a hectare planting. Seeds are tiny and there are 350 seeds per gram. Frequent watering is required to raise the seedlings. When the seedlings become ready 3-4 weeks after sowing, they are transplanted at the spacings of 45 x 30 cm. The plants are usually watered at transplanting to prevent wilting and early establishment.

Manures and Fertilizers

In general, FYM @10-12 t/ ha are mixed at the time of field preparation. Besides, 120 kg nitrogen, 60 kg P_2O_5 and 50 kg K_2O /ha are required to get optimum size of solid sprouts. One third of nitrogen should be given at the time of planting and rest of the nitrogen may be applied at an interval of 30 and 45 days after planting.The higher doses results into loose sprouts while sulphate based fertilizers and excessive K results in bitterness adversely affecting the quality of the sprouts (Chadha, 2000).

Irrigation and Interculture

First irrigation should be given soon after planting of seedlings for their early and better establishment. Thereafter irrigation is given at 10-15 days interval depending upon the moisture conditions of the soil. At the time of the maturity, irrigatiom may be avoided to prevent the loosening of the sprouts. Brussels sprouts are a long season crop and weeds can also be a serious problem. Pre-emergence application of Trifluralin @ 0.5 l/ha or Basalin @ 0.5 l/ha are also effective in suppressing the weeds. Black polythene mulch can also used to check the weed infestation. An earthing up before sprout formation helps in their proper formation.

Plant protection

Some of the more damaging insects on this crop are Harlequin bugs, cabbage loopers, diamond back moth, imported cabbage worm, cutworms, cabbage maggot, thrips and web worms. Aphids are especially difficult to control. A rigid control program will be necessary and the program must be started early. Brief description of these insect pests are given as under:

Cabbage Maggot (*Delia brassicae*)

The cabbage maggot is a serious pest in early direct-seeded or transplanted cole crops. Pupae overwinter in the soil. The adult flies emerge from the soil and lay their eggs on cole crop plants and related weeds near the soil surface or in the soil at the base of the plants. The short (1/4 inch), white maggots emerge a few days later and begin to eat and burrow through the soil into the plant stems and roots. The maggots feed for 3 to 4 weeks and then pupate. Adults emerge in 2 to 3 weeks. Young plants that are invaded by maggots usually wilt and die. Heavy infestation of maggots results in stunted growth of plants.

Cabbage worm (*Pieris rapae*)

Cabbage worms are the most common foliage pest of cole crops. Cabbageworm adults, the white butterflies often seen around cruciferous crops, overwinter as pupae and emerge in late April or early May and lay their yellow eggs singly on the leaves of cole crops and other cruciferous crops and weeds. The velvety green worms, which grow to over one inch in length, eat holes in leaves and leave large amount of green debris on the leaves. They tend to leave the plant to pupate in the soil. Damage from imported cabbage worms causes a loss in quality and yield. Fields should be monitored for adult activity and plants should be checked for eggs and larvae.

Cabbage looper (*Trichoplusia ni* (Hübner) Order: Lepidoptera)

The adults are about 1 to 1¼ inches across, gray-brown, fly and lay eggs mostly at night. The larvae are light green, with a white stripe on each side, about 1 inch long, and move by humping their back like an inch-worm, from which they get their name 'looper.' There may be 2 or 3 generations per year. As the larvae grow, they become more difficult to control. They cause foliar injury and can be a contaminant at harvest for cole crop. Plant damage and product contamination are similar to that of imported cabbage worm. Hosts of the cabbage looper include cole crops, celery, tomatoes and potatoes. Eggs are laid singly on the underside of the foliage.

Control

i) Monitor fields regularly for eggs, larvae, and damage

Diamond back moth (*Plutella xylostella*)

The gray adults can lay eggs in the spring and the small (1/3 inch) yellow green larvae emerge soon thereafter. The worms eat numerous small holes in the layers of leaves. Diamond back moths can cause foliar injury and contaminate the product.

Control

i) Natural enemies often effectively control diamond back moth. Applications of *Bacillus thuringiensis* is also very effective.

ii) Older plants are not usually seriously damaged. Destroy culls and mustard-type weeds several weeks before planting

Cabbage aphids

Cabbage aphids are small (1/16 inch), blue-gray insects that suck sap from the plants. They overwinter as eggs on cole crops residue. Heavy infestations cause leaves to cup and curl inward. The presence of live or dead ones makes the Brussels sprouts unmarketable.

Control

i) Natural enemies like parasitic wasps can be utilized

ii) Many predators also feed on aphids e.g.lady beetle adults and larvae, lacewing larvae and soldier beetles.

iii) Silver-colored reflective mulches have been successfully used to reduce transmission of aphid-borne viruses.

iv) Insecticidal soaps and oils are the best choices for most situations.

v) Spray formulations of malathion, permethrin, and acephate.

Harvesting and Yield

The sprouts begin maturing from the bottom upwards. In picking, the leaf below the sprout is broken away from the main stem. Harvesting should start before the lower leaves begin turning yellow. Often the central growing point is removed to hasten harvest. This is done when the sprouts are well formed. Sprouts are harvested when they attain 3-5 cm in diameter and are firm and dark green. Harvesting usually occurs 3 months after the planting. Sprouts begin to mature on the plant from the bottom upward. The sprouts can either be picked several times or the harvest can be delayed and the whole stalk taken at one time. One plant is capable to yield 750-1000 g of sprouts. The average yield varies from 100-150 quintals per hectare. Brussels sprouts can also be stored for period as long as 30 days if kept at 32°F and 90-95 per cent relative humidity. The storage condition with 10 per cent Co_2 and 2.5 per cent O_2 reduces yellowing without injury at 0° C. Sprouts can be stored in perforated plastic bags at 1-1.5° C .

Genetic Rcsources and Improvement

Germplasm collections of Brussels sprouts are available in several European genebanks, e.g. at the Centre for Genetic Resources (CGN), Wageningen, Netherlands. The possibility of inter-specific hybridization within the Brassicaceae family makes the genepool very wide.

The main breeding objectives include quality, yield, earliness, stress tolerance and resistance to diseases (e.g. clubroot). Seed companies focus exclusively on the production of single-cross F_1 hybrids. During the last forty years breeders have changed the shape of the plants from a pyramidal sprout setting, with big sprouts at the bottom and small ones at the top, to a plant type with a cylindrical sprout setting with all sprouts of the same size, suitable for once-over mechanical harvesting. As in the other *Brassica* vegetables, hybrids based on self incompatibility are now being replaced by hybrids based on cytoplasmic male sterility. Increasingly, inbred lines are being developed from anther and microspore cultures. DNA-markers are used for precise screening for resistance to diseases and other important traits (Tjeertes, 2004).

Embryo culture

In Brussels sprout, it has been possible to grow the material at all embryonic stages, ranging from the pre-heart to mature, into normal seedlings, however, pre-heart and heart-stage embryos require coconut milk for growth. After a short period of growth on a medium containing coconut milk, the embryos have to be transferred to a simpler medium. Growth substances as well as sucrose are needed for the growth of Torpedo and Walking-stick embryos. When the embryos have sufficiently advanced in differentiation (mature I and older) to germinate (i.e. to form roots) they can grow on a simple inorganic medium. The uniformity of the plants grown from *in vitro* embryo cultures and the percentage of success are generally high (Wilmar and Hellendoorn, 1968). Eight inbred lines of Brussels sprouts and ten F_1 hybrids derived from them were tested for their response to anther culture. From 5-19 plants per genotype were tested, and each plant was tested on 3-6 separate occasions. Results from the inbred lines were broadly similar to those from the F_1 hybrids, despite the inbreds producing fewer buds and having a higher frequency of anther deformities. The maximum embryo yield from an inbred line was 215 embryos per 100 anthers, and from a hybrid was 275. From estimation of the variance components it was calculated that, for both inbreds and hybrids, about half the total variation was genetic whereas variation due to plants within genotypes and to occasions within plants were each about 13% of the total. The narrow sense heritability of responsiveness to anther culture (estimated by the proportion of variation between inbred lines which was genetic) was 0.48, and there was partial dominance for this character. In three cases the hybrid outyielded the better inbred, and this heterosis may well be due to dispersed dominant genes (Ockendon and Sutherland, 1987). Incorporation of $AgNO_3$ (1-10 $mg l^{-1}$) into the culture medium of Brussels sprout callus significantly improved growth and allowed long-term callus culture. In the absence of $AgNO_3$, callus died shortly after removal from the hypocotyl explants. Regeneration of shoots from callus on low-hormone medium was also enhanced by $AgNO_3$. Significant differences in shoot production were found between the three genotypes examined. cv. Aries produced large numbers of shoots even in the absence of $AgNO_3$. Investigation of callus production from the inbred parent lines of cv. Aries indicated that tissue culturability may be determined genetically (Williams *et al.*, 1990).

References

Chadha, K.L. (2000). Hand Book of Horticulture. Indian Council of Agricultural Research, New Delhi.

Gopalan, G., Rama Sastri, B.V. and Balasubramanian, S.C. (2004). Nutritive Value of Indian Foods. National Institute of Nutrition, ICMR, Hyderabad, India.

Ockendon, D. J. and Sutherland, R. A. (1987). Genetic and non-genetic factors affecting anther culture of Brussels sprouts (*Brassica oleracea* var. gemmifera). Theoretical and Applied Genetics, 74 (5): 566-570.

Roger, G. F. and Griffiths, N.M. (1981). The identification of the goitrogen, (-)5-vinyloxazolidine-2-thione (goitrin), as a bitter principle of cooked Brussels sprouts (*Brassica oleracea* L. var. gemmifera). Zeitschrift für Lebensmitteluntersuchung und Forschung A.,172(2):90-92.

Tjeertes, P. (2004). *Brassica oleracea* L. (Brussels sprouts) In: Grubben, G.J.H. and Denton, O.A. (Eds.). PROTA 2: Vegetables/Légumes. Wageningen, Netherlands.

Williams, J., Pink , D. A. C. and Biddington, N. L. (1990). Effect of silver nitrate on long-term culture and regeneration of callus from *Brassica oleracea* var. gemmifera. Plant Cell, Tissue and Organ Culture, 21(1):61-66.

Wilmar, J. C. and Hellendoorn, M. (1968). Embryo culture of Brussels sprouts for breeding. Euphytica, 17(1):28-37.

14

WELSH ONION

Welsh onion or Japanese bunching onion is mainly grown for its green leaves. It is very similar to common onion. The major difference between common onion and Welsh onion is that the Welsh onion does not form dormant bulbs. The Welsh onion was historically the main *Allium* vegetable of China and Japan where it has been cultivated for more than 2600 years, and it remains very important there. China ranks first in the world in Welsh onion production. During 2011, the onion (including Welsh onion) has approximately 28 thousand hectares in cultivation and a total production of 979 thousand tons (FAO, 2011).

Green leaves of Welsh onion are used in salad or as herb to flavor soups and other dishes. Its young inflorescence is sometimes deep fried and eaten as snack. The odour of Welsh onion is mild as compared to common onion. This odour is mainly due to volatile allyl sulfides. Leaves of Welsh onion are rich in vitamin A, B_2 and C contents. As like other *Alliums*, Welsh onion also contains allicin a precursor which plays an important role in the intake of thiamin (Inden and Asahira, 1990). Ethanolic extracts of green leaves and white sheathe have inhibitory effects on bacteria and fungi (Fan and Chen, 1998). A number of therapeutic properties of Welsh onion have been mentioned in Chinese medicine. The consumption of Welsh onion is said to improve eye sight and the functioning of internal organs and the metabolism. It also prevents the common colds, headaches etc. The nutritive value of Welsh onion in per 100 g of edible portion is given in table (14.1).

Table (14.1) : Nutritive value of Welsh onion (per 100 g of edible portion)

Constituents	Pseudostem	Green leaves
Water	91.6 g	92.0 g
Proteins	1.1 g	1.7 g
Fat	0.1 g	0.2 g
Digestible Carbohydrates	6.7 g	5.4 g
Calcium	47.0 mg	80.0 mg
Iron	0.6 mg	1.0 mg
Sodium	1.0 mg	1.0 mg
Potassium	180.0 mg	200.0 mg
Phosphorous	20.0 mg	38.0 mg
Energy	27 kcal	35 kcal
Vitamin A	85 I.U.	480 I.U.
Vitamin B_1	40 µg	60 µg
Vitamin B_2	60 µg	100 µg
Niacin	300 µg	400 µg
Vitamin C	14 mg	33 mg

Source: Inden and Asahira (1990)

Origin and Distribution

The term Welsh derives from the German 'Welsche' meaning foreign. This was probably applied when it was initially introduced into Germany near the end of middle ages (Purseglove, 1972). Welsh onion probably originated in North Western China, although there is no authenticity about its progenitors. *Allium altaicum* is closely related to Welsh onion and it is still in cultivation in Siberia and Mongolia. The earliest description of the crop and its cultivation is found in Chinese book of 100 B.C. whereas in Japanese literature, its description was made in 720 A.D. Welsh onion is cultivated mainly in Japan, Korea, and China. In these countries, Welsh onion is in the list of very important vegetables.

Botanical Description

Perennial glabrous herb, growing in tufts, usually grown as an annual or biennial plant, up to 50(-100) cm tall, with indistinct, ovoid to oblongoid bulb up to 10 cm long, lateral bulbs few to several or virtually absent; tunic white to pale reddish brown. Leaves 4-12, distichously alternate, glaucous, with tubular sheath; blade cylindrical, hollow, 10-50(-100) cm × 0.5-2.5 cm, acute at apex. Inflorescence a spherical umbel 3-7 cm in diameter, on a long, erect, terete, hollow scape up to 50(-100) cm long and up to 2.5 cm in diameter; umbel composed either of flowers or of bulbils only; spathe 1, hyaline, persistent, up to 1 cm long, splitting into (1-)2-3 parts. Flowers bisexual, narrowly campanulate to urceolate; pedicel slender, up to 3 cm long; tepals 6, in 2 whorls, free, ovate-oblong to oblong-lanceolate, 6-10 mm long, white with greenish midvein; stamens 6, exceeding tepals, connate at base and adnate to tepals; ovary superior, 3 -celled, style slender, exceeding tepals. Fruit a globular capsule c. 5 mm in diameter, splitting loculicidally, few-seeded. Seeds 3-4 mm × 2-2.5 mm, black. The basic chromosome number of Welsh onion is 8 (Oyen and Soenoeadji, 1993).

Pollen Structure and Development

The ultrastructure of pollen development from microsporocyte to early 2-celled pollen study showed that at diakinesis intercellular substance exists between the primary walls outside the callose walls and within the intercellular spaces of the microsporocytes. This substance still partly remains at tetrad stage. Before meiosis the microsporocytes have lipids in cytoplasm. After microspore mitosis the lipids increase in both number and size. In anaphase the plastids in the microspore cytoplasm accumulated one to several starch grains. In early 2-celled pollen, some starch containing plastids in vegetative cytoplasm also contain lipids. At all developmental stages, free and polyribosomes are abundant, and a large amount of rough endoplasmic reticulum (RER), dictyosomes and vesicles and numerous mitochondria are present. These features indicate that synthesis and transportation of proteins, saccharides and other substances are active. The microspore lacks a large central vacuole. At anaphase of mitosis, the organelles in microspore cytoplasm are concentrated in the future vegetative cell pole. Microspore at cytokinesis some RER are close to or associated with the pollen plasma membrane. They may fuse with each other and enlarge the plasma membrane surface to adapt the growth of pollen grain (Xin Xiang Yuan, 2000).

Climate and Soil

Welsh onion grows well in area of 100 m or above elevation. It is very tolerant to cold weather and come over winter even in Siberia. Most cultivars of Welsh onion are tolerant to heavy rainfall and hot humid conditions. Welsh onion does not have a long day storage life like common onion and develops most rapidly in 12 hrs included with 4 hrs of blue light. So, it remains in vegetative growth and does not develop a real bulb (Cuvanova, 1970). For optimum growth neutral pH is required. Plants grown in acidic soil give poor growth. Geriya (1977) realized higher yield of green leaves from the crop in Republic of Georgia, USSR on the same plot for 4-5 year, however, crop can be grown as annual crop.

Raising the Seedlings and Planting

Welsh onion is grown by raising the seedlings in seed bed and there after plating the seedlings in main field. Crop can also be raised through direct seed sowing in the field. In the South East Asia, Welsh onion is propagated mainly using basal tillers. The optimum temperature for seed germination is between 15 and 25°C. Germination rate is low at temperature below 10°C. The optimal temperature for growth ranges from 15-20°C. For raising the seedling 2-4 kg seeds are enough whereas for one hectare direct plating, 8-10 kg seed/ ha is required. The seedlings are transplanted in to furrows about 15 cm deep with about 9 cm between plants in the row. Tillers are transplanted into raised beds or ridges, which are alternated with furrows for irrigation and drainage. Planting distance are maintained 25 cm × 20 cm. Before plating the tillers, their top one third portions are trimmed to reduce transpiration.

Manures and Fertilizers

Welsh onion does well in soil rich in phosphate. Application of ammonium nitrogen as compared to nitrate nitrogen gives better results. For raising one hectare crop, 200-300 kg N, 100-200 kg P_2O_5 and 150-200 kg K_2O are commonly recommended (Inden and Isahira, 1990). Nitrogen application should be given in three split doses. Use of *Arbuscular mycorrhizae* has been found beneficial, however, selection of suitable species and optimal phosphate application are must because of species differences in the rate of infection, increase uptake and growth of plants. Among three species of *Arbuscular mycorrhizae, Glomus mosseae, Glomus fasciculatum* and *Glomus caledonium*, response of *G. fasciculatum* was found to be most effective (Tawaraya *et al.*, 1996).

Inter-culture

Welsh onion is moist loving plant and needs frequent irrigation. For chemical weed control in Welsh onion, use of trifluralin as P.P.I. was found effective (Nishi, 1976). Effective control of annual weeds (75-95 %) was obtained with the application of Ramrod as P.E. This resulted in 10-15 per cent enhancement in yield without any phytotoxic effect (Kameneve and Penkov, 1971). For blanched pseudo-stem production of Welsh onion, field must be deeply prepared and mounted at the base of the plant, the first earthling 50 DAT, thereafter earthling may be given at 20 and 40 days before harvesting in summer and winter, respectively.

Intercropping

Welsh onion can be raised as a mix cropped/intercropped with other vegetables. Arie *et al.* (1987) noted little occurrence of *Fusarium* wilt disease in bottle gourd where Welsh onion had been mixed cropped. In allelopathic study, Choi (1993) isolated 4 deterrents inhibitory substance from Welsh onion. This substance inhibited the root growth of chrysanthemum cuttings but promoted the growth of rice seedlings.

Harvesting and Yield

Plant of Welsh onion are pulled by hand and packed in to bundles. Harvesting is labor as well as time consuming operations especially for pseudo-stem. Mechanized harvesting equipment has been developed in Japan. Mechanized harvester takes about 200 to 1000/hrs for digging and lifting mechanically whereas it takes 3000 to 5000 hours by hand (Naguchi, 1977). Its average yield in Japan and Korea are about 25t/ha.

Flowering, Pollination and Seed Production

In a study on effect of photoperiod under low temperature on the growth and bolting, Yamasaki and Miura (1995) concluded that flower initiation in Welsh onion is facultative with respect to short days and that the inhibition on the flower initiation by a long photoperiod can be overcome by low temperature. Among the most abundant (81-91%) flower visiting insect of Welsh onion is honeybee (*Aphis mellifera*). Honeybee foraging flowers at about 6.00h and continue up to 19.00 h. Foraging activity is significantly positively correlated with solar radiation intensity, and negative with R.H. (Choi, 1987). In Welsh onion, seed production is directly dependent on flower induction and bolting. The Welsh onion belongs to the green plant verbalization type, specific seedling characteristics and sufficient accumulated time at low temperature are indispensible for the completion of its verbalization process and when these conditions for verbalization are fulfilled, the plants bolt in the following year. Dong *et al.* (2013) suggested that for control bolting in Welsh onion production, it is required there should be proper selection of appropriate cultivar, sowing date and transplant location. Choosing a late bolting cultivar, such as cultivar XiaHei sowing around October, and transplanting in the open field can significantly delay bolting, while a sowing date in late August should be selected for seed production, and the seedlings should be transplanted in a plastic tunnel to accelerate development of the flower buds.

Major Pests

1. Beet army Worm (*Spodoptera exigua*)

The beet army worm is a serious pest in South East Asia. This insect has attacked the crop of Welsh onion in Japan since early 1980s and is now the most important pest of Welsh onion in south western Japan. The efficacy of most of the insecticides has been reduced. Taki and Wakamura (1995) observed the feasibility of synthetic sex pheromone by dispensing a 7:3 mixture of (Z,E)-9, 12-tetradecadienyl acetate and (Z)-9-tetradecen-1-olas a communication disruption agent for the control of the beet army worm.

2. Robin bulb mite

Mites feed on sap released after piercing plant calls, and some encircled the plants in a typical web. Soil incorporation of granules of pyraclofos carbosulfan may reduce mite population in the soil. Dipping young plants with a diluted solution of pyraclofos and protinofos resulted in the higher reduction of mite population (Kadono and Kato, 1993).

Major Diseases

1.Rust : This is one of the most important disease of Welsh onion caused by *Puccinia allii*. The outbreak of the disease varies from year to year; depending on climatic conditio. Infected plants have yellowish-orange uredial lesions on the leaves.

2.Purple Blotch : This disease is caused by *Alternaria porrii*. Infected plant show characteristic concentric spot on the leaves.

3.Downy Mildew : This is caused by *Pernospora destructor*. Downy mildew does not cause serious damage to Welsh onion. Symptoms appear as grey/purple felt sporing bodies, chlorosis and collapse of leaves and flower stocks. Debris of previous crop and infected seed is the main source of infection (Cheremushkina *et al.*, 1990).

4.Bacteria Leaf Spot : This disease is caused by *Pseudomonas syringe*. The bacteria is highly infectious on welsh onion and Chinese chives (Goto, 1972).

5. Leaf Blight : This disease is caused by *Botrytis squamosl*. Infested plants show small leaf spot, rot round pseudostem with grey mould on it. Welsh onion is highly resistant to this disease while common onion is highly susceptible.

6. White Rot Disease : This disease is caused by the fungi *Sclerotium cepivorum*. This disease does not normally infect the green part of the plant but continues to the white portion. The only known source and means of dissemination of *S. cepivorum* is *via* infected plant material or sclerotia.

Welsh Onion Yellow Stripe Potyvirus (WoYSV) Potyvirureses are the most important because they induce yellow striping which may result in large yield reduction. Transmission takes place by various aphid species. It causes similar mosaic type symptoms, including chlorotic mottling, streaking and stunting and distorted flattening of the leaves. Relative tolerance has been observed in cultivars in Kujyo Group (Oyen and Messiaen, 2004).

Cultivars

The Welsh onion varieties may be classified in two groups (I) one with a bulb neck: bulb diameters ratio of 0.49 and a bulb weight of 40-50 g and (II) with values of 0.60-0.70 and 6.67 and 10.0 g respectively (Veselovskii and Khireba, 1972). In Japan, 4 groups of Welsh onion are defined (Oyen and Soenoeadji, 1993). These groups are given as under.

1.Kaga: These groups of cultivar have dark green and thick leaves and pseudostem. This group of plant have low dormant in winter. It is mainly grown for pseudostems.

2.Kujyo: These groups of cultivar have tender green leaves and plants remain green during winter and their ability to stand in cold conditions is medium. Plant have strong tendency of tillering. It is mainly grown for culinary tops.

3.Senju: These groups of cultivars are considered intermediate between Kaga and Kujyo. Plants growth continues during winter season. It is mainly grown for pseudostem.

4.Yakura Negi (*A. fistulosum*): This group of plants have strong tendency of tillering and they do not flower but produce bulbils. Propagation it done by divisions of the basal cluster or by bulbils.

Genetic Resources and Improvement

Collections of Welsh onion germplasm are maintained in Japan, the United States, the United Kingdom, Germany and the former Soviet Union. IPGRI ranked *Allium fistulosum* second in importance in the genus *Allium* because of its disease resistance, ecological adaptability and close relationship to *Allium cepa* (Oyen and Messiaen, 2004).The major breeding objectives of Welsh onion are to breed cultivars/ hybrids having high yield potential along with excellent horticultural traits. Possessing greater resistance to bolting, tolerance to high temperature and other abiotic and biotic stress are other desirable attributes. Interspecific hybrids (2n = 16) between *Allium fistulosum* L. (2n = 16) and *Allium schoenoprasum* L. (2n = 16) were produced by reciprocal crossings through ovary culture, but the hybrids were much fewer in the combination using *A. schoenoprasum* as a seed plant. All the hybrids have eight long chromosomes originated from *A. fistulosum* and eight short chromosomes originated from *A. schoenoprasum*. In addition, the hybridity was confirmed by randomly amplified polymorphic DNA(RAPD) analysis and cleaved amplified polymorphic sequence (CAPS) analysis of the rDNA internal transcribed spacer (ITS) region. The interspecific hybrids showed a vigorous growth habit; their foliage was slightly bloomy and deep green. The hybrids did not form bulbs, but rather propagated vegetatively by tillering. Carotene contents of the hybrids and both parents were quantified using high-performance liquid chromatography (HPLC). The contents of all edible parts of the hybrids were approximately seven times higher than those of either parent. These results indicate that the hybrid is a new and carotene-rich vegetable of *Allium* species (Umehara *et al.*, 2006)

Micro Propagation

Vegetative propagation of Welsh onion was established by tissue culture using immature leaf segment. MS medium supplemented with 1.0 mg, 2.4 D/liter +0.5mg kinetin/liter+1.0 mg BA/liter resulted in the most rapid callus growth, with a callus induction rate of 89.9 per cent. MS medium supplemented with 0.88 mg 2, 4.D/liter +1.73mg BA/liter was suitable for shoot induction, with a maximum rate of 13.21 per cent. Rooting occurred readily on MS medium without hormones (Zhang and Zhang, 1995). To propagate male sterile plants of Welsh onion, shoot-tip explant were successfully cultured by Fujita *et al.*(1977) on MS medium containing 2 mg/liter kinetin and 0.5mg/liter NAA, kept at 2^0c for 30 days. Adventitious shoots were transferred to a medium containing no growth substance and rooted after another 30 days. About 40 plants were obtained from a single shoot tip after 3 months.

The ovary and flower bud culture of Welsh onion developed callus and regenerated plants on media A6 and A11, which contained 1-5mg BA, 0.5mg NAA and 1mg 1BA/liter (Ionescu and Popandron, 1995). In an another study flower buds cultured on MS media supplemented with different concentration of NAA and BA revealed that higher NAA and BA concentration were favorable for callus induction, with 4 mg NAA+5.5mg BA/liter resulting in the highest rate of (33.02%). Adventitious shoots grew vigorously on medium supplemented with 1 mg NAA + 1 mg BA/litre on transfer to MS medium with no growth regulator, 100% of shoot rooted and the regenerated plant had a 76.32% survival rate after transplanting (Zhang and Zhang , 1994).

References

Arie, T., Namba, S., Yamashita, S., Doi, Y. and Kijima, T. (1987). Biological control of *Fusarium* wilt of bottle gourd by mix cropping with Welsh onion inoculated with *Pseudomonas gladioli*. Annals of the Phytopathological Society of Japan, J3 (4):531-539.

Cheremushkina, N.P., Orekhoskaya, M.V. and Koptyaeva, T.F. (1990). In the causal agent of onion *Peronosporos* transmitted by seed? Zashchita Rasteniij, Moskva, 11:39.

Choi, S.T. (1993). Studies on the biologically active substances from *Allium fistulosum*. II. Allelopathic substances from *Allium fistulosum*. Journal of the Korean Society for Horticultural Science, 34(5):355-361.

Choi, S.Y. (1987). Foraging activity of insect on Welsh onion (*Allium fistulosum*). Korean Journal of Apiculture, 2(2):67-74.

Cuvanova, L.G. (1970). Characteristic of growth and development of some onion species in relation to the light regime. Shorn trud. Aspir. Molod, nauc, Sotrud, Leningrad, 15:377-381.

Dong, Y., Cheng, Z., Meng, H., Liu, H., Wu, C. and Khan, A.R. (2013). The effect of cultivar, sowing date and transplant location in field on bolting of Welsh onion (*Allium fistulosum* L.). Plant Biology, 2, 13:154.

Fan, J.J. and Chen, J.H. (1998). Antimicrobial activity of Welsh onion ethanol extract. Journal of the Chinese Agricultural Chemical Society, 36(1):12-20.

FAOSTAT (2011). Onion harvested area and production in 2011.

Fujita, K., Ando, Y. and Fujita, Y. (1977). Propagation of Welsh onion through shoot tip culture. Journal of the Faculty of Agriculture, Kushu, Univ., 22(1-2):89-98.

Geriya, R.I. (1977). Economic biological characteristic of Welsh onion. Bulletin Vsesoyuznogo-ordena Lenina Instituta Rasteniev odstva imeni N.I. Vavilova, 68:78-80.

Goto, M. (1972). Bacterial leaf spot of onion in Japan. Plant Disease Reporters, 56(6)490:493.

Inden, H. and Asahira, T. (1990). Japanese bunching onion (*Allium fistulosum* L.) In: Onions and Allied crops of III. Brewster, J. L. and Rabinowitch, H. D. (eds) CRC. Press, Boca Raton, Florida, United States, pp. 159-178.

Inoescu, A. and Popandron, N. (1995). Research on the *in vitro* culture of ovules, ovaries and flower buds of some *Allium* genotypes. Anale Institutul de Cercetari pentru-Legumicultures Floricultura vindra, 13:17-29.

Kadono, F. and Kato, T. (1993). Chemical control of the robin bulb mite (*Rhizoglyphus robini claparede*, Acari- Acaridae), in Welsh onion fields in China. Proceedings of the Kanto Toson Protection Society, 40:249-253.

Kameneve, E. A. and Penkov, L. A. (1971). Ramrod on onions, Zasc Rast, 16(4):26.

Naguchi, K. (1977). Digger for Japanese bunching onion, in seed production Technology of vegetable crop. (in Japanese) V of 2. Assoc. Farm. Tech. Sub. Tokyo. Eds. Sciebundoshinkousya, Tokyo 230.

Nishi, S. (1976). Evaluation of candidate pesticide (c-4) herbicides: vegetable fields. Japan Pesticide Information, 28:27-30.

Oyen, L.P.A. and Soenoeadji, (1993). *Allium fistulosum* L. In: Siemonsma, J.S. and Kasem Piluek (eds.). Plant Resources of South-East Asia No 8. Vegetables. Pudoc Scientific Publishers, Wageningen, the Netherlands. pp. 73-77.

Oyen, L.P.A. and Messiaen, C.M. (2004). *Allium fistulosum* L. [Internet] Record from PROTA4U. Grubben, G.J.H. & Denton, O.A. (Editors). PROTA (Plant Resources of Tropical Africa.

Purseglove, J.W. (1972). Tropical crops: monocotyledons. Longman, London, 52.

Takai, M. and Wakamura, S. (1995). Control of the beet armyworm *Spodoptera exigua* (Hubner), with synthetic sex pheromone. Technical Bulletin (Asian and Pacific Council. Food and Fertilizer Technology Center); 142. 8 p

Tawaraya, K., Kinebuchi, T., Watanabe, S., Wagatsuma, T. and Suzuki, M. (1996). Effect of Arbuscular mycorrhizal fungi *Glomus mosseae*, *Glomus fasciculatum* and *Glomus caledonium* on phosphorus uptake and growth of Welsh onion (*Allium fistulosum*), in Andosol. Japanese Journal of Soil Science and Plant Nutrition. 67(3):294-298.

Umehara, M., Takayuki Sueyoshi, T. Shimomura, K. Iwai, M. Masayoshi Shigyo, M., Hirashima, K. and Nakahara, T. (2006). Interspecific hybrids between *Allium fistulosum* and *Allium schoenoprasum* reveal carotene rich phenotype. Euphytica, 148:295-301.

Veselovskii, J.A. and Khireba, A. (1971). Disease forms of Welsh onion. Byulleten Glavnogo Botani Cheskogo Sada, 86:87.

Xin Xiang Yuan, (2000). Observations on ultrastructure of pollen development from microsporocyte to early two - celled pollen in Welsh onion. Acta Botanica Yunnanica, 22(2):161-165.

Yamasaki, A. and Miura, H.(1995). Effectof photoperiod and low temperature on growth and bolting of Japanese bunching onion (*Allium fistuloqum* L.).Journal of the Japanese Society for Horticultural Science, 63:805-810.

Zhang, S. and Zhang, Q. P. (1994). Study on plant regeneration *via* flower bud culture of Welsh onion (*Allium fistuloqum* L.). Journal of Shandong Agricultural Univ. 25(3): 277-281.

Zhang, S. and Zhang, Q. P. (1995). Studies on propagation *in vitro* using immature leaf of Welsh onion. Acta Horticulturae Sinica, 22(2):161-165.

15

LEEK

Leek resemble large onion plants with flat leaves. Unlike onion and garlic, leeks do not form bulbs or produce cloves. The edible portion is a pseudo stem consisting of the elongated base and lower blade parts of the foliage leaves. Leeks are eaten raw as well as cooked. They are used primarily for flavouring soups and stews in place of onion. The taste of leek is milder than that of either onion or garlic. Compared to onion, leek contains more proteins and minerals on fresh weight as well as dry weight basis. The energy value of 100 g of edible portion of leek is also higher than that of onion. Nutritional composition of the leek is given in table (15.1).

Table (15.1) : Nutritive value of leek

(per 100 g of edible portion)

Energy	61 kcal
Carbohydrates	14.2 g
Sugars	3.9 g
Dietary fiber	1.8 g
Fat saturated	0.3g
Monounsaturated	0.004 g
Polyunsaturated	0.166 g
Protein	1.5 g
Water	83 g
Vitamin A equiv.	83 µg
Thiamine	0.06 mg
Riboflavin	0.03 mg
Niacin	0.4 mg
Vitamin B_6	0.233 mg
Folate (vit. B_9)	64 µg
Vitamin C	12 mg
Vitamin E	0.92 mg
Vitamin K	47 µg
Calcium	59 mg
Iron	2.1 mg
Magnesium	28 mg
Phosphorus	35 mg
Potassium	180 mg
Sodium	20 mg
Zinc	0.12 mg

Source: USDA Nutrient Database

Origin and Distribution

Leek is an ancient crop and its mention has been made in the Bible. Most probably it was domesticated in the eastern Mediterranean where many of the related types (Turkish leek, Kurrat) and related *Allium* species occur. Leeks are mainly grown in northern Europe, the United States and Canada. It is especially important in North European countries such as Belgium, Denmark and the Netherlands. In our country, it is not commonly cultivated at larger scale.

Botanical Description

Leek is a member of the genus *Allium* (family Alliaceae), which consists of over 600 species (Hanelt, 1990; Jones, 1991). One characteristic of this family is the basal plate. The basal plate contains both shoot and root meristems in a circular disc. Some disagreement exists about the taxonomic position of leek. Some taxonomists use the taxonomic classification *Allium porrum* L. for leek (Mathew, 1996). Hanelt (1990) reviewed the *Allium* genera and distinguished 5 subgenera. Leek was placed in the *A. ampeloprasum* complex, section *Allium* of the subgenus *Allium*. Its chromosome number (2n) is = 32 (tetraploid).

Taxonomic position of the *Allium porrum* L.

Kingdom	Plantae
Subkingdom	Tracheobionta
Superdivision	Spermatophyta
Division	Magnoliophyta
Class	Liliopsida

Subclass	Liliidae
Order	Liliales
Family	Liliaceae
Genus	*Allium* L.
Species	*Allium porrum* L.

Morphological description of plant

Robust erect herb up to 150 cm tall; true stem consisting only of a basal plate or disk, with little or no bulb formation, or with bulbs producing two kinds of cloves, small stiped ones and larger sessile ones; roots adventitious. Leaves distichously alternate; sheath tubular, forming a pseudostem up to 50 cm long; blade linear, up to 50 cm × 7 cm, V-shaped in cross-section. Inflorescence a spherical umbel 4-12 cm in diameter, bearing several hundreds of flowers, on a solid, terete scape up to 150 cm long; umbel subtended by a single, long-pointed spathe shed at flowering. Flowers bisexual, campanulate; pedicel 1-5 cm long; tepals 6, in 2 whorls, free, ovate-oblong, 4-6 mm long, purple or white; stamens 6, the inner ones tricuspid; ovary superior, 3-celled. Fruit a depressed globose to ovoid capsule 2-4 mm in diameter, up to 6-seeded. Seed 2-3 mm × 2 mm, black. Seedling with epigeal germination (Messiaen and Rouamba, 2004).

Cytogenetics

Leek, shows extensive and frequent quadrivalent formation during prophase I of meiosis, and that despite the resolution of most of these into bivalents some quadrivalents persist to metaphase I; in addition, a substantial number of univalents are found at this stage. Irregularities of chromosome segregation are observed at anaphase of both meiotic divisions, giving rise to micronuclei in dyad and tetrad stage pollen mother cells. It is shown that micronucleus frequency in tetrads is correlated with univalent frequency at metaphase I, suggesting that micronuclei originate largely from missegregation of univalents. Seedling populations of the same four cultivars were screened cytologically and shown to contain aneuploids at frequencies ranging between 4.3% and 8.4%. It is proposed that the majority of aneuploid seedlings originate from meiotic irregularity, especially univalent missegregation. It is very likely that the presence of this level of aneuploidy in the seedling populations contributes to the problems of phenotypic non-uniformity experienced by the leek crop, although its relative importance is difficult to assess (Khazanehdari and Jones, 1997).

Varieties

In Himachal Pradesh, Musselburgh and Prizetaker are commonly grown.The cultivars of leek which are commonly available with seed man include the following:

Carina: It has heavy, thick, long white shanks

Titan: It is an extra long, early type has vigorous growth. The leaves are dark green with white stems

Alaska: It has dark blue green foliage

Pancho: It is considered an early variety. It has thick white shanks

Palam Paushtik: It has been developed from Himachal Pradesh Agricultural University, Palampur. It possesses long, broad flat garlic like blush green leaves and long white pseudostem thickening cylindrically to attain 3-4 cm diameter at harvesting. It takes about140-150days to reach marketable stage. It has mild digestive flavour. The average yield is 30-35 tonnes per hectare. Its nutritional composition is given in table (15.2)

Table (15.2): Nutritive value of leek cv. Palam Paustik (per 100g of edible portion)

Protein	1.5 %
Fat	0.3 %
Fibre	2.4 g
Calcium	65.2 mg
Iron	1.6 mg
Phosphorus	61.0 mg
Potassium	297.0 mg
Carotene	1.25 mg
Ascorbic acid	28.0 mg

Source: Kalia and Singh (2007)

Lincoln: This variety is marketed by Bejo Sheetal Company. It matures in 80 days after transplanting. It is early Bulgarian giant type leek, which gives uniform long and thick stems.

Cultural Practices

The best temperature for growing leek is 20-25°C. In the tropics, it is usually cultivated in the highlands at about 1000 m altitude. Leek is a cool season crop, temperature is more important than the length of day in seed stalk development. It is propagated through seed. Leek seed does not show dormancy and germinate epigeally. Optimum temperature for seed germination is 18-22°C. In northern parts of India, seeds are sown in September-October in the nursery beds. In Himachal Pradesh, seeds are sown from August to October in mid hills and from March to April in high hills. There are approximately 370 seeds per gram and 1.5 kg seeds are required for raising the seedlings for planting one hectare field. Seedlings are ready to plant when they attain a height of 15 cm. Leeks can be grown on a variety of soils but grows luxuriantly on medium soils that are rich in organic matter. The most favourable soil pH range for leek is 6-8. Plants of leek are larger than the onion, hence requirements of manure and fertilizer are higher. A yield of 30 tonnes per hectare removes 100 kg of nitrogen, 65 kg P_2O_5 and 130 kg of potash from the field. The diameter and length of bulbs are increased by nitrogen fertilization. For realizing the good yield, well rotten FYM @ 25 t/ ha should be applied at the time of field preparation. Besides, 200 kg N, 120 kg P_2O_5 and 100 kg K_2O /ha may be given. Nitrogen should be given in four splits at an interval of one month after transplanting.

Transplanting is done at the spacing of 60×45 cm in trenches which are 30-45 cm deep to make the plant blanch. Leeks require a large amount of water during growth. Hence, frequent irrigation should be given to maintain the proper soil moisture. Weed control at the proper stage plays an important role in the growth, yield and quality of leeks. Hence, field should be kept free from the seasonal weeds. In chemical herbicides, pendimethialin @ 1.0 kg a.i. per hectare may be applied as pre-emergence. Blanching may be done by covering the plants to certain height with soil to improve the quality of the crop. For this purpose, plants are sunk up to their center leaves in trenches or pits that are heavily manured to earth up soil as they grow. Care should be taken not to earth up soils early when the plants are young.

Harvesting and Yield

It takes about 28-30 weeks from sowing to harvest. Leek is harvested just like green onion and marketed in bunches. After harvesting, plants should be kept in cool conditions. The average yield varies from 300-400 quintals per hectare. The green crop can be stored for 1-3 months at a temperature of 0° C with a relative humidity of 85-90 per cent.

Major Diseases

Purple blotch: It is the most important leaf disease of leek caused by (*Alternaria porri*).

Control Measures

i) Remove and destroy the infected plants.

Rust : It is caused by *Puccinia spp.*

Control Measures

i) Apply foliar application of Kerathane @1 g per litre of water.

Major Pests: *Thrips tabaci* can cause serious losses in yield and quality. Leek moth (or onion leafminer), *Acrolepiopsis assectella* Zeller, is an invasive alien species of European origin that damages *Allium* spp. It was first identified in Eastern Ontario in 1993. The distribution of the pest includes Asia, Africa, Europe and Canada. The leek moth is a known pest of Alliums, including onion, garlic, leeks, chives, green onion, shallot, elephant garlic, wild garlic, etc. Larvae are yellowish-green with a pale brown head capsule and eight small grey spots on each abdominal segment Larvae can cause extensive damage by tunnelling mines and feeding on leaf tissue and occasionally on bulbs.

Control Measures

1. Follow long crop rotation.
2. Removal of old and infested leaves is essential .
3. Destroy pupae or larvae.
4. Destruction of plant debris following harvest.

Genetic Resources and Improvement

Major objective in leek improvement are yield, uniformity and disease resistance. Leek can be considered an allogamous autotetraploid with 20-30% self-pollination in seed-production fields. In general, maintaining open -pollinated cultivars under permanent mass selection pressure, or developing F_1 hybrids are in vogue. Plant vigour loss by selfing is about 30% for the first generation and 40% between S_1 and S_2. F_1 hybrids are based on male sterility. F_1 hybrid leek has 20% higher yields and is more uniform. The rationale behind this is the exclusion of progeny resulting from self-pollination. F_1 cultivars can have a different genetic background but they always have the male sterile parent in common. This line is vegetatively propagated *via* in-*vitro* multiplication and/or formation of bulbils, obtained in the umbels by early flower ablation and exposition to photoperiods of 20 hours. Cytoplasmic male sterility has recently been transferred from *Allium cepa* (Messiaen and Rouamba, 2004).

The cultivated leek is tetraploid (2n=4×=32) (Levan, 1940), which also makes breeding complicated. Opinion differs towards its ploidy nature. Most authors agree that leek is an autotetraploid, because for most of the genes leek shows tetrasomic inheritance (Levan, 1940; Kadry and Kamel, 1955; Berninger and Buret, 1967; Schweisguth, 1970). According to Koul and Gohil (1970) leek is a segmental allotetraploid with the genomic formula AAA'A", obtained from three closely related parental species. The fact that in leek the chiasmata are localized near to the centromeres has as the consequence that the amount of recombination is limited (Gohil 1984; Stack and Roelofs, 1996). Because leek has a narrow gene pool, the use of closely-related wild species in breeding programmes would be interesting. They may provide a possible source of desirable traits, such as disease resistance or cytoplasmic male sterility.

The most desirable way to produce hybrid seeds in leek would be a system based on *CMS*. Unfortunately, no source of *CMS* is currently available in leek. The search for spontaneously occurring cytoplasmic male sterile leek plants in large seed production fields has not been successful up till now (Silvertand, 1996). It is possible that *CMS* is present in leek, but due to the action of restorer genes not phenotypically expressed. Therefore, other approaches to introduce or create *CMS* in leek have been investigated. Silvertand (1996) investigated the possibility of inducing *CMS* in leek by using chemical mutagens. In other crops the use of chemical mutagens has proven to be effective in inducing mitochondrial mutations (Kaul, 1988). It was possible to induce male sterility in leek by this method, but this was not proven to be of a cytoplasmic origin. Attempts to transfer *CMS* from related *Allium* species *via* interspecific crosses, whereby the *CMS* donor is used as a female parent have failed, due to serious difficulties in obtaining fertile hybrids. A possibility that has not been explored by leek breeders and researchers yet, is the introduction of alien cytoplasms in leek by intraspecific crosses with members of the *A. ampeloprasum* complex, in order to create an alloplasmic type of *CMS*. The members of the *A. ampeloprasum* complex, which are interfertile with leek, provide a large pool of cytoplasmic genome variation (Kik *et al.*, 1997). A more intensive search within these close relatives of leek might even result in the identification of a sterile cytoplasm (Buiteveld, 1998).

Tissue culture

The experiment was conducted utilizing leek seed radical as graft to induce healing tissue for regeneration culture, and application of PGR effects over hormone matching for cultivation and regeneration of leek healing tissue. The result shows that MS+NAA (1mg/L) +6-B (A2mg/L) is suitable for radical to induce healing tissue and regenerative cultivation, and rational hormone matching is beneficial for leek to rapidly form healing tissue (Erjun *et al.*, 2008).

References

Berninger, E. and Buret, P. (1967). Étude des déficients chlorophylliens chez deuz espèces cultivées du genre *Allium*: l'oignon *A. cepa* L. et le poireau *A. porrum* L. Ann Amélior Plantes 17:175-19.

Buiteveld, J. (1998). Regeneration and interspecific somatic hybridization in *Allium* for transfer of cytoplasmic male sterility to leek ISBN 90-5485-813-3, (CPRO-DLO), P.O. Box 16, 6700 AA Wageningen, The Netherlands

Er-jun, Q.U. *et al.* (2008). Study on the effect of hormone on inducing leek callus [J]. Journal of Anhui Agricultural Sciences; 18.

Gohil, R.N. (1984). Extent of recombination possible in the cultivated leek. In: Van der Meer QP (ed.) Eucarpia 3rd *Allium* symposium, Wageningen, pp 99-105.

Hanelt, P. (1990). Taxonomy, Evolution and History In: Rabinowitch, H.D. and Brewster, J. L.(eds.) Onions and Allied crops, Vol. I. CRCPress, Inc, Boca Raton, Florida.

Jones, R.N. (1991). Cytogenetics of *Alliums*. In:Tsuchiya, T.and Gupta, P. K. (eds) Chromosome engineering in plants: Genetics, Breeding and Evolution, part B. Elsevier, New York, Oxford, pp 215-227.

Kadry, A.E.R. and Kamel, S.A. (1955). Cytological studies in the two tetraploid species *Allium kurrat* Schweinf. and *A. porrum* L. and their hybrid. Svensk Bot Tidskrift 49: 314-324.

Kalia, P. and Singh, Y. (1997). Palam Paustik- a new leek. Indian Horticulture, Jan.-Feb, pp.16-17.

Kaul, M.L.H.(1988). Male sterility in higher plants. In Frankel, F. Grossman, M. Linskens, H.F. Maliga, P., Riley, R. (eds.) Monographs on Thoretical and Applied Genetics. Springer Verlag, Berlin, Heidelberg, New York, London, Paris, Tokyo, 100pp Press, Inc, Boca Raton, Florida.

Khazanehdari, K.A. and Jones, G.H. (1997). The causes and consequences of meiotic irregularity in the leek (*Allium ampeloprasum* spp. porrum); implications for fertility, quality and uniformity. Euphytica, 93(3): 313-319.

Kik, C., Samoylov, A.M., Verbeek, W.H.J. and Van Raamsdonk, L.W.D. (1997). Mitochondrial DNA variation and crossability of leek (*Allium porrum*) and its wild relatives from the *Allium ampeloprasum* complex. Theor. Appl. Genet. 94: 464-471.

Koul ,A.K. and Gohil, R.N. (1970).Cytology of the tetraploid *Allium ampeloprasum* with chiasma localization.Chromosoma, 29:12-19.

Levan, A. (1940). Meiosis of *Allium porrum*, a tetraploid species with chiasma localisation. Heriditas, 26:454-462.

Mathew, B. (1996). A review of *Allium* section *Allium*. IPBGR, Royal Botanic Gardens, Kew, United Kingdom. 176 pp.

Messiaen, C.M. and Rouamba, A. (2004). *Allium ampeloprasum* L. In: Grubben, G.J.H. & Denton, O.A. (Eds). PROTA 2: Vegetables/Légumes.PROTA, Wageningen, Netherlands.

Schweisguth, B. (1970). Études préliminaires a l'amélioration du poireau *A. porrum* L. proposition d'une method d'amélioration. Ann Amélior Plantes, 20: 215.

Silvertand, B. (1996). Induction, maintenance and utilazation of male sterility in leek (*Allium ampeloprasum* L.). Ph.D. Thesis Agricultural University Wageningen, The Netherlands.

Stack, S.M. and Roelofs, D. (1996). Localized chiasmata and meiotic nodules in the tetraploid onion *Allium porrum*. Genome, 39: 770-783.

16

CHIVES

Chives is the smallest species of the ed ible onions, and is the most widely distributed. It is only *Allium* species that is native to both the New and the Old World, found in Europe, Asia, and North America and growing in Arctic regions up to 70°N latitude or in mountainous regions at lower latitudes (Brewster, 1994). The name of the species derives from the Greek skhoínos (rush or reed) and práson leek (Jaeger, 1944). The term "chives" is commonly used in Japan as the English name for *A. schoenoprasum* var. foliosum, the Japanese name of which is "asatsuki". "Chives" is therefore restricted to *Allium schoenoprasum* var. schoenoprasum (type species) and the term "asatsuki" is used for *A. schoenoprasum* var. foliosum to discriminate the two varieties.

Chives do not produce large bulbs, but it develops new axillary shoots after 2 or 3 leaves have formed, with shoots attached to short rhizomes (Brewster, 1994). The plant forms dense clumps with narrow, linear, hollow leaves, typically 15-30 cm tall. The flowers are packed in dense umbels and do not produce bulbils (small bulbs). Chive leaves and flowers are both edible. The leaves are widely used as a culinary herb and flowers are used in salads. The flavor is similar to but milder than other onions. Chives can be used fresh or dried. Chives produce sulfur compounds including methyl sulfides and disulfides. Nutritional composition of hive leaves is given in table (16.1).

Table (16.1) Nutritional composition of Chives (per 100 g of edible portion)

Energy	30 kcal
Carbohydrates	4.35 g
Sugars	1.85 g
Dietary fiber	2.5 g
Fat	0.73 g
Protein	3.27 g
Vitamin A equiv.	218 μg
Beta-carotene	2612 μg
Lutein and zeaxanthin	323 μg
Thiamine (vit. B_1)	0.078 mg
Riboflavin (vit. B_2)	7%
Niacin (vit. B_3)	0.647 mg
Pantothenic acid (B_5)	0.324 mg
Vitamin B_6	0.138 mg
Folate (vit. B_9)	105 μg
Vitamin C	58.1 mg
Vitamin E	0.21 mg
Vitamin K	212.7 μg
Calcium	92 mg
Iron	1.6 mg
Magnesium	42 mg
Manganese	0.373 mg
Phosphorus	58 mg
Potassium	296 mg
Zinc	0.56 mg

Origin and Distribution

Chives have been in cultivation in Europe at least since the 16th century; records of related *Alliums*, including onions, garlics, and leeks, are found in Egyptian and Mesopotamian records dating to 5000 B.C. The Romans believed chives could relieve the pain from sunburn or a sore throat, and that eating chives would increase blood pressure and acted as a diuretic.

Chives are primarily grown in home gardens and small farms. The total area commercially harvested globally in 1990 was 1000 hectares, with Denmark, New Zealand, and Germany were leading producers (Brewster, 1994).

Botanical Description

Chives (*Allium schoenoprasum*) possess attractive green foliage, and purple flowers which make them useful as ornamental edging plants or for growth in clumps among other herbs (Jones and Mann, 1963). The species is highly polymorphous, being widely distributed throughout the world (Levan, 1936) and many ecotypes and varieties exist (Poulsen, 1990 ; Stearn, 1978).

Chives are a bulb-forming herbaceous perennial plant, growing to 30-50 cm tall. The bulbs are slender, conical, 2-3 cm long and 1 cm broad, and grow in dense clusters from the roots. The scapes (or stems) are hollow and tubular, up to 50 cm long, and 2-3 mm in diameter, with a soft texture, although, prior to the emergence of a flower, they may appear stiffer than usual. The leaves, which are shorter than the scapes, are also hollow and tubular, or terete, (round in cross-section) which is distinguishing feature. The flowers are pale purple, and star-shaped with six petals, 1-2 cm wide, and produced in a dense inflorescence of 10-30 together; before opening, the inflorescence is surrounded by a papery bract. The seeds are produced in a small three valved capsule.

Climate and Soil

Chives thrive in well-drained soil, rich in organic matter, with a pH of 6 to 7 and full sun. Chives seeds germinate at a temperature of 15°C to 20°C. In cold regions, chives die back to the underground bulbs in winter and new leaves appear in early spring.

Propagation and Agro techniques

The most successful means of propagating chives is planting rooted clumps. Established plants usually need to be divided every 3 to 4 years. Its seedlings can be raised in polyhouses/ indoor conditions. Seedlings become ready for planting in four weeks. Chives can be grown as an annual or a perennial. Plants are spaced at 10 cm apart in rows 30 cm apart. Adequate nitrogen is important and before planting incorporation of composted manure is recommended with additional nitrogen applications after the first harvest every year. Terhi (2003) reported that chive need heavy fertilization for realizing the significant yield and it was observed that annual uptakes in crop yield were 185-200 kg ha $^{-1}$ for nitrogen, 17-20 kg ha $^{-1}$ for phosphorus, and 120-140 kg ha -1 for potassium in the most intensively fertilized treatment producing the highest yield. Further, he observed that black plastic mulch was effective in increasing yield, controlling weeds and maintaining soil moisture. Weed control is particularly important during the first 2 months of growth when the plants are growing slowly and cannot compete with weeds (Halva and Craker. 1996) . Keep the plants green and lustrous; it is essential to keep the soil moist throughout growing season.

Major Pests

Thrips : They are tiny, slender insects that feed on leaves. Leaves turn silver or grey, may twist and die off. Thrips hide near where the leaves and bulb join.

Control

1. Spray miticide like Kelthane
2. Predatory mites can be used to control infestations.

Root maggots

Root maggot is a white worm that feeds on seedlings, roots or bulbs.

Control

1. Select pest and disease resistant cultivars.
2. Apply diazinon granules at planting.

Major Diseases

1. Pink root: Pink root is fungal disease that changes roots into a pink colour. Roots eventually die off and yields are severely reduced.

Control

1. Use 5-year crop rotation.
2. Soil solarisation where the chives will be planted can be done.

2. Downey mildew

Downey mildew is a fungal disease that turns leaves into a light tan to brown colour. Furry growth, greyish violet in colour, may be visible on the surface of infected leaves during moist periods.

Control

1. Cultural practices that facilitate air movement and drying of leaves will reduce disease severity.
2. Apply *Trichoderma* for seedling treatments and soil application.

Harvesting

Harvesting is made by snipping off leaves from the base of the plant. Cut flower stalks off at the the base of the plant. Chives are most flavourful when used fresh. It can be frozen by chopping up prewashed leaves into small pieces and freezing these in plastic containers.

Genetic Resources and Improvement

Chives (*Allium schoenoprasum*) continuously produce leaves during spring and summer, and continue to develop leaves in winter, although growth is slow (evergreen). They do not produce well-defined bulbs, whereas asatsuki (*A. schoenoprasum* var. foliosum) forms bulbs with leaf desiccation in summer. Bulb formation of F_1 progenies of chives×asatsuki and BC_1 plants obtained by crosses with asatsuki was investigated. No F_1 progenies formed bulbs. The bulb formation of BC_1 was classified based on the bulbing ratio into two phenotypes, e.g. non bulb forming and bulb-forming types; however, the variation persisted in obvious bulb forming and non-bulb forming individuals. The index of the maximum thickness of scale leaves was an alternative to clearly segregate the BC_1 progenies into three types, e.g. no bulb formation, intermediate, and bulb-forming phenotypes. These results indicate that more than two recessive genes are involved in the bulb formation of chive (Xiao *et al.*, 2010).

Tissue Culture

In a comparative study on chive (*Allium schoenoprasum*) cultivated plant and (*Allium schoenoprasum*) tissue culture organs antioxidant status (antioxidant enzyme activities viz., superoxide dismutase, catalase, guaiacol peroxidase and glutathione peroxidase, reduced glutathione quantity, flavonoids and soluble protein contents and quantities of malonyldialdehyde and OH radical), Štajner *et al.* (2011) concluded that tissue culture plants exhibited the highest activities in the roots in contrast to the cultivated plants where highest activities were observed in the leaves.

References

Brewster, J.L. (1994). Crop production science in horticulture, Volume 15: Onions and other vegetable Alliums. Wallingford, Oxon, Great Britain: CABI Publishing. 2nd ed.

Halva, S. and Craker L.E.(1996). Manual for Northern Herb Growers. HSMP Press.

Jaeger, E.C. (1944). A source book of biological names and terms. Springfield, IL: Charles C. Thomas.

Jones, H.A and Mann, L.K. (1963). Onion and their allies. Leonard Hill, London.

Levan, A. (1936). Zytologische Studien an *Allium schoenoprasum*. Hereditas, 22: 1-126.

Poulsen, N. (1990). Chives, *Allium schoenoprasum* L. p. 231-250. In: J. L. Brewster and H. D. Rabinowitch (eds.). Onion and allied crops. III. CRC Press, Boca Raton, Florida.

Štajner, D., Popovic, B.M., Calic-Dragosavac, D., Malencic, D., Zdravkovic-Korac, S. (2011). Comparative study on *Allium schoenoprasum* cultivated plant and *Allium schoenoprasum* tissue culture organs antioxidant status. Phytotherapy Research, 25(11):1618-1622.

Stearn, W. T. (1978). European species of *Allium* and allied genera of Alliaceae a synonymic enumeration. Ann. Musei Goulandris, 4:83-198.

Terhi (2003) Yield potential of chive: effects of cultivar, plastic mulch and fertilisation. Agricultural and Food Science, 12 (2):95-105

Xiao, J., Ureshino, K. Hosoya, M., Okubo, H. and Akira Suzuki, A. (2010). Inheritance of bulb formation in *Allium schoenoprasum* L. J. Japan. Soc. Hort. Sci. 79(3):282-286.

17

CHINESE CHIVES

Chinese chives is grown for its edible garlic flavoured leaves and young inflorescence. In China, shoots are often harvested after being blanched by excluding light. Chinese chives are also grown as a decorative plant for its beautiful white flowers. China is leading producer of Chinese chives and it exports to USA where it is preferred for its characteristic flavor (Jia *et al.* 1996). In China, it is considered 3rd important crop of Allium group after garlic and Welsh onion (Wuteng *et al.*, 1997). Chinese chives are rich in vitamin and sugar content. The leaves contain 3 per cent sugar on a fresh weight basis and taste very sweet. The main sugars are fructose, sucrose and glucose (Saito,1990). Chinese chives possess several medicinal properties and are used against tumours and intestinal disorders. Seeds of Chinese chives are also source of steroidal oligo-glycosides (Ikeda and Nohora, 2000). In Thailand, the seed is used against toothache. Consumption of Chinese chives is considered effective. The major properties of Chinese chives are that its fiber content does not change during normal heat treatment (Takahashi *et al.*, 1989). Nutritional composition of Chinese chives are given in table (17.1)

Table (17.1) : Nutritional composition of Chinese chives (per 100g of edible portion)

Compositions	Fresh	Boiled
Protein	2.1g	2.3g
Fat	0.1g	0.0
Sugar	2.8g	5.2g
Energy	19 Kcal	28 Kcal
Water	93.1g	90.8g
Fiber	0.9	1.1g
Ash	1.0g	0.6g
Calcium	50mg	46mg
Iron	0.6mg	0.6mg
Phosphorus	32mg	23mg
Vitamins A	1800IU	2200 IU
Thiamin	60 µg	40 µg
Riboflavin	190 µg	110 µg
Niacin	600 µg	300 µg
Vitamin C	25 µg	10mg

Saito (1990)

Origin and Distribution

The Chinese Chives probably originated in China where it was grown in the 10th century and possibly even as early as 200 B.C. It grows wild over much in East Asia from Magnolia in the North to the Philippines in the South and from Japan to Thailand East-West. Since plant of Chinese chives easily naturalizes from cultivation, Brewester (1999) expressed uncertainty about its initial centre of origin. Chinese chives are crop of commercial importance in East-Asia, notably Japan, Korea and China.

Botanical Description

Chromosomes and Nucleic Acids

The diploid chromosome number of Chinese chives is =32. Chinese chives contained 4C DNA values 109.36 pg. The high DNA Value of the tetraploid *A. tuberosum* was not proportionately double the probable diploid progenitor *A. cepa.* Autotetraploidy in *A. tuberosum* has possibly been associated with lowering of DNA content through mutational and other changes, conferring species stability (Talukder and Sen, 1999).

Morphology

Chinese chives are a perennial herb forming dense clumps, 20-40 cm tall, with a prominently spreading rhizome from which thick long persistent roots appear. Bulks indistinct, narrowly ovoid, 15-20 mm × 15 mm with several protective brown bulb coat leaves broken up into netted fibres. Foliage leaves 4-9, distichous, linear, 13-45 cm × 2-10 mm, flat above, slightly keeled below, not folded lengthwise sub-erect to curved. Scape 1, compressed, with 2 longitudinal ribs, up to 50 cm long, solid. Inflorescence umbellate, many flowered, 3-5 cm in diameter, with cut bulbils; Spathe short, persistent, opening with 1-3 valves: pedicels sub equal, 14-35 mm long; flowers white, widely opened, star-like, slightly fragrant; tepals oblong to ovate, 6 mm × 3 mm; Stamens and pistils up to as long as the tepals. Fruit obovoid, 5-6 mm long and wide. Seed irregularly depressed globose, 3-4 mm long black (Van der Meer, 1993).

Climate and Soil

The optimum temperature for cultivation of Chinese chives is about 20°C. In Indonesia, it is grown in the highlands up to 2200 m altitude on fertile and loose soils. Under tropical conditions, growth is not interrupted by dormancy or by flowering. Cultivars which are grown in Malaysia and Thailand come in flowering and inflorescence are supplied for markets. Flowering can be induced by using incandescent light to create artificially long days. Ma Shubing *et al.* (1999) reported that annual leaf production of Chinese chives varicd between 31 and 52 leaves with a peak production in May-June. The mean interval between formations of successive leaves was 8.1 days. Mean leaf longevity varied from 38 to 41 days. Leaf blade growth peaked twice in the year in May and August, and the blade expansion took 17-19 days. Tillering peaked in July and there was lesser peak in October. Between 5^{0}C and 30^{0}C, the number of new leaves formed and the daily rate of expansion were positively related to temperature and the interval between the production of new leaves was inversely related to temperature. Chinese chives thrives well in moist soils. For realizing the high yield, adequate supply of manure and chemical fertilizer is essential.

Propagation

Chinese chives is propagated both vegetatively and from seeds. Seed is formed in abundance. In general, plants did not find differ in vitamin 'C' content between transplanted and plants grown from seeds.

Raising the Seedlings and Planting

New crop of Chinese chives is raised through seed sowing in nursery beds. Germination of seed is epigeal, cotyledons with a typical bend (knee). Seedlings are transplanted in a cluster of 10 or so, spacing is maintained 20 cm apart in rows and 30-40 cm between rows. Han (1988) obtained maximum yield at closest planting distance with 18 plants/hill at planting distance of 15×15 cm. For home consumption and small scale cultivation, plants of Chinese chives are raised in pots. In the tropics, Chinese chives is propagated vegetatively by division of clumps.

Manures and Fertilizers

Chinese chives is perennial crop, hence it needs adequate supply of various nutrients. Lin (1973) advocated each of N,P and K @ 60kg/ha/month. Further, he observed better quality with high level of N and low levels of P and K. Top growth of Chinese chives is highly correlated with ineral uptake during the growth period. Han (1988) obtained higher fresh weisht, number of tillers and N, P, K, C and Mg in plants grown in hydroponic than soil culture. Plants grown in hydroponic contained higher vitamin C (24.7-96.4 mg/100g) than those grown in soil (21.6-51.1 mg/100g).

Intercropping

Chinese chives can successfully be grown in combination with vegetables like tomato, amaranth and other *Brassica* greens. In an allelopathic study, it was found that Chinese chives had no detrimental effect on the growth of tomato plant but significantly delayed and suppressed the occurrence of bacterial wilt of tomato (Yu, 1999)

Growth and Development

Outside the tropics, short photoperiods induce dormancy of buds which is broken by low temperatures; long photoperiods induce flowering. Under tropical conditions, the growth is hardly or not interrupted by dormancy, and normally no flowering occurs (Van der meer, 1993). Gao *et al.* (1992) studied the photosynthetic characteristics of Chinese chives and reported that light saturation point was 40 klux and the light compensation point was 120 lux. The optimal temperature for photosynthesis was 23°C. The CO_2 saturation point ws 400 ppm and the compensation point was 42 ppm. The diurnal variation of photosynthetic rate showed double curve during the day.

Harvesting

Cutting of flowers is avoided during Ist year to build up of the rhizome. Cuttings start in the second season when the leaves are about 20 cm long. In Japan, quality is graded by leaf length, flavour and tenderness, leaves 23 cm to 28 cm long are considered as top quality, 18 to 22 cm long as good, and 15 to 17 cm as medium quality (Saito, 1990). The number of cuts per year varies from about 4 in north China to 8 in the south of the country (Brewster, 1999). The edible flower stalks are cut when 30-40 cm long while the flower buds are still green.

Packing and Storage

Both the leaves, particularly when blanched and the flowers are highly perishable and should dispose off quickly to avoid post harvest losses. Ping (1998) reported that flower opening rate of Chinese chives scape exceeded 50 per cent after 18 days storage at 10°C. Endogenous IAA levels decreased and IBA levels increased during flower opening and senescence in storage. The optimum temperature for (26 days) of Chinese chives scape was approximately 3^{0}C. Treatment of scape with 2, 4-D or GA_3 at harvest delayed flower opening, reducing flower opening rate to less than 50 per cent of that the control. These treatments also increased the endogenous concentrations of IAA, $GA_3$and 2, 4-D and reduced that of ABA compared to control. Film wrapping was effective in maintaining freshness of scape. Losses through decay and flower opening were reduced <1% after 26 days of storage at 3°C when polypropylene film wrapping was used. Packing of Chinese chives flower should be done in bundles of about 110g and these bundles may be sealed in polyethylene bags (10 bundles/bag) of about 30×45 cm in sized about 0.025 mm in thickness and 4 bags should be vertically placed in a corrugated box and cooled to 5°C within a day. The cooled produce may be transported in insulated containers.

Cultivars

A number of cultivars of Chinese chives have been reported from Japan, China and Taiwan. These cultivars differ in leaf size, dormancy period, tillering and hardiness. Chinese chives cultivars may be classified into 4 categories on the basis of their degree of dormancy. The cv. 'Jumbo Nira' possesses the longest period of dormancy from mid October to late January. Yabana (1983) was in favour to divide the Chinese chives cultivars mainly in two groups ie., (i) varieties suitable for edible leaves and (ii) varieties suitable for edible flowers. Recently a Chilgokbuchu cultivar has been developed and its detailed attributes are are given as under

Chilgokbuchu

It was developed at kyongbuk ARES, Taegu in 1997 from local cultivar collected at Chilgok in 1993. Chilgokbuchu has longer and narrower leaves and stronger flavor than cv. Green Belt. Its content of ascorbic acid in field and in plastic-house are also higher than those of Green Belt. It has more number of tillers/plant than Green Belt. Its growth and re-growing capacity after cutting of top leaves in field are vigorous. Flowers and seeds/umbel were fewer in Chilgokbuchu than in Green Belt. Its yield potential is 7 per cent higher than that of Green Belt (Choi-Kyungbae *et al.*, 1998) .

Genetic Resources and Improvement

The main breeding objective of Chinese chives is to develop cultivars with high yield potential and rich in nutritive value especially in vitamin C, carotene and sugar content. In China, some selection of plants for seed production has been made. The genebanks at Zenfralinstitut fur Genetik and Kulturpflanzen for schung, Gatersleben (Germany) and at the Institute of Horticultural Research, Wellsbourne (UK) holds small collection (Van der Meer, 1993).

Micropropagation

Embryogenic calli from shoot tip of two cultivars of Chinese chives were used to establish donor cell suspension culture to produce protoplast. Cell division and colony formation from protoplast proved to be best in agarose solidified B5 medium containing glucose. Protoplasts of cv. Wide Green produced micro calli but those of cv. Super Green Belt did not. Small colonies excised from agarose beads developed into callus on solidified MS medium supplemented with 1 mg, 2,4-D/l, and these calli produced somatic embryos on hormone free MS medium. However, no transferable shoots or plantlets were produced (Matsuda *et al.*, 1998). Halved shoot bases of *Allium tuberosum* Rottl. ex Spreng. proliferated both axillary and adventitious shoots on B5 medium (1968) supplemented with either 6-benzylaminopurine (0.5 mg/l) or 1-naphthalene acetic acid (0.1 mg/l) and 2-isopentenyladenine (0.5 mg11). *In vitro* shoots proliferated further numerous shoots upon subculture to fresh medium, and these shoots rooted spontaneously. Plantlets were transplanted successfully to soil and retained the diploid condition of the parents (Pandey *et al.*, 1992).

References :

Brewster, J.L.(1999). Onion and other Vegetables *Alliums*. CAB International, Wallingford Oxford Shire, OX10 8DE UK.

Choi-Kyungbae, Yoon-Jaetak, Choi-Boosull, Kim-Kyungmin and Shoh Jaekeun (1998). Selection of a new Allium tuberosum R. cultivar Chilgokbuchu with good quality and

high yield. RDA Journal of Horticulture Science, 40(2):154-157.

Gao, Z. K., Zhang, M.X. and Tan, J.J. (1992) Studies an photosynthetic characterstics of Chinese chives. Acta Horticulturae, Sinica, 19(3):240-244.

Han, K.Y. (1988). Effect of planting density on the growth and yield of Chines chives (*Allium tuberosum* R.) in non heating plastic house. Res. Report. Rural Develop. Admn. Upland and Industrial Crops, 33(2):79-83.

Ikeda, Tsumagari and Nohora, T.H. (2000). Steroidal oligoglycosides from the seeds of *Allium tuberosum*. Chemical and Pharmaceutical Bulletin, 48(3):362-365.

Jia, X.,Peter,C.,Uden,C.,Xing Zhang,X., Bruce, D. Q. G. and Sullivan, J.J. (1996) *Allium* chemistry: Natural abundance of organoselenium compounds from garlic, onion and related plants and in human garlic breath. Pure & Appl. Chem.68(4): 937-944.

Ma Shubing, Yang Wan Yu , Chen Jian Hua Wei Kuiying, Nieyuxia, Xiang Fa Ting (1999). Preliminary study on the leaf growth dynamics and tillering in Chinese chives. Journal of Henan Agricultural Sciences, 23-26.

Matsuda, Y., Hoffmann, F. and Adachi, T. (1998). Embryogenic callus formation from protoplast of Chinese chives (*Allium tuberosam* Rottl.) SABRAO Journal of Plant Breeding and Genetics, 30(2):91-96.

Pandey, R., Chandel, K. P. S.and Rao, S.R. (1992). *In vitro* propagation of *Allium tuberosum* Rottl. ex. Spreng. by shoot proliferation. Plant Cell Reports, 11(7):375-378.

Ping, W. (1998). Studies on post harvest physiology of Chinese chive scapes. Acta Horticulture, 467:379-386.

Saito, S. (1990) Chinese chives (*Allium tuberosum* Rott.). In :Onion and Allied Crops Vol. III. Biochemistry, Food Science and Minor crops (eds). Brewster, J.L and Rabinowich, HD. CRC. Press. Pp. 219-230.

Takahashi, R., Ogawa, H., Satoh, E. and Mori, B. (1989). Dietary fibre contents of vegetables and their changes during heat treatments. J.Nutr. 46,266-269.

Talukder, K. and Sen, S. (1999). *In situ* cytophotometric estimation of nuclear DNA in different cultivars of *Allium* species. Cytobios, 99(39):57-65.

Van der Meer, P.Q. (1993). Chines Chives. In: PROSEA. Plant Resources of South-East Asia. (Siemonsma, J.S. and Piluek, Kasem Eds.). Pudoc Scientific Publishers. Wageningen. pp. 80-83.

Wuteng, D., Xinhua, W., Wang, D.W. X.H. and Burba, J.L. (1997). Germplasm, production and strategies of garlic and onion in China. Proceedings of the 1st International Symposium on edible Alliaceae. Mendoza, Argentina March 14-18, 1994. Acta Horticulturae, 433:179-184.

Yabana, T. (1983). Varieties and cultivation of Chinese vegetables (4). Agriculture and Horticulture, 58(10):1301-1304.

Yu, J.Q. (1999). Allelopathic suppression of *Pseudomonos solanacearum* infection of tomato (*Lycopersicon esculentum*) in a tomato-Chinese chive (*Allium tuberosum*) cropping system. Journal of Chemical Ecology, 25(1):2409-2471.

18

RAKKYO

Rakkyo is mainly grown in Japan and China for production of small edible bulbs. Its bulbs are mainly used for pickles in Japan. In Jawa, it is used raw and fried, often mixed with other vegetables (van der Meer and Augstina, 1993). The edible portion of the rakkyo constitutes 30-40 percent of the full grown plant. Rakkyo is rich in beta -carotene content (Chen *et al.*, 1993). It is used in Chinese medicine to treat heart problems and stagnant blood (Peng *et al.*, 1996). A Chinese drug Xiebai prepared from the rakkyo is used for treatment of thoracic and diarrhoea (Matsuura *et al.*, 1989). Nutritive value of rakkyo in its per 100 g edible portion is given in table(18.1).

Table(18.1). Nutritive value of rakkyo (per 100 g of edible portion)

Constituents	Amount
Moisture	86.2 g
Iron	0.2 mg
Protein	0.6 g
Sodium	1.0 mg
Carbohydrates	12.6 g
Vitamin A	0
Fibre	0.3 g
Vitamin B_1	0.03mg
Ash	0.2 g
Vitamin B_2	0.02mg
Energy	52Kcal
Niacin	0.9 mg
Calcium	6.0 mg
Vitamin C	10.0 mg
Phosphorus	15.0 mg

Source: Toyama and Wakamiya (1990)

Origin and Distribution

The ancestor of rakkyo is not known (Meer and Burba, 1997). However, its wild form is reported to grow in mountain regions of Chinagsu and Chechiang Provinces of China (Brewster, 2008). South East Asia imports rakkyo from Japan.

Botanical Description

Cytology: The chromosome number in the root apex of rakkyo was counted as 2n = 16 (Kotayama, 1928). Most of the commercial varieties of rakkyo are tetraploid and infertile. In cytological study of tetraploid *Allium chinense* Gobil and Koul (1981) reported Chs no. (2n) = 32 in root tip cells, consisting of 8 groups of 4 chromosomes. The chromosome in each group were heteromorphic, 3 in each group being similar and one distinct. Male meiosis was highly variable.

Morphological Description of Plant

Rakkyo is member of family Alliaceae and its scientific name is (*Allium chinense* G. Don). It is an unusual *Allium* species in that the leaves are hollow and scape is solid. It is a biennial herb, upto 60 cm tall. Bulb ellipsoidal, 2-4 cm X 7-15 mm, at top gradually tapering into the leaf blades; protective bulb-coat leaves several membranous, white to purplish; after planting the bulb divides rapidly and forms a cluster of sprouting shoots which can divide again and finally produce new bulbs. Foliage leaves 3 (-5), distichous, conical-cylindrical, 20-40 (-60) cm X 1-5 mm, hollow, 3-5 ridged, D-shaped or nearly triangular in transverse section. Scape terete, solid, up to as long as the leaves and ca .2 mm wide. Inflorescence umbellate, 6-30 flowered, without bulbils; Apathe 2-lobed, persistent, hyaline; pedicel 1-3 cm long; flowers campamulate, purplish, tinged with red;

tepals 6 in whorls,4-5 mm long; stamens 6, much longer than the tepals; inner 3 filaments with broadened bases and each with 2 short teeth; pistil much longer than the tepals; ovules 2 per locule (van der Meer and Augustina, 1993).

Climate and Soil

Bulb development in rakkyo occurs in long photoperiod conditions and is favoured by temperature in range between 15 and 25°C (Toyama and Wakamiya, 1990). Photosynthesis, transpiration and water use efficiency in rakkyo have been studied by Toyama *et al.* (1985). They observed higher rate of photosynthesis and water use efficiency under low air temperature and high relative humidity conditions, but the transpiration rate was reverse. At 15°C, 70 % RH and 40 Klx photosynthesis occurred at a rate of 26 mg CO_2 dm^{-2} h^{-1}, transpiration occurred at 0.6 gH20 dm^{-2} h^{-1} and water use efficiency (Pn/Tr) was 38 mg CO_2/g H 20. Rakkyo possesses vigorous growth habit and it does well in almost all types of soils. Proper drainage should be given. In fertile soils such as clay loam or volcanic ash, the yield increases but the bulbs become too large and therefore have a lower market value. (Toyama and Wakamiya, 1990).

Planting

Rakkyo is propagated vegetatively by means of bulbs which are planted after a storage period of 1-2 months to overcome dormancy. Planting is done at a spacing of 10-15 cm X 10-15 cm. Rakkyo forms bulbs like onion but is smaller and clamps like welsh onion (Jones and Mann, 1963). In rakkyo bulb is formed by the thickened leaf-sheaths only and no bladeless scales are formed like those in the onion; consequently, no thin neck occurs (van der Meer and Augustina, 1993).

Flowering

Flowering bud differentiation in rakkyo is similar to that in other *Alliums* except that the leaf blades of the main axis wither and die in the summer before the inflorescence emerges. Newly developed leaves which appear in mid to late September, at the end of the dormancy period, sprout from lateral buds, but the flower stalk is terminal on the plant's old main axis. The scape therefore appears to be off centred hence, it is not surrounded by the current succession of leaves (Toyama and Wakamiya, 1990). Plant size is important for flowering, plants grown from mother bulbs weighing 3g or less do not flower and a mother bulb weigh of about 11 g results in 50 per cent flowering (Brewster, 2008).

Harvesting and Yield

Harvesting in rakkyo starts from 45-60 days after transplanting. After pulling the bulbs, they are tied in bundles. Its average yield ranges between 150-200 quintals per hectare.

Diseases and Pests:

1.Basal rot : It is caused by *Fusarium oxysporum* f. sp. *allii*. *Fusarium* basal rot occurs at two periods in the life of the crop. Sets which are planted in July-September and plants die within 30-40 days. The fleshy leaf tissue next the stem base tissue rots, become water soaked and the roots decay. A yellow brown rot spreads to the rest of the set which, in dry conditions, become mummified or, in wet conditions, forms a wet rot. This stage of the disease probably results from the use of infected planting material (Entwistle, 1990).

2. Anthracnose: This disease is caused by *Colletotrichum circinans*. Moderate temperature and high humidity are conducive for disease development. Irrigation and continuous cropping favor disease incidence. The yield losses estimated in irrigated and non irrigated fields are 20-50 and 10-25 per cent, respectively (Shi and Tong, 1997). The disease can be controlled by application of

50 per cent cabendazim W.P. or 75 % chlorothalonil W.P.

3. Soft rot: It is a bacterial disease caused by *Erwinia carotovara* pv. carotovora. The primary infection occurs through diseased seed bulbs and soil. The bacteria infect the host through wounds, resulting in rotting and death. Disease can be controlled if at early stage, chemicals including DT and Streptomycin are applied (Tang *et al.,* 1996).

4. Yellow Streak of rakkyo : It is a viral disease caused by Garlic Latent Virus (GLV) and Onion Yellow Dwarf Virus (OYDV). Symptoms appear as streak on the basal part of the newly developed leaves which become thin and curled (Sako *et al.*, 1990).

Cultivars

In Japan, following 3 cultivars are most popular (Toyama and Wakamiya, 1990).

1. **Tamma :** Plants of this cultivar have short, thin leaves and each plant produces 10-25 small (1.5-3.0 g), white bulbs that are suitable for high quality pickles.
2. **Rakuda:** Plants of this cultivar are tall, more robust and produces 6-9 bulbs in elongated shape weighing 4-10 g.
3. **Yatsufusa***:* Plants of this cultivar produces smaller bulbs with a narrow, firm neck. It matures and is lower yielding than cv. Rakuda. Its bulbs are suitable for pickling.

Genetic Improvement

The production of interspecific hybrids between *A. chinense* (female parent) and *A. fistulosum* (male parent) relied upon the removal of ovaries (embryo rescue) 5 days after pollination and their culture on MS medium containing 5 per cent sucrose at 25°C. Their hybridity was confirmed by DNA analysis (Nomura *et al.,* 1994). The interspecific hybrid plants obtained from *A. chinense* and *A. fistulosum* were similar to *A. chinense* in appearance. All the hybrids plants formed white coloured bulbs similar to those in *A. chinense*. Bulb weight varied widely, ranging from 1.6 to 49 g with an average value of 15.8 g. The average weight of the hybrid (15.8 g) was much greater than *A. chinense* (8.1 g). The average value of the total bulb weight per hybrid plant was 104.3 g, again significantly higher than the 68.5 g in *A. chinense*. Number of bulbs per plant (multiplication) in the hybrid plants also ranged widely from 1.5 to 21.5 and the average value of 6.6 in hybrid plants was intermediate between the parents. Flowering time in the hybrid was intermediate between both parents. The characteristics of dormancy; inherited from *A. chinense* was found in a third of hybrids.

In the crop of *A. chinense* and *A. tuberosum*, only one hybrid was obtained. Its growth was poor and its leaf blades were thinner and shorter than those in both parents. White bulbs much smaller than those in *A. chinense* were formed in the hybrid plant. It was not dormant in summer like *A. tuberosum.* Hybrid plants obtained from the crops of *A. chinense* and *A. cepa* were intermediate between parents in morphological appearance. The average value for multiplication and bulb weight per hybrid plant were much greater than in the parents. All the hybrid plants had pale red bulb colour from *A. cepa*. Flowering time in the hybrid plants was intermediate between the parents. Hybrid plants from the crops of *A. chinense* and *A. ampeloprasum* were intermediate between both parents. Several large white bulbs with many small bulblets were observed in the hybrid plants. The average value for multiplication and bulb weight per plant in the hybrid plants was greater than in both the parents. Flowering time in the hybrid plants was intermediate between the parents, and they were dormant in summer (Nomura and Makura, 1996).

References

Brewster, J.L. (2008). Onion and other vegetables *Alliums*, 2nd Edition, CAB International, Wallingford.

Chen, B.H. , Chuang, J.R, Lin, J.H. and Chiu, C.P. (1993). Quantification of pro vitamin A compounds in Chinese Vegetables by high performance liquid chromatography. Journal of Food Protection, 56(1):51-54.

Entwistle, A.R. (1990). Root Diseases: In: Onion and Allied Crops, Vol.II. Agronomy, biotic Interactions, Pathology, and the Crop Protection (Eds.). Rabinowich, H.D. and Breweter, J.L., CRC, Press.

Gobil, R.N. and Koul, A.K. (1981). Cytology of the tetraploid *Allium chinense* G. Don. Cytologia, 34(1):73-81.

Jones, H.A. and Mann, L.K. (1963). Onion, Their Allies. Interscience, New York,.

Kotayama, Y. (1928). The chromosome number in *Phaseolus* and *Allium*, and an observation on the size of stomata in different species of *Triticum*. J.Sci. Agric. Soc., 303, 562.

Matsuura, H., Ushiraguchi, T. Itakura, Y. and Fuwa, T. (1989). A furastanol glycoside from *Allium chinense* G. Don. Chemical and Pharmaceutical Bulletin, 37(5):1390-1391.

Meer, Q.P.Van-der and Burba, J.L. (1997). Old and new crops within edible *Allium.* Acta Horticulturae, 433:17-31.

Nomura, Y., Maeda, M. Tsuchia, T. and Makara, K. (1994). Efficient production of interspecific hybrids between *Allium chinense* and *Alliums sp.* through ovary culture and pollen storage. Breeding Science, 44(20):151-155.

Nomura, Y. and Makara, K. (1996). Morphological and agronomical characte-rstics in interspecific hybrid plants between rakkyo (*Aliium chinense* and other edible *Allium* Rakkyo species. Breeding Science, 46(1):17.22.

Peng, J.P., Yao, X. S., Tezuka,Y. and Kikuch, T. (1996). Fura stanol glycosides from bulbs of *Allium chinense*. Phytochemistry, 41(1):283-285.

Sako, I., Nakasone, W., Okada, K., Ohiki, S.T., Osaki, T. and Inouye, T. (1991). Yellow streak of rakkyo (*Allium chinense*. G. Don.) a newly disease caused by garlic virus and onion yellow dwarf virus. Annals of the Phytopathological Society of Japan, 57(1):65-69.

Shi, Z.L. and Tong, X.M. (1997). Occurrence and control of Scallion anthranose. China Vegetables, 1: 8-10.

Tang , X.N., You, C.P. and Qiu Sibang (1996). Study on soft rot of *Allium chinense*. Progress of research on plant protection in China. Proceedings of the third National Conference on Integrated Pest manage-ment, Beijing, China, Nov. 12-15, 1996, pp. 302-305.

Toyama, M. and Wakamiya, I. (1990). Rakkyo (*Allium chinense*, G. Don. In: Biochemistry, Food Science, and Minor Crops, Vol. III. Brewster, J.L. and Rabinowitch, H.D. (eds). CRC, Press Boca Raton, Florida, pp.197-218.

Toyama, M., Takeuchi, Y., Kroyanagi, N. and Sugimoto, K. (1985). Relationship between photosynthesis, transpiration and water efficiency in rakkyo (*Allium bakeri*, Regel) and air temperature, relative humidity and light intensity. Journal of the Japanese Society of Horticultural Science, 53(40):444-452.

Van der Meer, Q.P. and Agustina, L. (1993). *Allium chinense*, In: PROSEA, Plant Resources for South East Asia, 8 Vegetables, Siemonsma, J.S. and Piluek, K.(Eds). Pudoc Scientific Publishers Wageningen, the Netherlands.

19

PARSNIP

Parsnip is grown for its fleshy, aromatic and slightly mucilaginous root which is eaten as a cooked or fried vegetable. It is also used in soups and to add flavour to stews. It is especially popular in the United Kingdom. The seed, which tastes similar to dill, is occasionally used as a condiment. The leaves have diuretic properties. A poultice from the roots has been applied to sores and inflammations, and to treat skin diseases. The nutritional composition of the parsnip root is given in the table (19.1).

Table (19.1). Nnutritional composition of parsnip root (per 100 g of edible portion)

Proximate principles	Quantity
Water	79.5 g
Energy	750 kcal
Protein	1.2 g
Fat	0.3 g
Carbohydrate	18.0 g
Fibre	4.9 g
Calcium	362 mg
Phosphorus	71 mg
Iron	0.6 mg
Magnesium	29 mg
Zinc	0.6 mg
Vitamin A	0
Thiamin	0.09 mg
Riboflavin	0.05 mg
Niacin	0.70 mg
Folate	67 µg
Ascorbic acid	17.0 mg

Source: USDA, (2002)

All parts of parsnip contain essential oil. The essential oil from the fully-grown root is rich in myristicin and terpinolene and contains small amounts of (E)-β-farnesene, β-bisabolene, β-sesquiphellandrene and γ-palmitolactone. Furanocoumarins are present in the plant; these compounds may cause contact dermatitis. Parsnip contains minute amounts of the steroid 5α-androst-16-en-3-one or boar-pheromone which contributes to its characteristic fragrance. Wild parsnips contain three furocoumarins (psoralen, xanthotoxin, and bergapten). These chemicals are phototoxic, mutagenic and photo-carcinogenic. Psoralens, which are potent light-activated carcinogens not destroyed by cooking, are found in parsnip roots at concentrations of 40 ppm (Ivie *et al.*,1981).

Origin and Distribution

Parsnips are native to Europe and Asia and were used by the ancient Greeks and Romans for medicinal and food purposes. Parsnips were introduced into North America in the early 1600's and were grown by early colonists and Indians. Parsnips are biennial, but are grown commercially as an annual. The edible portion is the enlarged fleshy taproot. In Ontario, parsnips are grown on approximately 105 ha at a farm gate value of $730,000 with most of the production in the Bradford Marsh region. Although a majority of Ontario-grown parsnips are marketed within the province, some are also shipped into eastern Canada. It is cultivated mainly in temperate regions worldwide and only occasionally in cooler parts of the tropics, including East and southern Africa. It is mainly grown in home gardens and for specialty markets. It was already a popular crop during Roman times and still its wide diversity occurs in Moroccan and Algerian markets

Botanical Description

Parsnip belongs to the family Apiaceae. Its chromome number (2n) is 22. Apiaceae is a large family with about 300 genera and more than 3,000 species.

Pastinaca sativa is an biennial plant that is strongly-scented and grows to over 1.5 meters in height. The edible taproot is fleshy, long (can be over 80 centimeters at times), and white. During a period of low temperature, such as the first frost of the year, after growing in the summer, much of the starch in the root is converted to sugar, yielding a distinctive, sweet flavor. Being hardy, the taproot can remain in the ground during the winter, enduring the freezing of the soil. Taxonomic position of the genus *Pastinaca* is given as under.

Taxonomic position of the genus *Pastinaca*

Kingdom	Plantae
Subkingdom	Tracheobionta
Superdivision	Spermatophyta
Division	Magnoliophyta
Class	Magnoliopsida
Subclass	Rosidae
Order	Apiales
Family	Apiaceae
Genus	*Pastinaca* L.
Species	*Pastinaca sativa* L.

Other species under the genus *Pastinca*

P. acanthura · P. alpina · P. altissima · P. ambigua · P. anethum (Wild Parsnip) · P. angulosa · P. apula · P. aquila· P. argyrophylla · P. armena · P. armena dentata · P. arvensis · P. aspera · P. atropurpurea · P. aurantiaca · P. austriaca · P. barbata · P. biebersteinii · P. brevicauda · P. brevivittata · P. byzanthina · P. candolleana · P. capensis · P. caroli-kochii · P. carvifolia · P. caspica · P. chrysantha · P. clausii · P. colchica · P. collina · P. cordata · P. cyclocarpa · P. dasyantha · P. dasycarpa · P. dentata · P. denticulata · P. dissecta · P. divaricata · P. dorsalis · P. esculenta · P. ferulacea · P. fleischmanni · P. fleischmannii · P. freyniana · P. glacialis · P. glandulosa · P. glauca · P. granatensis · P. grandis · P. graveolens · P. grisea · Dasyatis centroura · P. hastata · P. heracleoides · P. hirsuta · P. hispida · P. hispidula · P. hookeriana · P. incana · P. insularis · P. intermedia · P. involucrata · P. jeinorui · P. kochii · P. laevis · P. lanata (Common Cowparsnip) · P. lanatum · P. lasiocarpa · P. latifolia · P. lebmaniana · P. ligusticifolia · P. lucida · P. lucids · P. lutea · P. macrocarpa · P. mantegaziana · P. minima · P. montana · P. nudicaulis · P. officinalis · P. olivacea · P. opaca · P. opopanax · P. orphanidis · P. orsinii · P. pachyrhiza · P. panacifolia · P. parvifolia · P. pastinacella · P. petteri · P. pimpinellifolia · P. polakii · P. pratensis · P. propinqua · P. pubescens · P. pumila · P. pyrenaica · P. rectistylis · P. rigens · P. rigida

Morphological Description of Pant

Parsnip is a glabrous to slightly hairy biennial herb up to 150 cm tall, with fusiform, fleshy white taproot; stem erect, hollow, grooved. Leaves alternate, pinnate, without stipules; petiole sheathed at base; leaflets sessile, ovate-oblong, often with some lobes at base, 2-13 cm × 1-5 cm, toothed. Inflorescence a terminal, compound umbel with unequal rays; involucral bracts 0-2, deciduous. Flowers bisexual, but male flowers present in addition to bisexual flowers, c. 2 mm in diameter, 5-merous; petals yellow; ovary inferior, 2-celled. Fruit a flattened ellipsoid schizocarp 5-7 mm long, ribbed, slightly winged. Seedling with epigeal germination; hypocotyl 0.5-1.5 cm long, epicotyl absent; cotyledons stalked, ovate-lanceolate, herbaceous. Parsnip is a slow - growing, deep-rooted plant. The flowers are pollinated by insects. Parsnip is self-fertile crop (Oyen, 2004).

The modern cultivated parsnip has developed a leaf-stalk 2 feet long, the first pair of leaflets being several inches above the sheath. The leaflets are oblong, about 2 inches across at the basal part and 4 1/2 inches in length (more than double the size of those of the wild plant), and are entirely smooth and somewhat paler in colour. The flowers in each case are yellow and in umbels at the ends of the stems, like the carrot, though the umbels do not contract in seeding, like those of the carrot. The flowers of the cultivated parsnip are of a deeper yellow colour than those of the wild plant. The fruit is flattened and of elliptical form, strongly furrowed. Parsnip 'seeds' as the fruit is commonly called, are pleasantly aromatic, and were formerly collected for their medicinal value. Seeds contain an essential oil that has property of curing intermittent fever. A strong decoction of the root is a good diuretic and assists in removing obstructions of the viscera. It has been employed as a remedy for jaundice and gravel.

Varieties

Avonresister

This variety possesses a high intrinsic resistance to the various forms of canker. Its roots are more uniform in shape and size than those of the commercial variety Offenham, but total yields are lower. The relative yield difference persists in both low and high yielding environments (Channon *et al.*, 1970).

Climate and Soil

Parsnip is a cool season crop. Optimum temperature for growth is 15-18°C. Roots produced under warm conditions do not have the strong and distinctive flavour of those grown under cooler conditions. Parsnip is biennial and requires vernalization for flower induction. It can grow in slightly shady localities or in full sunlight. In the tropics it can only be cultivated above 900 m altitude. The root is tolerant of hard frost. Under the influence of low temperatures starch in the root is converted to sugars. Parsnip requires a deep, light to medium -textured soil with good drainage. In clay soils germination and root growth are poor

Seed Rate and Spacing

The weight of 1000 seeds is about 3.5 g. Seed is difficult to preserve and often has a low germination rate. Parsnip seed usually retains viability for only two years; thus it is advisable to use fresh new disease-free seed each year from a reliable sources. Parsnip seed is sown *in situ*, thinly in rows 40-50 cm apart at a spacing of 1-2 cm in the row and covered with about 1 cm of fine soil. Parsnip seeds germinate optimally when soil temperatures range between 19 to 24°C. In general, germination is slow even under ideal field conditions and where sub-optimal conditions prevail emergence may require more than four weeks. Plants are thinned after 4-5 weeks to a spacing of 10-12 cm. Slight earthing-up after thinning is recommended.

Manures and Fertilizers

Nutritional recommendations are very scanty. Apply 10-15 tonnes /ha well rotten FYM in the field before a month of sowing. Fresh manure should be avoided as it causes plants to become hairy and branched. Among chemical fertilizers, application of NPK @ 60:40:40 kg/ha can be given. Half doze of nitrogen and full doze of phosphorus and potassium should be given at the time of sowing where as remaining half dose of nitrogen should be applied in two split dozes in form of top dressing 30 and 45 days after sowing.

Irrigation

Parsnip prefers a constantly moist soil. Erratic irrigation may cause roots to crack or become fibrous. Parsnip roots are ready for harvesting after 90-150 days. In temperate areas they are often left until after the first frost as their sweetness increases under low temperatures.

Harvesting and Yield

roots are pulled up with the help of spade and care should be taken that root should not get damage. A yield of 25 t/ha is considered good. Once parsnip roots have been harvested they lose water quickly. Sound roots may be kept in cold storage for 4 to 6 months at 0^0C and 98 to 100% relative humidity. The sweet nut-like flavour of parsnips is not developed until roots have been exposed to temperatures near freezing. At low temperatures the starches in the root are converted to sugars. Mature roots which have not been exposed to low field temperatures may be stored for two weeks at 0 to 1°C to develop flavour. Extremely low temperatures in storage may damage root tissues. Parsnips will freeze at -1.7°C. The centre of the core of frozen parsnips may become water soaked with reddish brown areas developing in the cambium layer. The root may become externally dark brown after freezing, subsequent thawing, and exposure to room temperature for a short period of time. Parsnips should not be stored with ethylene-producing crops (e.g. apples), since the possibility exists that the ethylene may impart a bitter flavour to the root tissues.

Plant Protection

Parsnips are subject to two major foliar and root diseases, *Itersonilia* canker and *Phoma* canker. An adequate spray program should be rigorously followed to control Itersonilia canker. Channon (1965) reported that cankers of parsnips growing in the East Anglian fens were caused by a species of *Centrospora*. Comparison of the morphology and pathogenicity of isolates from infected roots with those of isolates from carrot and celery and also with published information, indicated that the fungus from parsnips was *C. acerina*. Principle pests of parsnips in include the carrot rust fly, carrot weevil, cutworms, flea beetles, leaf hoppers, nematodes and spider mites. The seriousness of these insects will depend on the location and year.

Genetic Resources and Improvement

The old cultivar 'Hollow Crown' is mostly cultivated in East Africa; other well-known cultivars are 'Guernsey' and 'Offenham', both with shorter roots. No breeding programmes are known to exist. The North Central Regional PI Station, Ames, Iowa, United States maintains a small collection of *Pastinaca* germplasm.

References

Channon, A. G. (1965). Studies on parsnip canker IV. *Centrospora acerina* (Hartig) Newhall-a further cause of black canker. Annals of Applied Biology, 56(1): 119-128.

Channon, A. G., Dowker, B. D. and Holland, H. (1970). The breeding of Avonresister, a canker-resistnt parsnip. Journal of Horticultural Science, 45 : 249-56.

Ivie, G. W., Holt, D. L. and Ivey, M. C. (1981). Natural toxicants in human foods: Psoralens in raw and cooked parsnip root. Science, 213: 909-910.

Oyen, L.P.A. (2004). *Pastinaca sativa* L. In: Grubben, G.J.H. & Denton, O.A. (Editors). PROTA 2: Vegetables/ Légumes. PROTA, Wageningen, Netherlands.

20

HORSH RADISH

Horse radish is cultivated at very limited scale in Uttarakhand, however, it is very much popular among the people, where they use this root vegetable during lean season. When it is available in plenty during main season, its roots are dried by cutting the small pieces for use during off-season. Horse radish is grown for its pungent roots which are mixed with salt and vinegar and eaten as a relish, condiment, or appetizer with meat and other foods. The pungency is due to the presence of an allyl isothiocynate and butyl isothiocynate similar to the mustard oils occurring in combination with the glucoside sinigrin. Horse radish root contains approximately 0.6% of glucosinolates; the most abundant of these are sinigrin (0.2%) and gluconasturtiin (0.1%). As soon as intact cells are damaged, these isothiocyanates are enzymatically hydrolyzed to yield allyl isothiocyanate and 2-phenylethyl isothiocyanate, respectively. Further, glucosinolates in horse radish are glucobrassicanapin and the indol-derived glucobrassicin (along with some closely related compounds like 4-methoxy glucobrassicin, 4-hydroxy glucobrassicin). On hydrolysis, glucobrassicanapin yields 4-pentenyl isothiocyanate; yet the glucobrassicines have no corresponding stable isothiocyanates. Instead, they hydrolyze to 3-hydroxyindole derivatives and free isothiocyanate ions. Gilbert and Nursten (1972) reported at least 17 volatile compounds in horse radish root. Out of these, 8 were methy, ethyl, isopropyl, 2 butyl, allyl, 4- pentenyl and 2 phenethyl isothiocynates and allyl thiocynate. Among the non-volatile constituents, flavone glycosides (quercetine, kaempferol) and particularly ascorbic acid (06% in horse radish root). Horse radish roots also possess nematicidal properties (Kotova *et al.*, 1994). Dabrovska and Skapski (1976) observed that local horse radish were more pungent than the imported ones and wild types were more pungent than cultivated ones. Further, plants grown in the pseudopodsolic soil were more pungent than those in sand. This may be due to increse rate of sulphur which may influence root allele isothiocynide content and this caused increased root pungency. The boiled extract of its green leaves is medicinally useful to cure and reduce obesity in human beings. Nutritive value of root is given in table (20.1).

Table (20.1). Nutritional composition of horse radish root (per 100 g of edible portion)

Constituents	Amount
Water	74.6 g
Energy	87 k cal
Ash	2.2 g
Fat	0.3 g
Carbohydrate	19.7 g
Protein	3.2 g
Calcium	140.0 mg
Phosphorus	64.0 mg
Iron	1.4 mg
Sodium	8.0 mg
Potassium	564 mg
Thiamine	0.07 mg
Ascorbic acid	81.0 mg

Origin and Distribution

The plant is thought to be of Mediterranean or Eastern European origin and is now widely cultivated in Central and Eastern Europe. Horse radish has been cultivated in Denmark since the 17th century and it is also found there in growing wild state (Heneriksen and Bjorn, 2004). In China, it is mistaken with wasabi in some areas because both of them possess a special aroma and pungent flavor, here, horse radish is called western wasabi, and can be used as a substitute for wasabi (Yang and Zha, 2000).

Botanical Description

Horse radish belongs to the family Brassicaceae and genus *Armoracia*. Its systematic position is given as under:

Taxonomic position of the genus *Armoracia*.

Kingdom	Plantae
Subkingdom	Tracheobionta
Superdivision	Spermatophyta
Division	Magnoliophyta
Class	Magnoliopsida
Subclass	Dilleniidae
Order	Capparales
Family	Brassicaceae
Genus	*Armoracia*
Species	*Armoracia rusticana*

G. Gaertn., B. Mey. & Scherb

Horse radish root is a perennial, thick, tapering, white, long, acrid, and from which arise many large leaves. From the center a round or angular, smooth, erect, branching stem arises, about 50-60 cm in height; those branches which flower are corymbose, smooth, and angular. The radical leaves are nearly 25-30 cm long, half as wide, oblong, crenately-toothed, waved, sometimes pinnatifid, of a dark-green colour, and upon long, channeled petioles; the cauline leaves are smaller, lanceolate, dentate, or incised, sessile, sometimes entire, and without footstalks; the lower ones are often pinnatifid. The flowers are numerous, small, white, peduncled and borne in terminal corymbose racemes. Calyx ovate, spreading, and equal at the base; sepals four, concave. Petals obovate, obtuse, entire, claw-like. Stamens without teeth, the length of the calyx; anthers cordate; silicle sessile, oblong, or ovoid-globose, and compressed; dissepiment thin; valves ventricose, thickish; cells many seeded; seeds not bordered and cotyledons flat and accumbent.

Varieties

Two varieties of horse radish are grown i.e., common horse radish, which has broad, crinkled leaves and produces roots of high quality, and other one is Bohemian.

Common horse radish

It is a broad-leaved, hollow-petiole type and it might have originated in Eastern Europe. It covers maximum acreage planted under horse radish. It produces a root crop of good quality in spite of being very susceptible to white rust and mosaic diseases.

Bohemian horse radish

It is also known as the Maliner Kren or Bayersdorf strain, is a smooth, narrow leaved, solid petioled variety and is believed to have originated in southern Europe. It lacks root quality and yield but appears to be highly field resistant to both the white rust and horse radish mosaic diseases.

Cultural Practices

The temperate regions at high attitude are considered ideal for its cultivation. In Uttarakhand, it is sown during the month of August-September. Horse radish can be grown in any good soil, but does best in a deep, rich, moist loam well supplemented with organic matter. It can be raised by dibbing the seeds, however, vegetative propagation through root cutting is more beneficial. Generally chemical fertilizers are not applied for raising the crop. Application of organic form of manure is preferred by local growers for realizing the original taste and flavour of the produce. Crop responds well to NPK application. Application of N as $(NH_4)2\ SO_4$, $Co(NH_2)2$ or $NH_4\ NO_3$ resulted in yield increases of 4.4, 3.7 and 3.2% respectively. The best treatment was a single application of N @ 240kg/ha. This treatement reduced the cellulose contain and increased the production of marketable roots by upto 8.3 per cent. Further, application of N in form of urea gave the best results (Pavule, 1972).

Weed Management

Crop of horse radish is infested with a number of annual and perennial weeds. Hopen *et al.* (1982) observed that oxyfluorfen had good weed control in horse radish and it did not exert any phyto-toxicity. They further, concluded that fluazifop butyl and sethoxydin were good post emergence in conjunction with oxyflorfen as pre emergence. Habling (1983) obtained 29.7 per cent weed control with the application of atrazine as pre emergence while nitrofen+desmetryne, Trizilin+ alachlor and chloroponsodium+ desmetryne gave more than 80 per cent control when applied as post-emergence. Dobrzanski *et al.* (1971) reported that simazine gave best control of weeds and up to 1.5 kg / ha could safely be applied.

Plant Protection

Crop of horse radish is attacked by a few pathogen and insects. Root rots are the most serious disease of horse radish. The most serious pest of horse radish is flea beetle. Brittle root is caused by MLO (mycoplasma like organism). Infected plant develops dark discoloration of root phloem (Raju *et al.*, 1981). It is transmitted by leaf hopper (Fletcher *et al.*, 1981). Horse radish is also infested by TpMv (Wang, 1978). The virus induces chlorotic to yellow vein and ring spot, confined to the older leaves and most distinct in the autumn. Infected plants expressed 37 per cent reduced yield (Paludan, 1973).

Harvesting and Yield

Three or four days prior to harvesting, the tops should be removed as near to the ground as possible. In Japan, mechanical horse radish harvester has been designated (Kudo *et al.*, 1985). After harvesting, the lateral roots must be trimmed. The average yield of horse radish is 25-30 t/ha. In European countries, for preparing horse radish commercially for table use, the roots are scraped and peeled and all defects are removed. Then, the root is grated directly into white wine vinegar or distilled vinegar (4.5-5.0% strength). Cider vinegar should not be used because it turns the grated horse radish discolour within short time

Seed Production

Completely pollen-sterile, clone of 'Common types' horse radish was obtained which produced ovaries in which about 0.5 per cent of the ovules contained normal appearing gametophytes; and two clones of Bohemian matured some functional pollen. The inflorescence of Bohemian horse radish is a terminal or axillary raceme of short-pedicelled, small white flowers. It can be induced to blossom profusely under greenhouse conditions. Generally, 'Common types' produced no any functional pollen. The pollen grains show signs of abortion in the early stages of development. Two clones of Bohemian produced some functional pollen. In one about 25 per cent and in the other about 50 per cent of the grains were normal in appearance. It is possible to effect fruit development on several of the pollen sterile 'common' clones by applying pollen from one of the pollen producing Bohemian clones to the receptive stigmas. Fruits develop very rarely following selfing of the pollen producing clones. Further, 30-40 per cent of the ovules contained normal appearing female gametophytes. The fruits develop rapidly for a short period following pollination and then development stopped in a high percentage of the cases, with usually only a few continuing to grow. The developing fruits were more or less distorted in appearance. Usually only one seed developed within a fruit, with two seeds rarely being present. The fruits matured in three to four weeks, at which time they dehisce, leaving the seed attached to the placenta wall or permitting it to drop. Dehiscence usually occurs while the fruits were still green in color.

The cross between the two types of horse radish proved to be the most effecttive means of securing seed. The seed harvested varied from fairly well developed to badly shrunken kernels. The selfed seeds were smaller and usually not as well developed as those resulting from the cross. Generally the mature kernels were brown to gray in colour. Further, obtained seedlings represented crosses between the 'common' and the Bohemian types of horse radish. The twelfth plant was an offspring of the pollen producing clone. These plants varied widely. The selfed offspring had petioles with solid piths while the crosses had hollow petioles. The leaf blades of the various plants differed in size, shape, smoothness, and colour. The offspring were vigorous and showed no evidence of the presence of virus. Preliminary tests indicated that the F_1 plants, of the cross between the white rust and virus-susceptible "common" clone and the resistant Bohemian clones, were segregating for these characters. In cytological studies, the somatic number of chromosomes as determined by acetocarmine root tip smears was 32 in both the "common" and Bohemian varieties of horse radish. Irregularities occurred in the premeiotic and meiotic divisions in microsporogenesis. The gametes were formed in the pollen grains prior to dehiscence of the anthers. The ovary contained 16 to 20 ovules. Meiosis was likewise irregular in mega sporogenesis. Oftentimes the spore failed to continue growth or the gametophyte started to develop and breaks down prior to maturity so that only a small percentage of the ovules contained female gametophytes that appeared normal. There was the occasional accumulation of food materials in the nucellar regions of non functional ovules following pollination. The production of fertile lines of horse radish through cross pollination and selection would greatly stimulate improvement work (Weber, 1949).

Micro-propagation

Virus free plant material obtained by meristem cultures of seven cultivated Polish types of horse radish was used as a source of leaf and root explants. Shoot regeneration from leaf explants on MS medium supplemented with 0.5 mg l^{-1} naphtaleneacetic acid (NAA) and 0.1 mg l^{-1} benzyladenine (BA) was more effective than on LS medium with 0.8 mg l^{-1} indole-3-acetic acid (IAA) and 4 mg l^{-1} kinetin. Root explants afforded possibilities for better shoot regeneration than leaf explants. Concentration of BA (0, 0.05, 0.1, 0.2, and 0.4 mg l^{-1}) in MS medium did not affect the number of shoots regenerated from root explants, but it affected shoot weight. The optimal BA concentration appeared to be 0.1 mg l^{-1}. Higher BA concentration resulted in leaf deformations and inhibition of rooting. The shoots regenerated on MS medium were characterised by higher weight than those obtained on ½ MS (containing half of the original concentration of macronutrients). Storage of isolated roots at 5°C for 4 weeks stimulated their capacity for shoot regeneration (Pawelczak *et al.*, 2006). The highest mean number of shoots, 41.5, from 1 cm^2 leaf fragment was obtained on Linsmaier and Skoog medium containing IAA at 0.8 + kinetin at 4.0 mg/liter, this compared with 1.2 shoots obtainied on a medium recommended by Meyer for horse radish *in vitro* propogation. Rooting of 100 per cent of the shoots in 2 weeks was obtained on medium containing auxins only. After transplanting 96.5% of the plantlets survived (Gorecka, 1987).

Genetic Resources and Improvement

Selection in horse radish is made on the basis of root shape and taste. It was found that variety Krenox exceeded the local variety by 60.9 per cent in yield and quality (Moravec *et al.*, 1974). Hansen (1974) observed significant variation in European clones with respect to content of isothiocynates which varied form 12.2 mg/g freeze dried root in the Danish Spangbjerg 12.68 to 20.4 mg/g in the German Frieslander.

References

Dabrowska, B. and Skapski, H. (1976). Some factors affecting the pungency in horse radish. Zeszyty Naukowe Szkoly glonej Gospodarstwa Wiejskiego akademii Rolniczej w Waszawie Ogrodnictwo, 9: 101-118.

Dobrzanski, A. Daszewski and Fajkowska, H. (1971). The effect of herbicide on weed infestation of horse radish and on the crop. Biuletyn warzywniczy, 12:171-184.

Fletcher, J., Schultz, G.A., Davis, R.E., Eastman, C.E. and Goodman, R.M. (1981). Brittle root disease of horse radish : evidence for an etiological role of *Spiroplasma citri*. Phtopatho-logy, 71(10):1073-1080.

Gilbert, J. and Nursten , H.E. (1972). Volatile constituents of horse radish. Journal of the Science of Food and Agriculture, 23(40):527-539.

Goerecka, K. (1987). *In vitro* propagation of horse radish *Cochlearia armora-cia* L. Acta Horticulture, 212 (2):671-674.

Habling, W. (1983). Controlling weeds in horse radish (*Armoracia rusticana* G.M. Sch.). Nachrichtenblatt fur den Pflanzenschutz in der DDR, 37(1): 22-24.

Hansen, H. (1974). Content of glucosin-olates in horse radish. Tidsskrift for Planteavel, 78(3):408-410.

Heneriksen, K. and Bjorn, G. (2004). Cultivation of horse radish. Gron Viden(1982). Pre and post emergence hercbicides for weed control in horse radish . Proceedings, North Central Weed Control Conference, 1982, pp94-95.

Kotova, V.V., A.A. Shestperov., R.F.Kononkov and K.G. Koza. (1994). Nematicidal properties of plants. Zashchita Rastenii Moskva.9: 25.

Havebrug, 160:6.

Hopen, H.J., Doll, C.C. and Binning, L.K.

Kudo, M., Aray, K., Meakawa, T., Komatsy, M., Nobuta, Y. and Sakai, T. (1985). Studies of horse radish harvester. Journal of Hokkaido College Senshu University, 18:115-122.

Moravec, J, Antoszewski, R., Harission, L. and Zych. C.C. (1974). The possibility of horse radish. Proceedings of XIX International Horticulture Congress. IsectionVII. Vegetables, pp 761-764.

Paludan, N. (1973). Turnip mosaic virus in horse radish (*Armoracia lapathifolia* Gilib.) Tidsskrift for Planteavl, 77(2): 161-169.

Pavule, A. (1972). The effect of different nitrogen fertilizer of yield and quality of horse radish. Trudy- Lativiiskoi Selkhozyaistvennoi Akademii. No. 53:38-47.

Pawelczak A., Majewska, A. Geszprych, A., and Dabrowska, B. (2006). Micropropagation of horse radish (*Armoracia rusticana*): V. International Symposium on *In Vitro* Culture and Horticultural Breeding. ISHS Acta Horti culturae No. 725.

Raju, B. C., Nyland, G., Backus, E.A. and McLean, D.L. (1981). Association of a spiroplasma with brittle root of horse radish. Phytopathology, 71(10):1067-1072.

Wang, G.C. (1978). Viruses causing rape mosaic in Zhejaing Province. Acta Microbiologica Sinica, 18(4):298-309.

Weber, W. W. (1949). Seed production in horse radish. J. Hered. 40: 223-227 .

Wu, H., Wang, C.J., Bian. X.W., Zeng, S.Y., Lin, K.C., Bo Wu., Guo-Anhang and Zhang, X. (2011). Nematicidal efficacy of isothiocyanates against root knot nematode Meloidogyne javanica in cucumber. Crop Protection, 30(1): 33-37.

Yang, R.Y. and Zha, D.S. (2000). Studies on resource of vegetable varieties in Shanghai (VI- distinction of several easily confused vegetables. Acta Agriculture Shanghai, 16 (1): 46-48.

21

JERUSALEM ARTICHOKE

The tubers of Jerusalem artichoke may be eaten raw or boiled; they are also pickled, made in to chips or ground into flour. Inulin is the storage carbohydrate of Jerusalem artichoke, whereas starch is the storage carbohydrate in the majority of plants. Only a small numbers of plant accumulate inulin in amount sufficient for cost effective extraction, with Chicory (*Cichoriumintybus* L.) and Jerusalem artichoke being most important inulin storing species. Inulin has value in the diet of persons suffering from diabetes. The crude inulin extracted from Jerusalem artichoke tubers were given in diet for diabetic rats on different levels of inulin (10 and 15%) had significantly lower in total cholestrol, triglyceride and total lipids in comparing to positive diabetic rats fed on control diet. It was observed that, HDL level increased significantly after fed on 10 and 15% inulin. On the other hand, LDL and VHDL levels were decreased significantly after fed on (10 and 15%) inulin in comparing to positive group rats fed on control diet (Gaffar *et al.*, 2010).

Table (12.1). Nutritive value of Jerusalem artichoke (per 100 g of edible portion)

Constituents	Amount
Energy	373 kcal
Carbohydrates	17.44 g
Sugars	9.6 g
Dietary fiber	1.6 g
Fat	0.01 g
Protein	2 g
Thiamine (vit. B_1)	0.2 mg
Riboflavin (vit. B_2)	0.06 mg
Niacin (vit. B_3)	1.3 mg
Pantothenic acid (B_5)	0.397 mg
Vitamin B_6	0.077 mg
Folate (vit. B_9)	13 μg
Vitamin C	4 mg
Calcium	14 mg
Iron	3.4 mg
Magnesium	17 mg
Phosphorus	78 mg
Potassium	429 mg

Botanical Description

Jerusalem artichoke is commonly known as Sunchoke, Sunroot, Earth Apple, Girasole and Topinambour. Jerusalem artichoke (*Helianthus tuberosus* L.) belongs to the family Asteraceae. The genus *Helianthus* has around 50 species. The most important species in commercial terms is the cultivated sunflower (*Helianthus annus* L.), grown mainly for its oilseed. In contrast, *Helianthus tuberosus* is distinguished by its large tubers, which have been used for their food value. Taxonomic position of the genus *Helianthus* is given as under:

Taxonomic position of the genus Helianthus

Kingdom	Plantae
Subkingdom	Tracheobionta
Superdivision	Spermatophyta
Division	Magnoliophyta
Class	Magnoliopsida
Subclass	Asteridae
Order	Asterales
Family	Asteraceae
Genus	*Helianthus* L.
Species	*Helianthus tuberosus* L.

Morphological Description of Plant

No other species of *Helianthus* is cultivated to a significant extent, although several have value as ornamentals. Jerusalem artichoke is an herbaceous perennial plant growing to 1.5-3 m tall with opposite leaves on the lower part of the stem becoming alternate higher up.The leaves have a rough, hairy texture and the larger leaves on the lower stem are broad ovoid-acute and can be up to 30 cm long and the higher leaves smaller and narrower.The flowers are yellow, produced in capitate flower heads which are 5-10 cm diameter, with 10-20 ray florets. The fruit is an achene and generally few are formed. The tubers are elongated and uneven, typically 7.5-10 cm long and 3-5 cm thick, and vaguely resembling ginger root, with a crisp texture when raw.They vary in colour from pale brown to white, red or purple. Makowetzky's (1929) studyof the chromosomes of *Helianthus tuberosus* shows the haploid number of Jerusalem artichoke to be about 51. Description of species under the different botanical group is given in table (21.2).

Table (21.2). Description of species under the different botanical group

Section	Series	Species b
Helianthus'	*Corona-Solis B*	*H. niveu2*
		H. deserticola2
		H. paradoxus2
		H. anomalu2
		H. debili2
		H. praecox2
		H. petiolaris2
		H. neglect us 2
		H. annuus2
		H. argophyllus 2
		H. bolanderi2
		H. exilis2
Divaricati	*Corona-Solis B*	*H. mollis2*
		H. divaricatus
		H. tuberosus6
		H. hirsutu4
		H. resinosus6
		H. strumosu4
		H. grosesserratus2
		H. nuttalli2
		Microcephali2
		Angust $olii'
		Atrorubentes"
		Ciliares PumiliX
		Ciliares c
		Agreste"
		H. giganteus2
		H. decapetalus x4
		H. eggertii6
		H. californicus6
		H. schweinit&
		H. salicifolius '
		H. maximiliani
	Microcephali'	*H. microcephalu?*
		H. glaucophyllus2
		H. laevigatus4
		H. smithii'
		H. porteri
	Angust $olii'	*H. longifolius2*
		H. carnosus'
		H. angust $olius'
		H. simulans'
		H. floridanus
		H. heterophyllus2
		H. radula'
	Atrorubentes"	*H. occidentalis'*
		H. silphioide?
		H. atrorubens2
		H. rigidus
Ciliares	*PumiliX*	*H. gracilentu?*
		H. pumilu?
		H. cusickii2
	Ciliares c	*H. arizonensi?*
		H. ciliaris4
		H. laciniatus 2
Agreste"		*H. agrestis2*

b 2,4, and 6 denote diploid (2n = 34). tetraploid (2n = 68), and hexaploid (2n = 102) species, respectively.

Origin and Distribution

Jerusalem artichoke is a native of North America and wild plants are distributed in north central region and eastern coast of USA and adjoining areas of

Canada. Jerusalem artichokes were first cultivated by the Native Americans long before the arrival of the Europeans; this extensive cultivation obscures the exact native range of the species. The French explorer Samuel de Champlain found domestically grown plants at Cape Cod in 1605. He then brought the plant back with him to France. By the mid-1600s, the Jerusalem artichoke had become a very common vegetable for human consumption in Europe and the Americas and was also used for livestock feed in Europe and colonial America. The French in particular were especially fond of the vegetable, which reached its peak popularity at the turn of the nineteenth century. The Jerusalem artichoke was titled 'best soup vegetable' in the 2002 Nice Festival for the Heritage of the French Cuisine. In India, it is grown in West Bengal, Assam, Uttar Pradesh, Maharashtra and Andhra Pradesh for its edible subterranean tuberous stem.

Climate and Soil

Jerusalem artichoke requires a growing season of at least 125 frost-free days. Optimum yields are obtained where temperatures range from 18 to 26°C. Rainfall of 127cm. or less is required. In dry areas irrigation may be necessary to begin germination. Tuber and top yields are limited if soil moisture is less than 30% of field capacity during the tuber formation period. Jerusalem artichoke can grow in poor soils, but thrive best in areas rich in organic matter. Loam and loose soil, sandy or clayish is best for its cultivation. It does not tolerate water logging, which promote disease development and hinder the growth of the tubers. It can successfully grow in soil with pH between 4.5 and 8.2.

Soil Preparation

Jerusalem artichokes are very tolerant of soil conditions, and minimal soil preparationis needed. They grow very well under a mulch of rotted hay or straw. Once they have started growing, they compete very successfully with weeds.The soil should be tilled well. It must be free of perennial weeds and grasses. Additions of compost or manure promote good fertility for cropgrowth.

Varieties

Veitch Improved Long White, Sutton New White, Mammoth French White and French White Improved.

Planting

The recommended planting time is spring season (March-May). To plant, cut the tuber into two or three sections, each one withan "eye"; cover the tubers with soil to a depth of 10 cm. Plant in rows 70 m apart, 25 cm between plants in full sun and mulch well. Jerusalem artichokes need a good supply of potassium. Application of heavy dose of nitrogen should be avoided as it favours vegetative growth on the cost of tuber formation. Select tubers about 25 to 35 mm in diameter. Larger tubers should be cut into similar sized pieces with two eyes, or buds, per set. Tubers sprout approximately 10 to 17 days after planting, but soil temperatures must be at least 7°C before germination begins.

Manure and Fertilizer

About 15-20 tonnes per hectare of FYM or compost are applied to the soil at field preparation. Application of fertilizer NPK is at the rate of 100-120 kg N, 50kg P and 150-200 kg K per hectare for

Jerusalem artichoke. The total quantity of P and half quantity each of N and K are given as basal. The rest half quantities of N and K are applied at time of intercultural operation.The maximum tuber dry matter yield was 18.4 Mg ha^{-1} (65.6 Mg ha^{-1}, fresh weight basis) and it was recorded where 100 kg N ha^{-1}, 2 plants m^{-2} and seed-tubers were planted. The best planting spacing was 2 plants m^{-2} in irrigated as well as in rain-fed conditions. Nitrogen significantly increased tuber yield only whenseed-tubers were used. Averaged across N fertilization rates and planting densities mean tuber dry matter yields were 12.8 and 6.9 Mg ha^{-1} for seed-tuber and botanical-seed, respectively. In general crops raised through seeds give poor yield as compared to vegetatively propagated through tubers (Rodrigues *et al.*, 2007).

Irrigation and Weeding

Four to five irrigations may be needed for optimum yields. Regular watering gives the best tuber production, but Jerusalem artichokes can tolerate dry periods. Jerusalem artichoke plants are extremely vigorous and will compete strongly with weeds. Early season cultivation is recommended to reduce emerging weeds, with a subsequent tillage operation to improve the hilling of rows.

Harvesting and Yield

Tubers can be harvested four to six weeks after flowering. Even though the flowers are pretty, yields will be better if the flower buds are pinched off as soon as they appear. In cooler areas with well-drained soils it is better to dig up the tubers only as you use them. In subtropical areas and poorly drained soils the tubers may rot if left in the ground once the tops die back, so it is better to dig up the entire crop in one operation. Tubers do not store well out of the ground. Carefully store them in slightly damp sawdust or sand in a dark place; or store them in a perforated plastic bag in the bottom part of the fridge.The average tuber yield is 20-30 tonnes per hectare. Tuber can be stored for several months at 0^0C temperature and 85-95% relative humidity.

Plant Protection

In general pests have not been serious problems. Sometimes, stalk borers have been observed, but they usually cause limited damage.Few diseases are reported to affect Jerusalem artichokes. The primary disease is *Sclerotinia* (white mould), which can cause early wilt, stalk rotand degradation of the tubers. This pathogen also can cause severe yield reductions in dry edible beans, sunflowers, and soya beans. If possible, susceptible crops should be rotated with small grains or maize. Avoid close rotations with dry, edible beans, sunflowers, and safflower, mustard and soya beans. Diseases such as downy mildew, rust and southern stem blight have been reported but have not been of economic importance.

Genetic Resources and Improvement

The major emphasis in Jerusalem artichoke breeding is to evolve varity having high fructose content. Analysis involving 37 cultivars originating from France or overseas showed that among the early cultivars studied, 'Grando' appear to be the richest with 193 mg/g fresh wt of dry matter (DM) and 508 mg/g DM of fructose (as fructan). For fructose production, among the late white cultivars analysed, 'Kharkov' gave the best performance with 179 mg/g fresh wt of DM, 531 mg/g DM of fructose, and 710 mg/g DM of total carbohydrates. The French cultivars 'Miello', 'Dub', and 'Rico' are considered in their group as the most productive with 549, 545, and 540 mg/g DM of fructose, respectively (Ben Chekroun *et al.*, 1996). The hexaploid species of *Helianthus tuberosus* is a potential source of resistance to several sunflower pathogens. But crossed with cultivated sunflower, it produces F_1, hybrids which have low values of fertility or even full sterility. Pollen viability and meiotic features were studied in 17 populations of the species *H. tuberosus* and in F_1 hybrids. Significant differences in pollen viability existed between populations (47.1-98.8%). In the 15 F_1 hybrids, plant fertility ranged from 0 to 100 per combination, while pollen viability ranged from 12.4 to 57.1. Meiosis was almost normal in the analyzed species, and irregular in the F_1, hybrids. The highest percentage of meiocytes was with bivalents (85.9), but univalents (0.3) and multivalents (13.8) occurred as well. In metaphase and anaphase, the percentage of meiocytes with fast and lagging chromosomes was high. In anaphase, chromosome bridges were detected in 9.9 of the meiocytes (Atlagic *et al.*, 1993).

References

Atlagić, J., Dozet, B. and ŠKorić, D. (1993). Meiosis and pollen viability in *Helianthus tuberosus* L. and its hybrids with cultivated sunflower. Plant Breeding, 111: 318-324.

Ben Chekroun, M.J. Amzile, J., A. Mokhtari, A. , El Haloui, N.E., J. Prevost, J. R. Fontanillas, R. (1996). Comparison of fructose production by 37 cultivars of Jerusalem artichoke (*Helianthus tuberosus* L.). New Zealand Journal of Crop and Horticultural Science, 24(1):115-120.

Gaafar, A.M., Serag, M.F. El-Din, E., Boudy, A. and El-Gazar, H. H (2010). Extraction conditions of inulin from Jerusalem artichoke tubers and its effects on blood glucose and lipid profile in diabetic rats. Journal of American Science, 6(5):36-43.

Makowetzky, M. Trud (1929.). S. silsk, Gost, Botan. Charkiw, 2:211 D.

Rodrigues, M.A., Sousa, L., Cabanas, J.E. and Arrobas, M. (2007). Tuber yield and leaf mineral composition of Jerusalem artichoke (*Helianthus tuberosus* L.) grown under different cropping practices. Journal of Agricultural Research, 5(4): 545-553.

22

ENDIVE

Endive is most commonly used as a fresh in salads, for which curly leaved forms are preferred. Plants for salads are blanched to reduce bitterness. The main area of production of endive is the European Community followed by North America. Plant contains bitter milky juice, producing a shortened stem with a rosette of large leaves. Rosette leaves are sessile, light to dark green or yellow in colour, appear alternate on stem. Leafy blade broadened, slightly crumpled, margin entire or dentate in curly leaved types, leaf blade reduced, very narrow, deeply pinnatifid and strongly curled. Endive is generally self pollinated crop but some cross pollination occurs due to insects. Endive is rich in many vitamins and minerals, especially in folate and vitamins A and K, and is rich in fiber. It also contains fair amount of other minerals like phosphorus, potassium and iron. Endive contains inulin and intybin which cause the typical bitter taste and stimulate appetite. Recently three hydroxycinnamic derivatives (5-O-caffeoylquinic acid, 3, 4-di-O-caffeoylquinic acid, and 5-O-feruloylquinic acid) were found in chicories, while 3, 5-di-O-caffeoylquinic acid was obtained from var. crispum and cis-caftaric acid of var. latifolium (Papetti *et al.*, 2008). Nutritive value of endive is given in table (22.1).

Table (22.1). Nutritive value of endive leaves (per 100 g of edible portion)

Constituents	Amount
Water	93 per cent
Energy	20 K cal
Protein	1.7 g
Fat	0.1 g
Carbohydrates	4.1 g
Calcium	81.0 mg
Phosphorus	54.0 mg
Iron	1.7 mg
Sodium	14.0 mg
Potassium	294.0 mg
Vitamin 'A'	3300 IU
Thiamine	0.07 mg
Riboflavin	0.14 mg
Niacin	0.5 mg
Ascorbic acid	10.0 mg

Koudela and Petrikova (2007) observed that the cultivars of *C. endivia* var. crispum contained significantly higher quantities of dietary fibre and dry matter than those of *C. endivia* L. var. latifolium. Nitrate levels were significantly higher in the leaves of *C. endivia* L. var. latifolium cultivars. Chemical composition varies among different species as well as varieties of *Cichorium endivia*. Chemical components of some of the varieties are given in table (22.2).

Botanical Description

Endive is also a common name for some types of chicory (*Cichorium intybus*). There is considerable confusion between *Cichorium endivia* and *Cichorium intybus*. Endive belongs to the *Cichorium* genus, which includes several similar bitter leafed vegetables. Species include endive (*Cichorium endivia*), (*Cichorium pumilum)* and common chicory (*Cichorium intybus*). Common chicory includes chicory types such as radicchio, puntarelle and Belgian endive. Its chromosome number (2n) is = 18 (36). Taxonomic position of the genus *Cichorium* is given as under:

Taxonomic position of the genus *Cichorium*

Kingdom	Plantae
Subkingdom	Tracheobionta
Superdivision	Spermatophyta
Division	Magnoliophyta

Table (22.2) : Chemical components of endive varieties

Constituents	Endive varieties					
	Maral	Manal	Nuance	Markant	Protos	Midori
Vitamin C (mg/kg)	218	205	213	282	145	255
Chlorophyll (mg/kg)	430	533	456	496	391	505
Potassium (mg/kg	3117	3302	3037	3682	2711	2483
Sodium (mg/kg)	258	237	199	256	213	162
Calcium (mg/kg	392	449	342	640	332	378
Magnesium (mg/kg)	148	158	126	139	147	133
Dietary fibre (g/ kg)	6.74	6.66	6.23	9.80	7.88	10.53
Dry matter (g/kg)	75	72	65	112	79	108
Nitrates (mg/kg	952	1030	982	797	265	190

Source: Koudela and PetRíková (2007)

Class	Magnoliopsida
Subclass	Asteridae
Order	Asterales
Family	Asteraceae
Genus	*Cichorium* L.

Endive can be classified into two general groups:

i.) The curled or fringed leaved culti vars

ii.) Broad leaved cultivars

There are two main varieties of cultivated endive:

1. **Curly endive (var. crispum)**: This type has narrow, green, curly outer leaves. It is sometimes called chicory in the United States and is called chicorée frisée in French.
2. **Escarole, or broad-leaved endive (var. latifolia):** It has broad, pale green leaves and is less bitter than the other varieties. Varieties or names include broad-leaved endive, Bavarian endive, Batavian endive, grumolo, scarola, and scarole. It is eaten like other greens, chopped into soups or as part of a green salad.

Varieties

Under the curly endive, common varieties are Lorca, Ruffec (resistant to cold and wet conditions) and Salad King. The cultivars representing the broad-leaved class are Broad Leaved Batavian, Full Heart Batavion or Escarole and Florida Deep Heart.

Cultural Practices

Endive can be grown on a wide range of soil types, however, loose fertile loams are best. Soils should have good water holding capacity and drainage. Optimum pH for endive is 6.5 and above. Endive is easy to cultivate during winter season in northern plains. It is more tolerant of night temperatures than lettuce. The mean daily optimum temperature for growth is 15-18°C. Cultivation of endive is similar to that of lettuce. For raising the crop, seedlings are produced by sowing the seeds in nursery bed in the month of October-November. For a hectare planting, 150-200 g seeds are required. The 1000 seed weight is 1.3-1.6 g. One month old seedlings are transplanted at the planting distance of 20×30 cm for narrow leaved and 30×45 cm for broad leaved cultivars. Nutritional requirement of endive is moderate. Heavy application of nitrogen should be avoided as it promotes nitrate content in the leaves. In general, a crop yield 12 t/ha removes about 20 kg P_2O_5 and 45 kg K_2O. Mg deficiency may occur on acidic soils which may be corrected by spraying of magnesium sulphate @ 2 per cent. For maintaining the optimum moisture in the field, timely irrigation should be given. However, 1st irrigation just after transplanting is essential.

Harvesting and Yield

Crops become ready for harvest in 60-90 days from sowing. About a week before harvesting, heads are often tied up in order to blanch them and to moderate the bitterness of the leaves. Approximate 15-20 tonnes yield are obtained from one hectare field. Endive is leafy salad green, not adapted to long storage. Even at 0°C which is considered to be the best storage temperature it cannot be expected to keep satisfactorily for more than 2-3 weeks. Vacuum cooling can help maintain their fresh appearance. Endive should keep somewhat longer if stored with cracked ice in or around the packages. The relative humidity of storage should be maintained above 95 per cent to prevent wilting.

***In -vitro* propagation**

An efficient method for totipotent callus formation and whole plant regeneration has been developed for chicory. Totipotent calli of chicory were induced from cotyledon, leaf, hypocotyl and root explants on MS medium supplemented with different concentrations of IAA, IBA, NAA and 2,4-D at 0.5-10 μM in combination with BAP (2 μM). These calli were transferred to shoot regeneration medium containing MS basal medium with different concentrations and combinations of BAP, KIN and IAA. Maximum number of shoots was obtained on MS medium with BAP (4 μM) + IAA (1 μM). The shoots were rooted on MS medium supplemented with IAA, IBA and NAA. Rooted plantlets were successfully established in the field after hardening (Velayutham *et al.*, 2006).

References

Koudela, M. and Petrikova, K. (2007). Nutritional composition and yield of endive cultivars *Cichorium endivia* L. Hort. Sci. (Prague), 34 (1): 6-10.

Papetti, A., Daglia, M., Aceti, C., Sordelli, B., Spini, V., Carazzone, C. and Gazzani, G. (2008). Hydroxycinnamic acid derivatives occurring in *Cichorium endivia* vegetables. J. Pharm. Biomed. Anal., 48(2):472-476.

Velayutham, P., Ranjithakumari, B.D. and Baskaran, P. (2006). An efficient *in vitro* plant regeneration system for *Cichorium intybus* L. an important medicinal plant. Journal of Agricultural Technology, 2(2): 287-298.

23

LETTUCE

Lettuce is named after milky juice (latex) that it produces. It is an herbaceous plant, grown during cool days. Lettuce occupies the largest area of the salad crop worldwide. It is a pleasure food with low nutrient density. It has a crisp texture, a large surface to volume ratio and serves as source of bulk for diet conscious consumers. It is one of the few vegetable crops used extensively as a fresh raw product. Lettuce is generally not canned, dried or frozen and rarely cooked. The loose lettuce and Romaine types are more nutritious than the head type lettuce, mainly because of its high vitamin A and vitamin C values. It is also a good source of calcium and phosphorus. Lettuce is generally low in calories and helps in the digestion of other foods eaten with it. Nutritive value of lettuce is given in table (23.1).

Origin and Distribution

The origin of lettuce is in Turkey and the Caucasus or the Middle East. The ancestor is probably the European prickly lettuce (*Lactuca serriola* L.), that crosses easily with the cultivated forms. Lettuce was known as a vegetable in the Mediterranean as early as 4500 BC; it was depicted in Egyptian tombs in 2500 BC and cultivated by the Greeks and Romans as a popular vegetable. In Western Europe, headed types have been known since the 14th century but leafy types have been known for much longer. At present lettuce, especially the headed types, is the world's most important salad crop. Lettuce is a major salad crop in North America, Europe, Australia, New Zealand and South America (Grubben, 2004). Its cultivation in our country is

Table (23.1). Nutritive value of different type of Lettuce (per 100 g of edible portion)

Constituents	Lettuce butterhead	Lettuce Romaine	Lettuce crisphead	Lettuce loose leaf
Water (%)	95	94	96	94
Food energy (K cal	14	18	13	18
Protein (g)	1.2	1.3	0.9	1.3
Fat (g)	0.2	0.3	0.1	0.3
Carbohydrates (g)	2.5	3.5	2.9	3.5
Calcium (mg)	35	68	20	68
Phosphorus (mg)	26	25	22	25
Iron (mg)	2.0	1.4	0.5	1.4
Sodium (mg)	9	9	9	9
Potassium (mg)	264	264	175	264
Vitamin A (IU)	970	1900	330	1900
Thiamine (mg)	0.06	0.05	0.06	0.05
Riboflavin (mg)	0.06	0.08	0.06	0.08
Niacin (mg)	0.3	0.4	0.3	0.4
Ascorbic acid (mg)	8	18	6	18

restricted to Himachal Pradesh and Jammu and Kashmir. However, it can successfully be grown during winter season in plains of India.

Botanical Description

Lettuce is most common leafy salad crop of family Asteraceae (Compostae). Its chromosome number (2n) is=18. Taxonomic position of the genus Lactuca is given as under:

Taxonomic Position of the Genus *Lactuca*

Kingdom	Plantae
Subkingdom	Viridaeplantae
Infrakingdom	Streptophyta
Division	Tracheophyta
Subdivision	Spermatophytina
Infradivision	Angiospermae
Class	Magnoliopsida
Superorder	Asteranae
Order	Asterales
Family	Asteraceae
Genus	*Lactuca* L.
Species	*Lactuca sativa* L

Morphological Description of Plant

Glabrous, annual herb up to 100 cm tall, containing latex, forming a dense basal rosette and later a tall, branched, flowering stem; root system shallow, but with a strong, fleshy tap-root. Rosette leaves loose or arranged in more or less compact heads; petiole short; lamina undivided to sawtoothed or pinnatifid, sometimes curly and fringed, pale to dark green or with red or brown anthocyanin pigment; stem leaves arranged spirally, sessile, progressively smaller, ovate to orbicular in outline, entire, cordate-amplexicaul. Inflorescence a head; many heads arranged in a dense, corymbose, flat-topped panicle; involucre 10-15 mm long, consisting of 3-4 rows of lanceolate or ovate bracts. Flowers 7-35 per head, bisexual; corolla ligulate, yellow; stamens 5, with connate anthers; ovary inferior, 1-celled, style bifid. Fruit a narrowly obovate achene 3-8 mm long, compressed, ribbed, white, yellowish, grey, brown or black, with narrow beak, surmounted by a white pappus of 2 equal rows of soft hairs. Seedling with epigeal germination (Grubben, 2004).

Classification of Lettuce

There are several distinct types of lettuce. The main division is between 'headed' and 'non-hearting' types.

A) Headed lettuce

1. Butter head: These types of lettuce have soft, rather buttery leaves. The best of the modern cultivars have good firm hearts and resistance to the diseases. Butterhead lettuce produces a smaller and less compact head than the crisphead type. Leaves are broad, crumpled, relatively thin, and tender with a soft oily texture. The colour of outer leaves is lighter than most crisphead lettuces, and the inner colour is yellowish. Some cultivars have red pigmentation on the outer leaves. Taste varies from bland to relatively sweet.

2. Crisp head: The heads are usually larger than butter heads, and the leaves crisper and wrinkled in texture. Some have an excellent flavour, others are notable more for their crunchy texture than their flavour. The early leaves are elongated at the rosette stage, and gradually increase in width with each successive leaf until they are broader than long at maturity. After 10-12 leaves, leaves change to cup-shaped and begin overlapping each other and enclosing later leaves to form a head structure. New leaves continue to appear and expand from inside to fill the head, which becomes large and firm. Over mature heads become hard, and leaves may rupture and taste bitter. The head may then burst with the elongation of the seed stalk. Outer leaves are bright green or dull green, and the interior colour changes progressively from lighter green to whitish or creamy yellow towards the centre. The tightly folded inner leaves are rugose, brittle, and crispy with a mild taste. Included in this type is a crisp subtype called Batavia that forms a less dense and softer head and weighs about 500 g when mature. A cultivar named 'Iceberg' is actually a Batavia type of lettuce.

3. Cos: These are upright, thicked leaves lettuces, with a fairly loose heart in most cases. They are widely considered to have an excellent flavour. It has elongated, coarse, and relatively crispy textured leaves with prominent broad mid veins. Plants tend to have an upright stature and form a loaf-shaped head after the rosette stage. The heads are either closed or relatively open at the top. The closed type is used for "romaine heart" production, in which only the closed inner leaves are harvested. Outer leaves are usually light to dark green, and interior leaves are yellowish. Romaine heads may weigh as much as 750 g. They have a sweeter and stronger taste than that of crisp head lettuces. Some cultivars have red leaves, which are usually harvested while young and small to produce a value-added product (Mou, 2008)

B) Non-hearting lettuce

The non-hearting type are essentially loose-leaved forming only very insignificant, if any, hearts.

1. Salad Bowl: It forms rosette of leaves rather than a heart. In this type of variety, harvesting is done by removing the individual leaves of the plants can be cut across the head and left to resprout.

2. Cutting lettuces: These are old European cultivars of the loose-leaf type which were traditionally grown for a cut and allow for seed crop.

Varieties

Varieties s of different type of lettuce have been grouped as given in table (23.2).

Table (23.2). Important varieties of lettuce under different groups

Type	Varieties
Crisp head	Great Lakes, Imperial 44, Imperial 459, Red Coach, Vanguard
Butter head	White Boston, Big Boston
Cos or romaine	Parris Island, Dark Green
Leaf	Black seeded Simpson, Early Curled Simpson, Slobolt

Cultural Practices

Lettuce is a cool season crop, requires a mean air temperature of 10-20°C. Cool nights are essential for good quality lettuce. It is successfully grown in hilly parts of the country. In Himachal Pradesh, sowing is done from September to November in lower hills, September to October in mid hills and March to July in higher hills. In plains of Northern India, its sowing is done in month of September-October in nursery bed for raising the seedlings. Approximately 350-500 g seeds are required for one hectare planting. A well drained fertile soil with a pH of about 6.0 is considered most desirable for its growth. The growth and yield of lettuce is greatly influenced by application of fertilizers. For its good crop, 10-12 tonnes of farm yard manure supplemented with 65 kg nitrogen, 40 kg P_2O_5 and 40 kg K_2O per hectare should be applied. About one third of the nitrogen should be given at the time of planting and the rest after a interval of one month from the date of planting. Seedlings become ready for transplant within 3-4 weeks from the date of sowing. Plant seedlings at the four to five leaf stage with the seed leaves just above soil level. Lettuce should never be planted deeply. Small lettuce can be planted 15-20 cm apart in each direction. Standard Butter Head cultivars can be 25-30 cm apart between and within the rows. Crisp Head needs little bit wider spacing i.e. 40×30 cm or 45×30 cm to produce bigger heads. Frequent and light irrigations have been found to be more effective in achieving high yield and quality production. After planting, lettuce should be irrigated at 8-10 days of intervals.

Harvesting and Yield

Harvesting of lettuce commences as plants reach acceptable size and firmness and should be completed before the leaves become tough and bitter and before plants start to bolt. Head lettuce is harvested when the heads are well formed and solid (Ryall and Lipton, 1979). Maturity is based on head compactness, and the firmness of the head is related to its susceptibility to certain postharvest disorders. Soft heads are easily damaged while fairly firm heads have a higher respiration rates. Firm heads have maximal storage-life, while hard and extra-hard heads are more prone to develop russet spotting, pink rib and other physiological disorders. It is usually cut with a long handled sharp knife. The average yield varies from 70-80 quintals per hectare. Lettuce is a highly perishable crop as it has very high water content. Ideally, lettuce may be pre-cooled to about 1°C and held at this temperature at high humidity i.e. 95-97 per cent for storage or transport. It can be stored in good quality for 10-14 days under the ideal temperature conditions. Deterioration starts with increasing temperatures

Grading and Packaging: Head lettuce is graded by size and firmness, while leafy types are graded by size (Hardenburg *et al.*, 1986). Lettuces, as with other leafy vegetables, must be kept clean and free of soil and mud. This is easier when grown on mineral than on organic soils. A strong bitter taste and toughness develops if harvest is delayed or if over-mature, and then the product becomes unmarketable. Crisphead or iceberg lettuce is usually packaged in 20 to 22 kg, 24-count, cartons. Cos or romaine lettuce is commonly packaged in 24-count cartons. Leaf lettuce is usually packaged in 9 to 11 kg or 24-count cartons. Butterhead or Boston lettuce is usually packaged in 9 kg cartons. Lettuce harvested for processing is placed in large bulk bins for transportation to the pre-cooling or processing facility. The package is made from special films that are selected to maintain a desired lower O_2 and higher CO_2 concentration than in air.

Pre-Cooling: Vacuum-cooling is the preferred method for pre-cooling all lettuces (Hardenburg *et al.*, 1986; Ryall and Lipton, 1979). For effective vacuum-cooling, containers and film wraps are perforated. To aid cooling, clean water is sprinkled on the heads of lettuce prior to carton closure if they are dry and warmer than 25°C. A modification called hydro-vacuum reduces water loss during cooling. Hydro-cooling is effective for non-heading lettuce types, but should not be used with head lettuce since the water retained in the head fosters decay. In general, head types are better adapted to prolong storage than are the other types, .Film liners or individual polyethylene head wraps are desirable for attaining high RH; however, they should be perforated or be permeable to maintain a non-injurious atmosphere and to avoid 100% RH on removal from storage. Lettuce is easily damaged by freezing, so all parts of the storage room must be kept above the highest freezing point of lettuce of -0.2°C .

Controlled Atmosphere (CA) Storage: Lettuce, especially crisp head and fresh-cut responds favorably to CA (Saltveit, 1997a). Levels of 1 to 3% O_2 at temperatures of 0 to 5°C (32 to 41°F) reduce russet spotting in susceptible lots. Intact heads do not benefit from elevated CO_2, and injury, ie., brown stain, may develop when lettuce is transferred from storage in > 2% CO_2 to air at 10°C (50°F) (Ke and Saltveit, 1989). A 2 to 5% O_2 atmosphere maintains appearance of lettuce and inhibits pink rib and butt discoloration compared to air. Brown stain is intensified when O_2 is reduced to 2 to 3%, but the effect differs with cultivar. Romaine and leaf lettuce appear to tolerate a slightly higher CO_2 level when packaged than head lettuce. Browning is a major problem with fresh-cut lettuce, and is controlled by packaging in <1% O_2 and 10% CO_2 atmospheres (Lopez-Galvez *et al.*, 1996; Smyth *et al.*, 1998). The elevated level of CO_2 is more effective at reducing browning of the cut surfaces than it is at inducing brown stain.

Flowering, Pollination and Seed Production

Lettuce is a quantitative long-day plant and for seed production plants require vernalization Wien (1997; Prince, 1980). However, vernalization requirements differ with species as well as cultivars (George, 2009). Waycott (1993) observed the effect of photoperiod on the transition to flowering in lettuce genotypes. The inflorescence of lettuce, which is called a capitulum, contains approximately 24 florets. These are highly developed in favour of self-pollination and the crop is therefore largely self-fertilized. However, some cross-pollination can take place between lettuce cultivars and also between cultivated lettuce and some wild Lactuca species. Common lettuce (Lactuca sativa) is cross-compatible with its wild species Lactuca serriola, which grow as a weed in Mediterranean and parts of Asia. Hence it is utmost essential to avoid contamination of the lettuce seed crop with seeds of other wild Lactuca species. There are several mechanical methods have been used to enhance the bolting in lettuce. An alternative is to manually peel off the leaves around the heart, although this is only really practical when relatively few plants are to be used for seed production. The timing of the mechanical or manual operations is important; if left too late, the extended flowering shoot inside the heart will be damaged (George, 2009). Growth regulating chemicals have been used to promote bolting in lettuce. Gibberellic acid has been applied before hearting to promote early bolting, thus avoiding the problems associated with flower shoot emergence. The application of aqueous solutions of gibberellic acid at concentrations between 20 and 500 ppm have been used for the butter head types (Harrington, 1960). However, the application of chemicals to lettuce before the hearts have been completely formed in order to assist flower-stem emergence does not enable the seed producer to confirm the plant's morphological characters at normal market maturity and can therefore adversely affect the selection or roguing efficiency. Wurr *et al.* (1986) advocated the use of gibberelin 4 plus 7 in their studies of the effects of seed production techniques on seed characteristics, subsequent seedling growth and crop performance of the crisp lettuce cultivar 'Pennlake'. They concluded that the major differences between commercial seed lots of lettuce result from differences in the environment rather than the seed production technique.

The succession of inflorescences during anthesis subsequently provides a steady sequence of ripe seed. The length of time from flowering to ripe seed produced on an individual capitulum is 12-21 days, depending on environment. High temperatures increase the rate of development and ripening. The influence of post-flowering temperature on seed development and subsequent performance in crisp lettuce, cv. Saladin, was investigated by Gray *et al.* (1988b). They reported that seed yields per plant increased from a mean of 15-27 g, and reduced to 20 g with gradual increase in temperature (in 16 h day and 8 h night) of 20/10°C, 25/15°C and 30/20°C; they also included evaluations of resulting seed quality. In general, seed produced at 25/15°C exhibited a greater variation in numbers of seeds per floret, cell numerical volume density, seed weight, times of seedling emergence and seedling and mature head weight than seed produced at the lower temperatures. To avoid the loss of seeds by shattering growers should harvest the plant when an estimated 50% of seed heads are ready. The stage of ripeness at which the pappus is fully developed and dry is referred to as 'feathering'. There is less shattering if the plants are cut with the dew on them. The cut material is normally left in windrows for up to 5 days, but in very arid areas the seed can be extracted the same day as cutting. A satisfactory seed yield for hearting lettuce under good conditions is between 0.5 and 1 t/ha. The 1000 grain weight of lettuce is from 0.6 to 1.0 g according to the cultivar cultivars (George, 2009).

Important Pests

1. Beet armyworm: Green or black larva, up to 30 mm long; three pairs of legs near head; five pairs of fleshy prolegs; three lightly coloured stripes running length of body; black spot on each side of the second segment behind the head. It damages bud and young leaves of lettuce plant.

2. Cabbage looper: Green caterpillar with longitudinal white stripes; body up to 30 mm long, tapers toward the head; three pairs of legs near head; three pairs of fleshy prolegs, young larva on underside of leaf; mature larva deep within head; It causes damage to tender leaf tissue, leaving most veins intact.

3. Cutworms: Fat, basically gray, brown, or black caterpillars 40 to 50 mm long when fully grown; three pairs of legs near head; five pairs of fleshy prolegs; young larva on underside of leaf; mature larva deep within head. It feeds tender leaf tissue, leaving most veins intact.

4. Imported cabbage worm: Velvety green caterpillar up to 32 mm long; yellow stripe down back; row of yellow spots down each side; three pairs of legs near head; five pairs of prolegs. It feeds deeper in plant and more likely to eat small veins than the cabbage looper; leaves wet, greenish-brown excrement deep among leaves.

Major Diseases

1. Bottom Rot (*Rhizoctonia solani*)

This is a fungus that attacks mature lettuce in poorly-draining soil. Affected plants show symptomsof necrosis on both the stems and on lower leaves. Plants will also become wilted and start to decay.

Control Measure

1. Ensure proper drainage in the field and avoid over watering.

2. Downy Mildew (*Bremialactucae*)

It is also a fungal disease. Once infected, the lettuce will eventually turn brown and die. Moist conditions exacerbate the problem.

Control measure

1. Select disease resistant varieties of lettuce.

Genetic Resources and Improvement

Large germplasm collections of lettuce and wild *Lactuca* species are kept at institutes in the Netherlands (Centre for Genetic Resources, Wageningen), Russia (Vavilov Institute of Plant Industry, Petersburg), United Kingdom (Institute of Horticultural Research, Wellesbourne) and the United States (National Seed Storage Laboratory, Fort Collins, Colorado). There is a great risk of genetic erosion of this material, since seed companies promote improved cultivars. Many hundreds of cultivars have been bred in temperate countries (Europe, North America, Japan) with a large variation of very specific characters. Resistance to mosaic and downy mildew is common, and cultivars with resistance to aphids, tip burn and bottom rot occur. Important selection criteria for tropical headed lettuce cultivars are heat resistance, a short growing period, slow bolting, and compact heads that are not easily damaged during transport (Grubben, 2004).

Tissue Culture

Explants (2-3 mm long) from axillary buds were successfully grown on MS+1.0 or 2.0 mg litre-1 kinetin and 6.4 mg litre-1 IAA to promote shoot growth. Concentrations of 0.5 and 4.0 mg litre-1 kinetin gave poor shoot growth. The cultures were successfully rooted after 3-4 wk on MS+6.4 mg litres-1 IAA after transfer from MS+1.0 mg litre-1 kinetin and on MS+4.8 mg litre-1 IAA after transfer from MS+2.0 mg litre-1 kinetin. Concentrations of 3.2 and 8.0 mg litre-1 IAA gave poor root initiation. Root initiation was more successful when cultures were grown at 40 Wm-2 than in cultures grown at 5 Wm-2. Rooted cultures were established in compost with a 90-95% success rate and the regenerated plants flowered c. 18 wk after the cultures were initiated (Pink and Carter,1987). An efficient method for the regeneration of shoots directly from cell suspensions of three commercial cultivars of lettuce i.e. Great Lakes 659-700, Salad Bowl, and Prize Head) was standardized by Teng *et al*. (1992). Cell suspensions were prepared byosterizing cotyledon-derived callus for 60 seconds. The effects of callus quality, lightintensity, carbohydrate type and concentration, auxins, and cytokinins on cell growthand differentiation in the suspension culture were examined. Among variousfactors,callus quality and carbohydrates were the most critical. The optimal medium for regenerationof shoots in suspension culture was SH (Schenk and Hilderbrandt) basalmedium containing 1000 mg *myo*-inositol/liter, 1.5% glucose, 0.44 µM BA, and 0.54µM NAA. The pH of the medium was adjusted to 5.8. Under such condition, hundredsof shoots could be produced from 50 to 55 mg (dry wt) of cell aggregates within 2 weeks.

Biotechnology

Among the leafy vegetables lettuce has witnessed more advances in development of maps and mapping aspects than any other leafy vegetable. Lettuce has all important linkage maps developed from intra- and interspecific crosses. The first intraspecific genetic map of lettuce using molecular markers was reported in 1987 (Landry *et al*., 1987) from an F_2 population. Linkage map had 53 genetic markers with 41 RFLP loci, 5 downy mildew resistance genes, 4 isozyme loci and 3 morphological markers. The detailed genetic map later published by Kesseli *et al*. (1994). The map comprised of 319 loci covering 1,950 Cm with 13 major and 14 minor linkage groups. In addition three interspecific maps have been developed by different research groups (Jeuken *et al*., 2001; Syed *et al*. 2005). Recently, integrated map consisting of 2744 AFLP, SSR and EST markers on 9 linkage groups was created using seven intra and interspecific populations (Truce *et al*., 2007). Many genes and quantitative trait loci for downy mildew resistance, plant growth and seed germination traits have been mapped on different maps. A total of 13 major QTLs for root characters and water use efficiency were mapped on different linkage groups using F2:3 population of *L. Serriola* X *L sativa* (Johnson *et al*., 2000). For seed germination characteristics and seedling quality, seventeen QTLs were mapped on map derived from permanent mapping population (RILs) developed from *L. sativa* X *L. serriola*. Further, the Compositae Genome Project has developed expressed sequence data over many years for *L. sativa* and *L. serriola* (http://cgpdb.ucdavis.edu). To aid cloning of important genes bacterial artificial chromosome library have also been developed (Frijters *et al*., 1997). Recently, introgressed lines have been developed for *L. saligna* in the background of *L. sativa* (Jeuken and Lindhout, 2004). These developments in the field of lettuce genomics are likely to aid conventional breeding and increase the efficiency of breeding.

References

Frijters, A.C.J., Zhang, Z., van Damme, M., Wnage, G.L., Ronald, P.C., Michelmore, R.W. (1997). Construction of a bacterial artificial chromosome library containing large Eco RI and HindIII genomic fragments of lettuce.Theor. Appl. Genet. 94:390-399.

George, R. A.T. (2009). Vegetable seed Production, 3rd edition CABI Head Office Nosworthy Way 875 Wallingford Oxford shire OX10 8DE UK.

Gray, D., Wurr, D.C.E., Ward, J.A. and Fellows, J.R. (1988). Influence of post-flowering temperature on seed development and subsequent performance of crisp lettuce. Annals of Applied Biology, 113:391-402.

Grubben, G.J.H. (2004). *Lactuca sativa* L. In: Grubben, G.J.H. & Denton, O.A. (Editors). PROTA 2: Vegetables/ Légumes. PROTA, Wageningen, Netherlands.

Hardenburg, R.E., Watada, A.E. and Wang., C.Y. (1986). The Commercial Storage of Fruits, Vegetables, and Florist and Nursery Stocks. USDA Agric. Hndbk No. 66, pp.136.

Harrington, J.F. (1960). The use of gibberellic acid to induce bolting and increase seed yield of tight-heading lettuce. Proceedings of the American Society of Horticultural Science, 75: 476-479.

Jenuken, M., van Wijk, R., Peleman, J. and Lindhout, P. (2001). An integrated interspecific AFLP map of lettuce (*Lactuca*) based on two *L. sativa* X *L. saligna* F_2 populations. Theor. Appl. Genet. 103:638-647.

Jeuken, M.J.W. and Lindhout, P. (2004). The development of lettuce backcross inbred lines (BILs) for exploitation of the *Lactuca saligna* (wild lettuce) germplasm. Theor. Appl. Genet. 109: 394-401.

Johnson, W.C., Jackson, L. E., Ochoa, O., van Wijk R., Peleman, J., St Clair, D. A. and Michelmore, R.W. (2000). Lettuce, a shallow rooted crop, and *Lactuca serriola*, its wild progenitor, differ at QTL determining root architecture and deep soil water exploitation. Theor. Appl. Genet. 101:1066-1073.

Ke, D. and Saltveit, M.E. (1989). Carbon dioxide-induced brown stain development as related to phenolic metabolism in iceberg lettuce. J. Amer. Soc. Hort. Sci., 114(5):789-794.

Kesseli, R.V., Paran, I. and Michelmore, R. W. (1994). Analysis of a detailed genetic map of *Lactuca sativa* (lettuce) constructed from RFLP and RAPD markers. Genetics, 136:1435-1446.

Landry, B.S., Kesseli, R.V., Farrara ,B. and Michelmore, R. W. (1987). A genetic map of lettuce (*Lactuca sativa* L.) with restriction fragment length polymorphism, isozyme, disease resistance and morphological markers. Genetics, 116:331-337.

Lopez-Galvez, G., Saltveit, M.E. and Cantwell,M. (1996). The visual quality of minimally processed lettuce stored in air or controlled atmospheres with emphasis on romaine and iceberg types. Postharv. Biol. Technol. 8:179-190.

Mou, B. (2008). Lettuce In: Handbook of Plant Breeding Vol. 1. Asteraceae, Brassicaceae, Chenopodicaceae and Cucurbitaceae(Eds.) Fernando Nuez and Jaime Prohen,s Springer Science+Business Media, LLC, 233 Spring Street, New York, NY, pp. 85-126.

Pink, D.A.C. and Carter, P. J. (1987). Propagation of lettuce (*Lactuca sativa*) breeding material by tissue culture. Annals of Applied Biology,110(3): 611-616.

Prince, S.D. (1980).Vernalization and seed production in lettuce. In: Hebblethwaite, P.D. (ed.) Seed Production.Butterworths, London, pp. 485-499.

Ryall, A.L. and Lipton, W.J. (1979). Handling, transportation and storage of fruits and vegetables. Vol. 1, 2nd edition, Vegetables and Melons. AVI Pub. Co, Westport CT. ISBN 0-87055-115-9.

Saltveit, M.E. (1997a). A summary of CA and MA requirements and recommendations for harvested vegetables. In: 7th Int. Controlled Atmos. Res. Conf.,

Vol. 4, Vegetables and Ornamentals. Univ. Calif., Davis, Postharvest Hort. Series,18: 98-117.

Saltveit, M.E. (1997b). Postharvest Diseases. In: Compendium of Lettuce Diseases. R.M. Davis, K.V. Subbarao, R.N. Raid, E.A. Kurtz (eds) Amer. Phytopath. Soc. Press, pp. 57-59.

Smyth, A.B., Song, J. and Cameron, A.C. (1998). Modified atmosphere packaged cut iceberg lettuce: effect of temperature and O_2 partial pressure on respiration and quality. J. Agric. Food Chem., 46: 4556-456.

Syed, N. H., Sorensen, A. P., Anonise, R., van de Wiel, C., van der Linden, C.G., van t Westende, W., Hoofman, D.A.P., den Nijs, H.C.M., Flavell, A.J. (2005). A detailed linkage map of lettuce based on SSAP, AFLP and NBS markers. Theor. Appl. Genet., 112: 517-527.

Teng, W.L., Liu,Y.J. and Tai-Sen Soong T.S. (1992). Rapid regeneration of lettuce from suspension culture. Hort. Science, 27(9):1030-1032.

Truce, M.J., Antonise, R., Lavelle, D., Ochoa, O., Kozik, A., Witsenboer, H., Fort, S.B. Jeuken, M.J.W., Kesseli, R.V., Lindhout, P., Michelmore, R.W. andPeleman. J. (2007). A high-density, integrated genetic linkage map of lettuce (*Lactuca* spp.). Theor. Appl. Genet., 115: 735-746.

Waycott, W. (1993). Transition to flowering in lettuce: effect of photoperiod. Hort. Science, 28, 530.

Wien, H.C. (1997). Lettuce. In: Wien, H.C. (ed.) The Physiology of Vegetable Crops. CAB International, Wallingford, UK, pp. 479-509.

Wurr, D.C.E., Jane, R., Fellows, D.Gray, and Joyce, R.A. Steckel. (1986). The effects of seed production techniques on seed characteristics, seedling growth and performance of crisp lettuce. Annals of Applied Biology, 108, 135-144.

24

SUGAR LOAF CHICORY

The plants of sugar loaf chicory resembles cos or romaine lettuce in its growth being upright in nature. The heads are erect, leaves are tightly packed and so naturally blanched and sweettened. The foliage has slightly bitter taste. While studying the change in carbohydrate content during the growth stage in chicory, Michel *et al*. (1995) reported that the levels of sucrose and fructose in chicory roots remain about the same over the growing season while levels of total non-structural carbohydrates (TNC) and fructans increase. During low temperatures and especially in cold storage, sucrose and fructose contents increase. Concurrently, a second fructan series that contains no glucose accumulates, while at the same time, fructans (inulin) with a high degree of polymerization (DP) and TNC decrease. Chicory leaves contain low concentrations of fructans. Only trace amounts are present in the leaf lamina but leaf petioles, especially in the basal region, contains higher levels. The inulonose series up to DP 4 (inulotetraose) is also detectable in leaves. Sugar loaf chicory leaves is rich in vitamin A and mineral contents. Its nutritive value is given in table (24.1).

Table (1). Nutritive value of sugar loaf chicory (per 100 g of edible portion)

Constituents	Amount
Water	94.0 g
Protein	1.2 g
Fat	0.2 g
Carbohydrates	1.5 g
Energy	12 Kcal
Calcium	42.0 mg
Iron	2.0 mg
Magnesium	20 mg
Phosphorus	30 mg
Vitamin 'A'	1600 IU
Thiamine	0.09 mg
Riboflavin	0.07 mg
Niacin	0.4 mg
Vitamin 'C'	5.0 mg

Botanical Description

Sugar loaf chicory is botanically known as *Cichorium intybus* var. foliosum. It has elongated, blanched leaves and is usually referred to as Belgium endive. In Belgium, it is known as witloof. Taxonomic position of the genus *Cichorium* is given as under:

Taxonomic position of the genus *Cichorium*

Kingdom	Plantae
Subkingdom	Tracheobionta
Superdivision	Spermatophyta
Division	Magnoliophyta
Class	Magnoliopsida
Subclass	Asteridae
Order	Asterales
Family	Asteraceae
Genus	*Cichorium* L.
Species	*Cichorium intybus* L.
CV. group	*Cichorium intybus* L. foliosum

Varieties

Binaca di Milano, Pain de Sucre, Sugar Loaf, Biondissisma de Trieste, Poncho, Snowflake are some of the important cultivars of sugar loaf chicory popular in European countries.

Pluto: It is medium early, cylindrical dark green heads with a good height 25 to 40 cm. It is bitter in taste and has good nutritive value. In India, it is marketed by Bejo Sheetal Seeds Pvt. Ltd.

Cultural Practices

Sugar loaf chicory can successfully be grown as a winter salad vegetable in plains of north India. For raising the crop, seeds are sown in nursery beds and approximately 20,000 to 25,000 seeds are required for a hectare planting. When seedlings attain 4-6 leaf stage, they are transplanted at 15-25 cm apart in rows spaced 45-60 cm apart. For realizing the good yield, apply FYM @10-12 t/ha in the soil at least 1 month before transplanting and at the time of planting 80 kg N, 40 kg P_2O_5 and 40 kg K_2O may be given. Nitrogen should be given in two split doses. Half dose at the time of planting and rest is in form of top dressing after a month of transplanting. A light irrigation just after transplanting is essential. Clean culture and frequent but light irrigation at definite interval promote early growth of plant.

Harvesting and Yield

Plants of sugar loaf chicory become ready for harvest in 75 days after planting. Harvesting should be done 2.5 cm above the ground, leaving the stump to resprout. Approximately 200-250 quintals per hectare yield are obtained. After harvest, damaged and diseased leaves are trimmed. The trimmed heads are washed and may be cooled by vacuum cooling or hydro cooling to 0°C. Heads of sugar loaf chicory can be stored for a longer period in cold storage.

Genetic Resources and Improvement

In general, Belgian endive is not a commercially cultivated vegetable in the world. Recently, its cultivation has started in many European countries. To meet the standards of different markets and preference of consumers, breeding chicory with specific characteristics are major objective of its improvement. Having molecular markers for specific features of chicory, it will be possible to stimulate and support breeder's efforts for developing new cultivars in a more efficient way (Maurice de Proft *et al.*, 2003). A genetic linkage map based on an intraspecific cross between two inbred lines of witloof chicory has been constructed. In total, 129 RAPD markers were scored in 565 F_2 plants. Grouping of these markers at a LOD of threshold 4.0 resulted in nine linkage groups, which was equal to the chicory haploid genome. The nine linkage groups covered 609.6 cM. All 129 RAPD markers were linked to one of the nine groups. Three RAPD markers could not be mapped. Out of the 126 remaining RAPD markers, 18 showed segregation distortion with significance value of $P < 0.01$ (Stallen *et al.*, 2003)

References

Maurice de Proft, Nico van Stallen and Noten Veerle (2003). Breeding and cultivar identification of *Cichorium intybus* L. var. foliosum Hegi, Eucarpia Leafy Vegetables 2003 (eds. Th.J.L. van Hintum, A. Lebeda, D. Pink, J.W. Schut), pp 83-90.

Michael Ernst, Jerry Chatterton, N. and Philip A. Harrison (1995). Carbohydrate changes in chicory (*Cichorium intybus* L. var. foliosum) during growth and storage. Scientia Horticulturare,63(3-4):251-261.

Stallen, N. Van Vandenbussche, B. Verdoodt, V. and De Proft, M. (2003). Construction of a genetic linkage map for witloof (*Cichorium intybus* L. var. foliosum Hegi). Plant Breeding,121(6):523-525.

25

RED CHICORY

Red chicory is also known as Radicchio. The radicchios are types of chicory that grow in tight heads, like cabbage. Most radicchios are wine-red with white veins and about the size of an orange or small grapefruit. They often look like little red cabbages. It is a popular salad vegetable and has a distinctively bitter flavour. Plant is characterized by its red and in some cases strikingly red/variegated colouring. The leaves are slightly bitter like endive but also sweeten slightly with cooler day temperatures. Red chicory is generally used as raw or lightly grilled or roasted and added to salads. It is also used to serve as a colour garnish. Red chicory develops a small crisp heart with inner leaves. Presently improved cultivars develop tighter heads. Water soluble compounds of *Cichorium intybus* var. silvestre (Treviso red chicory) possesses anti- and pro-oxidant activity (Adele Papetti *et al.*, 2002). Chicory leaves are good source of vitamin A (carotene content) and calcium. Nutritive value of chicory leaves is given in table(25.1).

Table (25.1). Nutritive value of chicory (per 100 g of edible portion)

Constituents	Amount
Water	92 g
Energy	13 K cal
Protein	1.7 g
Fat	0.3 g
Carbohydrates	1.1 g
Calcium	100 mg
Iron	0.9 mg
Magnesium	30 mg
Phosphorus	47 mg
Vitamin A	4000 IU
Vitamin B_1	0.06 mg
Vitamin B_2	0.10 mg
Niacin	0.5 mg
Vitamin C	24 mg

Origin and Distribution

Chicory is originally from the Mediterranean region. Chicories and radicchios are especially popular in Italy, where the leaves and stems are eaten raw or cooked, and the roots are boiled or roasted. The wild form of chicory has become naturalized as a weed in many parts of the world, and is especially common in North America where it grows in old fields and along roads and powerlines.

Botanical Description

Red chicory is a plant of family of Asteraceae and it has been place under the genus *Cichorium* and species *intybus* L.

Varieties

The varieties of radicchio are named after the Italian regions where they originate. The most ubiquitous variety in the United States is radicchio di Chioggia, which is maroon, round, and about the size of a grapefruit. Somewhat less common in the States is the radicchio di Treviso, which resembles a large Belgian endive. Other varieties include Tardivo, and the white-colored radicchio di Castelfranco, both of which resemble flowers and are only available in the winter months, as well as Gorizia, Trieste (biondissima) and Witloof/Bruxelles (also known as Belgian lettuce). Alouvette, Cesare, Red Devil, Red Treviso, Verona Palla Rossa are other important varieties of red chicory.

Agro Techniques

Red chicory is grown much like head lettuce and needs a long cool growing period. In the northern plains of India, red chicory can successfully be grown during winter season. Seed are sown in nursery bed during the months of October-November and seedlings are transplanted in the fully prepared bed at the distance of 30 cm between the row and 20 cm from plant to plant. Nutritional requirement of chicory is like endive. Excess nitrogen fertilization of leafy vegetables may cause undesirable accumulation of nitrates and a decrease in essential amino acids, resulting in a decrease in protein nutritional quality (Mirjana Herak *et al.*, 2009). Frequent irrigation should be given to maintain the optimum moisture content in the soil.

Harvesting and Yield

Harvesting can be done by picking the individual leaves or by cutting the hearts. Harvesting should not be delayed otherwise long days/or high temperatures can cause to bolt, increase in bitterness and develop tipburn.

Postharvest

While studying the changes in the quality of minimally processed red chicory storage in a polyvinyl chloride (PVC) film and in a mono-oriented polypropylene (OPP) film at 4°C, Lavelli *et al.* (2009) observed the maximum limit of 5×107 CFU g^{-1} for mesophilic aerobic bacterial count after 4 days using the PVC film and after 4.5 days using the OPP film. The loss of fruity aroma occurred after 3 days in the products stored in PVC but not in those stored in OPP. In spite of microbial growth, no off-odour was observed. Storage in PVC slightly decreased the contents of cyanidin 3-O-malonyl glucoside, total phenolics and antioxidant activity, whereas these parameters increased upon storage in OPP. Red chicory leaf was evaluated as a residue natural substitute for synthetic antioxidants for the food and feed industry. After lyophilization, a red chicory extract (RC) was characterized for its phenolic profile and its oxidative stability as compared to BHT. RC was shown to reduce lipid peroxidation of different oils. In addition, the antioxidant property of RC was studied in a model system by evaluating the saccharomyces cerevisiae response to oxidative stress by means of gene expression. In this analysis, the RC extract, added to the yeast culture prior to oxidative stress induction, exhibited a pleiotropic protective effect on stress responsive genes (Anna Lante *et al.*, 2011).

Genetic Resources and Improvement

Eight red chicory (*Cichorium intybus* L. var. silvestre Bisch.) cultivars were screened for plant characteristics, yield, resistance to *Erysiphe cichoracearum* and their persistence of bolting over a two-year period. The highest total and marketable weight of heads (i.e. total weight - weight of removed damaged leaves) were obtained from cv. 'Averto', followed by cv. 'Mesola' and 'Rubino'. The most compact heads in both years, were produced by cv. 'Palla rossa-super'. Less productive cultivars of red chicory were on average the most infected with the powdery mildew caused by *E. cichoracearum*. Among eight cultivars studied through two years, cv. 'Castel Franco', 'Averto' and 'Mesola' were free from incidence of bolters (Dragan *et al.*, 2004).

Italian red chicory possesses sporophytic incompatibility system. As a result of an incompatibility mechanism which prevents both self-fertilization and intermating between plants with identical genotypes, inbreeding does not occur in *Cichorium intybus.* This self-incompatibility system complicates the production of inbred lines and the generation of F_1 hybrids in Italian red chicory. Histological observations on pollinated stigmas show that the incompatibility reaction in chicory is very rapid and either prevents pollen adhering to the stigmatic papillae or determines the arrest of pollen-tube growth on the stigmatic surface; incompatible pollen cannot reach the transmitting tissue of the style (Varotto *et al.*, 1995). Castano *et al.* (1997) reported that the cv. Carolus was highly self-incompatible, cv. Flash and cv. Pax could be self-compatible or self-incompatible and the cv. Sigma and cv. Focus are self-compatible. All the tested cultivars were intra-compatible (pollinations within the same cultivar) and cross-compatible among them. There was difference in percentage of viable seeds when cultivars were self-pollinated and intra-pollinated. This difference in percentages of viable seeds can be used as a numerical criterion to determine genotypic identity of a cultivar. The higher this difference value, the less uniform the behaviour of the cultivars in the field and during hydroponic forcing for production of the chicory head was observed. The results led to the conclusion that some of the commercial cultivars are not genotypic identical which has unfavorable implications in Belgian endive production and commercialization. There was equal reaction to incompatibility in any sense of crossing (plants used as females compared with the result when plants were used as males in cross-pollinations).

***In vitro* propagation**

Best storage condition (95% of shoot survival after 9 months) was obtained with the maintenance of 'Treviso precoce' at 4°C in darkness. As for cryopreservation, shoot tips survived satisfactorily when they were loaded in PVS2, prior to direct plunging into liquid nitrogen)(Previati *et al.*, 2005). Red chicory shoot tips survived and regrew satisfactorily from the storage at -196°C when a procedure of "vitrification/one-step freezing", consisting in loading the explants in the PVS2 vitrification solution was applied (Lambardi *et al.*, 2006). Embryos collected at the heart or early torpedo stage of development (three days after pollination) produced green plantlets after two weeks of *in vitro* culture and thus they could be transferred to the greenhouse for acclimation. Culture of zygotic embryos *in vitro* has shown to be a suitable technique to accelerate the breeding process in biennial cultivated Italian red chicory. In fact it is possible to obtain plantlets ready for performing the selection for the desired agronomic traits one month after pollination. Since flowering occurs after vernalization, the selection can be performed in the correct season before flowering (Serena Varotto *et al.*, 2000).

References

Adele Papetti, Maria Daglia and Gabriella Gazzani (2002). Anti- and pro-oxidant activity of water soluble compounds in *Cichorium intybus* var. silvestre (Treviso red chicory). Journal of Pharmaceutical and Biomedical Analysis, 30(4):939-945.

Anna Lante, Tiziana Nardi, Federico Zocca, Alessio Giacomini, and Viviana Corich (2011). Evaluation of red chicory extract as a natural antioxidant by pure lipid oxidation and yeast oxidative stress response as model systems. J. Agric. Food Chem., 59(10):5318-5324.

Castaño, C. I., Demeulemeester, M. A. C. and Proft, M. P. De (1997). Incompatibility reactions and genotypic identity status of five commercial chicory (*Cichorium intybus* L.) hybrids .Scientia Horticulturae, 72(1):1-9.

Dragan , Z, Jo•e, O. and Stanislav, T. (2004). Plant characteristics for distinction of red chicory (*Cichorium intybus* L. var. silvestre Bisch.) cultivars grown in Central Slovenia. Acta agriculturae slovenica, 83 - 2, November 2004 str. 251 - 260.

Lambardi, M., Benelli, C., De Carlo, A., Previati, A., Da Re, F. and Giannini, M. (2006). Biotechnologies for the preservation of selected red chicory (*Cichorium intybus* L.). Acta Hort. (ISHS) 725:311-318

Lavelli, V., Pagliarini, E., Ambrosoli, R. and Zanoni, B., (2009). Quality of minimally processed red chicory (*Cichorium intybus* L.) evaluated by anthocyanin content, radical scavenging activity, sensory descriptors and microbial indices. International Journal of Food Science & Tech., 44(5):994-1001.

Mirjana Herak Usti , Marija Horvati and Marija Pecina (2009). Nitrogen fertilization influences protein nutritional quality in red head chicory. Journal of Plant Nutrition, 32(4):598-609.

Previati A., Benelli C., Da Re F., De Carlo A., Vettori, C. and Lambardi, M. (2005). *In vitro* propagation and conservation of red chicory- the role of biotechnoly Villa Gualino, Turin, Italy - 5-7 March, 2005

Serena Varotto, Margherita Lucchin and Paolo Parrini (2000). Immature embryos culture in Italian red chicory. Plant Cell tissue and Organ culture, 62:75-77.

Varotto, S., Pizzoli, L., Lucchin, M. and Parrini, P. (1995). The incompatibility system in Italian red chicory (*Cichorium intybus* L). Mendeley, 114 (6):535-538.

26

GLOBE ARTICHOKE

Globe artichoke (*Cynara scolymus* L.) is an herbaceous perennial that is grown for its tender, edible, immature flower buds. The globe artichoke should not be confused with Jerusalem artichoke, another member of the compositeae family native to North America, which is grown for its fleshy tubers. Globe artichoke plants can become large: four to five feet tall and wide, with long, heavily serrated silvery green leaves. Unopened flower buds are used for vegetable purpose.The buds are harvested at an immature stage before they open and expose the flower. The base of each bract and the large fleshy base or receptacle (artichoke "heart") on which the flower and bracts are borne are fleshy and edible. If the buds are allowed to mature and open, the resulting flowers are quite attractive, large, and fragrant. Nutritional composition is significantly affected by genotype, head fraction, location, season and genotypes. 'Blanc Hyerois', Harmony F_1', 'Madrigal F_1' and 'Violetto di Provenza' showed high levels of both macro- and micromineral content. The globe artichoke contains a high level of K and as compared to other vegetables, it posseses low Na/K ratio (Pandino *et al.*, 2011). Nutrition value of globe artichoke is given in table (26.1).

Table (26.1). Nutrition value of globe artichoke (per 100 g of edible portion)

Principle Nutrients	Value
Energy	47 Kcal
Carbohydrates	10.51 g
Protein	3.27 g
Total Fat	0.15 g
Cholesterol	0 mg
Dietary Fiber	5.4 g
Folates	68 µg
Niacin	1.046 mg
Pantothenic acid	0.338 mg
Pyridoxine	0.116 mg
Riboflavin	0.066 mg
Thiamin	0.072 mg
Vitamin C	11.7 mg
Vitamin A	13 IU
Vitamin E	0.19 mg
Vitamin K	14.8 µg
Electrolytes Sodium	94 mg
Potassium	370 mg
Calcium	44 mg
Copper	0.231 mg
Magnesium	60 mg
Manganese	0.256 mg
Phosphorus	90 mg
Selenium	0.2 µg
Zinc	0.49
Carotene- ß	8 µg
Iron	1.28 mg
Crypto-xanthin	0 µg
Lutein-zeaxanthin	464 µg
ORAC value 7904	

Source: USDA National Nutrient data base.

Origin and Distribution

Globe artichokes are native to the Mediterranean region, but have been grown in this country since colonial times. Thomas Jefferson grew artichokes in his extensive food gardens at his central Virginia Monticello plantation as early as 1767, successfully overwintering the plants using protective coverings. Today, in this country, over 12,000 acres of artichokes are grown commercially in California's central coast area. The mild West Coast climate allows year-round perennial culture and quality bud production. In recent years, annual winter production systems using newer cultivars (cultivated varieties) have been developed in the southern desert areas of California and Arizona where growers target off-season markets.

The evolutionary history of artichoke and cultivated cardoon and their relationships to wild allies of the genus *Cynara* are not fully understood yet. To try resolve the evolutionary patterns leading to the domestication of these two crops, a study of molecular evolution was undertaken. The species *C. cardunculus*, including artichoke, cultivated and wild cardoon, together with four wild *Cynara* species were taken into consideration. Internal (ITS) and external (ETS) rDNA transcribed spacers were used as markers of nuclear genome, the psbA-trnH spacer as a marker of chloroplast genome. Sequences were analysed using phylogenetic analysis packages. Molecular data indicate that the whole genus is quite recent and that the domestication of artichoke and cultivated cardoon, crops diverging for reproduction system and use, are independent events which diverge in time and space. As for wild *Cynara* species, an evolutionary pattern consistent with their present geographical distribution was hypothesized in relation to the climatic changes occurring in the Mediterranean during the last 20 millennia: *C. humilis* and *C. cornigera* appeared to have differentiated first, *C. syriaca* and *C. baetica* were differentiated in a second period, while *C. cardunculus* showed to be the most recent and plastic species. The high plasticity of *C. cardunculus* has not only allowed its nowadays wide distribution, but has also given the potential for domestication (Sonnante *et al.*, 2007).

Botanical Description

Eight species and four subspecies are recognized under the genus *Cynara*, viz., *C. algarbiensis*, *C. auranitica*, *C. baetica* including subsp. *baetica* and subsp. *maroccana* (formerly known as *C. hystrix*), *C. cardunculus* including subsp. *cardunculus* and subsp. *flavescens*, *C. cornigera*, *C. cyrenaica*, *C. humilis* (formerly sometimes in the genus *Bourgaea*) and *C. syriaca*. The cultivated artichoke (formerly *C. scolymus*) and cardoon are both included in *C. cardunculus*. One species, *C. tournefortii*, is excluded from *Cynara*. Wiklund (1992) has made a detailed study of the genus and its morphology, anatomy and phytogeography. All species of the genus *Cynara* have a diploid number of chromosomes (2n=2x=34).

Climate and Soil

Artichokes have narrow preferences for climatic conditions. Considered a cool season crop, they grow best at a 75°F day time temperature mean with 55°F night time temperatures. They have an effective adaptive range of 45° to 85°F. In order to form buds, new artichoke plants require 'vernalization' or 'chilling.' This occurs just after planting, with seedling exposure to cool temperatures (eight to ten days or 190 to 240 hours of 50°F or less) required for plants to initiate buds. Thus early spring planting is needed to meet this requirement. Artichokes are deep-rooted plants adapted to a wide range of soil types, but will perform best in well-drained, fertile, and deep soil. The extremes of heavy clay and light sandy soils should be avoided. Raised-bed culture is recommended where drainage is inadequate

Propagation and Planting

Globe artichoke is propagated both by seeds as well as by vegetative means.

Seeding

Seeds for transplants should be sown six to seven weeks ahead of anticipated transplant date. Seedlings will germinate in seven to ten days at greenhouse temperatures of 75° to 85°F days and 60° to 65°F nights. Beginning at the first true-leaf stage, a once weekly soluble fertilizer (100 to 200 ppm N) should be applied to keep seedlings vigorous. Plants should be hardened outside of the greenhouse in a cold frame or other protected area for seven to ten days prior to planting.

Planting

Use a measuring tape or other means to mark plant spacing. A hole 5- 7 cm diam. should be made in the plastic mulch and soil. A sharp bulb planter, pointed trowel, or other hole punch device works well for this. Plants should be set 60-70 cm apart (2,906 plants per acre at 3 m between-row spacing) in the row. Use of closer spacings may result in less secondary bud production per plant, but also more plants per acre. A 60 cm in-row spacing (3,630 plants per acre) has been used with some success. Use of a twin row configuration on the bed with 30 to 40 cm between rows, and with plants staggered 1.0 m apart (4,148 plants per acre) between rows, may also be practical, and increase plants per acre. Irrigation rates will need to be adjusted to match planting density. Double-row configurations may require two drip tapes on each bed, depending on water movement patterns in the soil type. Installing a soil moisture-monitoring device, such as a tensiometer, is recommended.

Transplanting should not be too deep as seedlings are sensitive to depth of planting . Root balls should be lightly covered with soil. A transplant solution (300 to 400 ppm N) may be applied to water in each plant individually after planting. Within a few days, the irrigation should be given for establishment of seedlings. It is not unusual to observe very slow early transplant growth; plants may sit for ten to 14 days with little noticeable growth. A small percentage of plants may die early; these and any weak plants should be replaced. This is likely a consequence of planting into adverse weather condition speed growth.

Artichokes are tolerant of low temperatures (approximately 30° to 32°F); however, given the early planting date, protection of newly developed young seedlings using a row cover or straw mulch advocated.

Irrigation

Once seedlings are established, artichokes require frequent irrigation during the growing season. This is particularly true as young plants begin to develop some size. Low soil moisture and plant stress during bud formation will result in loosely formed, tough, poor-quality buds that do not size well. Tensiometer can be used to monitor soil moisture and irrigation should be given accordingly to keep soil moisture in the available range at 15 to 25 centibar tension, depending on soil type. Over-irrigation and long-term saturation should be avoided, especially on heavier soils. The amount of normal rainfall must be taken into consideration, especially in bare-ground plantings.

Use of PGR

The application of the plant growth regulator (PGR) giberellic acid (GA) has been found effective to enhance yield earliness and uniformity, as well as improve the percentage of marketable buds in perennial artichoke systems. Application of GA @ 20 ppm induced bud development. The first application should be made five to eight weeks after planting. This treatment is repeated two more times at two-week intervals.

Interculture

Weeds can be managed through a combination of both pre- application of herbicides. Plasticulture production will require weed management between the beds, while bare-ground production requires weed management over the entire field. Several effective herbicides are registered for use in artichokes. Weeds can be mechanically managed by tillage as well. Once plants reach a certain size, the large leaves will shade the soil, discouraging weed growth.

Major Pests

Aphids are major concern, congregating on the undersides of leaves and in bud bract. Cutworms and slugs can cause considerable damage to new seedlings. Earwigs also inhabit the crowns of plants, but cause little damage, except leaving frass deposits. Plants should be kept free from two-spotted spider mites during hot weather.

Major Diseases

Powdery mildew and *Ramularia* leaf spot can attack the bud bracts and foliage in artichoke, causing premature leaf drop and unmarketable buds. Young seedlings can damp off (*Pythium*) following planting. This occurs in poorly drained soils. *Verticillum* wilt, a vascular disease, can also affect artichoke.

Viral Diseases

Artichoke is very much vulnerable to infestation of a number of of virus disease. Some of the important virus diseases are described as under:

Potyviruses:

Artichoke latent virus (ArLV) was first recovered from symptomless globe artichoke plants by Marrou and Mehani (1964) in Tunisia, then by Fischer and Lockhart (1974b) and Foddai *et al.* (1977) in Morocco and Sardinia (Italy), respectively. ArLV is a typical potyvirus with particles ca 730 nm long, that can be transmitted experimentally in non-persitent manner to artichoke seedlings by *Myzus persicae*, *Brachycaudus cardui*, and *Aphis fabae* (Rana *et al.*, 1982). Transmission efficiency is very high, as shown by the quick re-infection rate (over 75% in two years) registered in a field trial carried out with virus-free plants in Sardinia (Foddai *et al.*, 1985). The host ranges of ArLV isolates from eight different countries were studied by Rana *et al.* (1982), who also characterized an Italian isolate of the virus, proving that it is serologically unrelated to the following potyviruses: Bean common mosaic virus (BCMV), Bean yellow mosaic virus (BYMV), Clover yellow vein virus (CYVV), Beet mosaic virus (BMV), Lettuce mosaic virus (LMV), Zucchini yellow fleck virus (ZYFV), Papaya ringspot virus (PRSV), Watermelon mosaic virus (WMV), Turnip mosaic virus (TuMV), Potato virus Y (PVY), Soybean mosaic virus (SMV), and Dasheen mosaic virus (DsMV). Originally, ArLV was thought not to have any detrimental effect on infected plants. However, a number of trials conducted in France and Italy have demonstrated that it has a distinct depressive effect on the crop. Recent studies have shown that accessions of *C. cardunculus* and *C. syriaca* are susceptible to ArLV, whereas an interspecific hybrid between these two species has a high level of resistance (Manzanares *et al.*, 1995).

Bean yellow mosaic virus (BYMV): This virus was initially recorded from globe artichoke by Russo and Rana (1978) in Apulia (southern Italy) and later in Greece (Rana and Kyriakopoulou, 1980). Apulian artichokes were also infected by BBWV and ArLV and showed vein yellowing, yellow flecking and line patterns on the leaves. These symptoms closely resembled those characterizing the aetiologically undetermined "mosaic" described by Gigante (1951), and the yellow mottle from Campania and Apulia with which BBWV was associated (Rana *et al.*, 1987b). Turnip mosaic virus (TuMV) was isolated in southern Italy (Sardinia and Apulia) from symptomless artichoke plants of cvs Spinoso sardo (Foddai *et al.*, 1993) .

Carlaviruses

Artichoke latent virus M (ArLVM) was found in Apulia in mixed infection with one or two of the following viruses: AILV, AMCV, ArLV (Di Franco *et al.*, 1989). Virus particles in leaf dip preparations from naturally infected globe artichoke leaves measured 697 nm in length and contained a single species of RNA with an estimated size of ca 7.5 kb and a coat protein with an Mr of 31,000. ArLVM was not associated with any particular symptom in the field and was not mechanically transmitted to a number of hosts (*C. quinoa*, *C. amaranticolor*, *Cucumis sativus*, *Cucurbita pepo*, *C. scolymus*, *G. globosa*, *N. benthamiana*, *N. clevelandii*, *N. megalosiphon*, *N. tabacum*) . In immuno-elec-tron microscopy tests, ArLVM was weakly decorated by antisera to Carnation latent virus (CarLV) and Poplar mosaic virus (PopMV), both members of the genus Carlavirus. The cytopathology of globe artichoke cells infected by ArLVM did not differ from that induced by other carlaviruses in their hosts. The cells contained complex cytoplasmic inclusions composed of deranged organelles, lipid droplets and accumulations of membranes (Di Franco *et al.*, 1989). ArLVM is a tentative species in the genus Carlavirus (Adams *et al.*, 2004). Artichoke latent virus S (ArLVS) was probably identified for the first time in symptomless Californian artichoke plants (Costa *et al.*, 1959). Two viruses causing the same symptoms as those induced by ArLVS in three herbaceous hosts (*C. amaranticolor*, *N. clevelandii*, and *Zinnia elegans*) were later found in symptomless globe artichoke plants in Apulia (Majorana and Rana, 1970b) and Morocco (Fischer and Lockhart, 1974b). The particle lengths of these viruses were 678 and 674 nm, respectively. There is no evidence that a virus found in Brazil in symptomless artichoke plants (Costa and Camargo, 1969; Kitajima *et al.*, 1970) was the same as ArLVS as infected plants contained the cylindrical inclusions (pinwheels and laminated aggregates) typical of potyviral infections. However, degenerated artichoke plants from Spain (Peña-Iglesias and Ayuso-Gonzales, 1972), were possibly infected by ArLVS (or another carlavirus) and a potyvirus because infected cells had cytopathological modifications suggestive of the simultaneous presence of a carlavirus and a potyvirus. No antiserum to ArLVS is available. ArLVS is a tentative species in the genus Carlavirus (Adams *et al.*, 2004).

Unnamed putative carlavirus: A virus with filamentous particles was isolated in Apulia from symptomless globe artichoke plants of cvs Terom and Romanesco obtained by *in vitro* meristem tip culture. Its biological, physico-chemical, and cytopathological properties resembled those of members of the genus Carlavirus (Rana *et al.*, 1989). This virus differed from ArLVM because of its sap-transmissibility to herbaceous hosts, genome size (7.7 kb versus 7.5 kb), coat protein subunit size (29,000 versus 31,000 Da), and particle length (664 nm versus 697 nm). Like ArLVM, it was serologically related to PopMV. However, the relationship was distant enough (SDI = 4-5), to support its identification as an artichoke isolate of PopMV, as suggested by Rana *et al.* (1989).

Potexviruses: Potato virus X (PVX) was isolated in Tunisia by Chabbouh (1989) from artichoke plants showing mosaic, leaf deformation and a strong reductionin size of the leaf blade. The virus was mechanically transmitted to a number of herbaceous hosts and identified serologically. Cytopathological alterations were those expected in cells infected by PVX. Artichoke curly dwarf virus (ACDV) is currently classified as a tentative species in the genus Potexvirus (Adams *et al.*, 2004). It was recovered by Leach and Oswald (1950) and Morton (1961) in California (USA) from stunted artichoke plants with distorted leaves, variously extended vein necrosis, delayed flower head development, and low quality crop. Mechanically inoculated artichoke and cardoon seedlings reproduced the field syndrome (Leach and Oswald, 1950). ACDV is poorly characterized; its particles are filamentous, 582 x 15 nm in size. Its epidemiology is unknown. In California, this virus was reported to occur consistently in mixed infections with another virus with longer filamentous particles (Morton, 1961) that may be the same as the putative carlavirus ArLVS. Artichoke degeneration virus (ADV) is a virus with filamentous particles that infects globe artichokes in Spain. It was identified by Peña-Iglesias and Ayuso- Gonzales (1972) in dwarfed plants of cv Tudela with mottling, curling and crinkling of the leaves. Virus particles from partially purified preparations had an estimated length of 585 nm. The cytopathology of infected cells resembled that induced by potexviruses. Attempted transmissions by aphids were unsuccessful. According to Peña-Iglesias and Ayuso-Gonzales (1972), ADV and the putative potexvirus ACDV are not the same. This, however, was not ascertained experimentally.

Criniviruses: In California, natural infections by Tomato infectious chlorosis virus (TICV) were detected in artichoke and weeds (*Nicotiana glauca*, *Picris echioides*) that were host to high populations of Trialeurodes vaporariorum. Known in the USA since 1993, TICV elicits interveinal yellowing and necrosis of the leaves of tomato plants. Symptoms in artichoke are undetermined. In Italy, TICV was recorded from tomato plants in Liguria, Sardinia, Latium, and Campania. The virus is transmitted by *T. vaporariorum* but not by *Bemisia tabaci*, and is included in the EPPO alert list as it is regarded as an emerging pathogen with a high potential to have an economic impact.

***In vitro* production**

Side shoots excised from underground dormant buds of globe artichoke were used as primary explants to establish *in vitro* cultures. The highest rate of axillary shoots was induced on Murashige and Skoog agar medium supplemented with 0 mg NAA/liter and 1.0 mg BA/liter (4.44 μM). Other cytokinins tested (kinetin, zeatin, and 2-isopentenyl-adenine were less effective than BA in inducing axillary shoot growth. Up to 60% of elongated microshoots rooted after 5 weeks on 1/2 MS agar medium supplemented with 2 mg/liter (11.42 μM) indole-3-acetic acid (IAA). Seventy per cent of rooted plantlets were transferred successfully into soil. Plants are under evaluation for their genetic uniformity and clonal fidelity (Anonymous, 2010.).

References

Adams, M.J., Antoniw, J.F., Bar-Joseph, M., Brunt, A.A., Candresse T., Foster, G.D., Martelli, G.P., Milne, R.G. and Fauquet, C.M. (2004). The new plant virus family Flexiviridae and assessment of molecular criteria for species demarcation. Archives of Virology, 149:1045-1060.

Anonymous (2010). Micropropagation of globe artichoke (*Cynara scolymus* L.) from underground dormant buds (ovoli). Biomedical and Life Sciences, 32(4):249-252.

Chabbouh, N. (1989). Mise en évidence de quatre virus présents sur l'artichaut en Tunisie. Annales de l'Institut National de la Recherche Agronomique de Tunisie, 62:3-14

Costa, A.S., Duffus, J.E., Morton, D., Yarwood, C.E. and Bardin R. (1959). A latent virus of California artichokes. Phytopathology, 49:49-53.

Costa, A.S. and Camargo, I.G.B. (1969). Occorência do virus latente da alcachofra en São Paulo. Revista da Sociedade Brasileira de Fitopatologia, 3:53-54.

Di Franco A., Gallitelli, D., Vovlas, C. and Martelli G.P. (1989). Partial characterization of Artichoke virus M. Journal of Phytopathology 127: 265-273.

Fischer, H.V. and Lockhart, B.E. (1974a). Occurrence in Morocco of a virus disease of artichoke related to artichoke mottled crinkle. Plant Disease Reporter, 58:1117-1120.

Fischer, H.V. and Lockhart B.E. (1974b). A Moroccan isolate of artichoke latent virus. Plant Disease Reporter, 58: 1123-1126.

Foddai A., Marras F., Idini G. (1977). Presenza di un "Potyvirus" sul Carciofo (Cynara scolymus L.) in Sardegna. Annali della Facoltà di Agraria dell'Università di Sassari 25:398-407.

Foddai A., Marras F., Idini G., (1992). Il saggio immunoenzimatico (ELISA) per la diagnosi del Potyvirus latente delFoddai A., Cugusi M., Idini G., 1993. Artichoke (*Cynara scolymus* L.): a new host for turnip mosaic virus. Phytopathologia Mediterranea, 32: 247-248.

Foddai A., Marras, F. and Idini, G. (1985). Indagine sulla reinfezione naturale del carciofo "Spinoso sardo" risanato da ALV. Informatore Fitopatologico, 35: 57-58.

Foddai, A., Cugusi M. Idini G., (1993). Artichoke (*Cynara scolymus* L.): a new host for turnip mosaic virus. Phytopathologia Mediterranea 32: 247-248.

Gigante, R. (1951). Il mosaico del carciofo. Bollettino della Stazione di Patologia Vegetale di Roma, 7:177-181.

Kitajima, E.W., Camargo, I.G.B. and Costa A.S. (1970). Microscopia elettronica de particulas do virus latente da alcachofra e das inclusoes citoplasmicas associadas a sua infeccao. In: Proceedings of the 1st "Corgreso Brasileiro de Microscopia Elettronica", Riberao Preto, 1970, 97-98.

Leach, L.D. and Oswald, J.W. (1950). Curly dwarf, a virus disease of globe artichoke. Phytopathology, 40: 967-968.

Majorana, G. and Rana, G.L. (1970b). A latent virus of artichoke belonging to potato virus S group. Phytopathologia Mediterranea, 9: 200-202.

Majorana, G. and Rana, G.L. (1970a). Un nuovo virus latente isolato da carciofo in Puglia. Phytopathologia Mediterranea, 9:193-196.

Manzanares, M.J., Corre, J. and Hervé, Y. (1995). Evaluation of globe artichoke and related germplasm for resistance to artichoke latent virus. Euphytica, 84: 219-228.

Marrou, J. and Mehani, S. (1964). Etude d'un virus parasite de l'artichaut. Comptes Rendues de l'Académie d'Agriculture de France, 50:1051-1064.

Morton, D.J. (1961). Host range and properties of the globe artichoke curly dwarf virus. Phytopathology 51:731-734.

Nazarenko, A., Bhatnagar, S.K. and Hohman R.J.,(1997). A close tube format for amplification and detection of DNA based on energy transfer. Nucleic Acids Research, 25:2516-2521.

Pandino, G., Lombardo, S. and Mauromicale, G. (2011). Mineral profile in globe artichoke as affected by genotype, head part and environment. J. Sci. Food Agric., 91(2):302-308.

Peña Iglesias, A. and Ayuso Gonzales, P. (1972). Degeneration of Spanish globe artichoke (*Cynara scolymus* L.) plants. Virus isolation, host range, purification and ultrastructure of infected hosts. Annales de l'Instituto Nacional de Investigaciones Agronomicas, 2:89-122.

Rana, G.L., Migliori A. and Ragozzino, A. (1987b). La maculatura gialla del carciofo in Campania e Puglia. Informatore Fitopatologico, 37:41-43.

Rana, G.L., Piazzolla, P., Lafortezza, R. and Greco, N. (1989). Characterization of a latent elongated virus from globe artichoke (*Cynara scolymus* L.) in Italy. Journal of Phytopathology, 125:289-298.

Rana, G.L. and Kyriakopoulou, P.E. (1982). Artichoke Italian latent and artichoke mottled crinkle viruses in artichoke in Greece. Phytopathologia Mediterranea, 22: 46-48.

Russo M. and Rana G.L. (1978). Occurrence of two legume viruses in Artichoke. Phytopathologia Mediterranea, 17: 212-216.

Rana, G.L., Russo, M., Gallitelli, D., Martelli, G.P. (1982). Artichoke latent virus: characterization, ultrastructure, and geographical distribution. Annals of Applied Biology 101: 179-188.

Rana, G.L. and Cherif, C. (1981). Occurrence of artichoke mottled crinkle virus in Tunisia. Phytopathologia Mediterranea, 20: 179-180.

Rana, G.L. and Kyriakopoulou, P.E. (1980). Bean yellow mosaic virus in artichokes in Greece. In: Proceedings 5th Congress of the Mediterranean Phytopathological Union, Patras, 1980, 38-40.

Sonnante, G., Carluccio, A.V., Vilatersana, R., and Pignone, D. (2007). On the origin of artichoke and cardoon from the *Cynara* gene pool as revealed by rDNA sequence variation. Genetic Resources and Crop Evolution, 54(3):483-495.

Wiklund, A. (1992). The genus *Cynara* L. (Asteraceae-Cardueae). Botanical Journal of the Linnean Society, 109: 75-123.

27

PARSLEY

Both dried and fresh leaves of parsley have numerous uses. The fresh leaves of curly-leaved cultivars are commonly used for garnishing and seasoning, while those of flat-leaved cultivars are mainly used to flavour dishes as like coriander leaves. Dried leaves of parsley, known as parsley flakes are used to flavour soups, meat and fish dishes, vegetables and salads. Essential oils from the fruits and the herb are used as a seasoning agent in all kinds of food products, especially meats and processed products. The maximum permitted level of parsley leaf oil is about 1.5% in processed vegetables and of parsley oleoresin about 0.05% in condiments. The oil from the fruits is occasionally used in perfumery. It possesses diuretic, carminative, emmenagogue and antipyretic properties, and is effectively used to control hypertension. Crushed leaves of parsley are used as a poultice for sore eyes and can be applied to insect bites and stings. It is rich in calcium, iron, folate, β-carotene, thiamin, riboflavin, vitamins C and E content as compared to other vegetables (Athar *et al.*, 1999). A high proportion of the carotene ie., 9-cis β-carotene, possibly active against cancer and cardiovascular disease has also been reported in parsley (Benamotz and Fishler, 1998). Nutritional composition of the fresh parsley leaves is given in table (27.1).

Table (27.1). Nutritional composition of fresh leaves of parsley (per 100 g of edible portion)

Constitutents	Amount
Energy	36 kcal
Carbohydrates	6.33 g
Sugars	0.85 g
Dietary fiber	3.3 g
Fat	0.79 g
Protein	2.97 g
Vitamin A equiv.	421 µg
β–carotene	5054 µg
Lutein and zeaxanthin	5561 µg
Thiamine	0.086 mg
Riboflavin	0.09 mg
Niacin	1.313 mg
Pantothenic acid	0.4 mg
Vitamin B_6	0.09 mg
Folate	152 µg
Vitamin C	133 mg
Vitamin E	0.75 mg
Vitamin K	1640 µg
Calcium	138 mg
Iron	6.2 mg
Magnesium	50 mg
Manganese	0.16 mg
Phosphorus	58 mg
Potassium	554 mg
Sodium	56 mg
Zinc	1.07 mg

The aerial parts of parsley are reported to contain furanocoumarins (mainly bergapten) and glucosides (derivatives of glucose and apiose with apigenin and luteolin as aglucon parts). The essential oil constituent apiol has been used to treat kidney ailments and menstrual problems. It is also an abortifacient, but large doses may cause paralysis and death. Apiol in minute doses helps cure epileptic fits. Some people get dermatitis from working regularly with parsley.

Origin and Distribution

Parsley probably originated in the western Mediterranean. It occurs naturally or naturalized in most Mediterranean (including northern Africa) countries and in many temperate countries. Parsley is an old crop, which was already well known in classical Greece and Rome. It is widely grown for its leaves in most Mediterranean countries, Europe and North America. In the tropics, including South-East Asia, it is cultivated on a small scale at higher elevations. Forms with a thickened, edible taproot are of recent origin and were probably developed around 1500 AD in Northern Germany. Their cultivation is concentrated in North-Western and Eastern Europe and North America.

Botanical Description

Parsley (*Petroselinum crispum* (Mill.) Nyman ex A.W. Hill) is a member of the family Apiaceae There is a subspecies, *P. crispum* subsp. tuberosum, Hamburg parsley, with an edible root. Taxonomic position of the genus *Petroselinum* is given as under:

Taxonomic position of the genus *Petroselinum*

Kingdom	Plantae
unranked	Angiosperms
unranked	Eudicots
unranked	Asterids
Order	Apiales
Family	Apiaceae
Genus	*Petroselinum*

Morphological Description of Plant

Plant erect, copiously branched, biennial to perennial herb, 30-100 cm tall, aromatic in all parts, glabrous. Root system slender, fibrous, taproot up to 1 m long, sometimes thickened, with a radical rosette of leaves when young. Stem terete, sulcate, hollow. Leaves alternate, 1-3-pinnately compound, dark green, glossy, flat or curled, sheathing at base; petiole longest in lower leaves; pinnae long stalked, leaflets obovate-cuneate to finely linear, divided into acute segments, higher leaves gradually less divided, topmost one consisting of a few acute segments only. Inflorescence a terminal or axillary compound umbel; bracts 1-3-foliolate, rather short; peduncle up to 12 cm long; primary rays 5-20, 1-5 cm long; bracteoles 3-8-foliolate; secondary rays (pedicels) 3-15, 2-5 mm long; flowers small, yellow-green, bisexual; calyx obscure; corolla consisting of 5 petals, each one suborbicular to obovate, up to 1 mm x 0.5 mm, sub-emarginate with an inflexed apical lobe; stamens 5; pistil with inferior, 2-carpelled ovary, each carpel with a thickened stylopodium, a style and a small globose stigma. Fruit a schizocarp, ovoid, 2-3 mm long, at maturity splitting into 2 mericarps, each one with 5 narrow ribs. The 1000-seed weight is about 1.5 g.

Climate and Soil

Parsley tolerates temperatures of 3°C to over 24°C, but grows best between 7-16°C. In the tropics, it is generally grown above 600 m altitude, its growth rate being reduced at lower altitudes. It is cultivated in Malaysia at 200 m and in the Philippines at 600 m altitude. Adequate moisture supply is most important during germination and early growth. Crop thrives well in areas receiving precipitation in range of 1000-1200 mm. Parsley prefers well drained fertile loams rich in organic matter with pH ranging from 4.9-8.2.

Seed Sowing and Planting

Parsley is propagated by seed, which can be sown directly in the field. Curly-leaved Parsley and Flat-leaved Parsley can also be sown first in a nursery and then transplanted. Seeds germinate very slowly, but the process can be enhanced by soaking the seeds in warm water overnight. Slight freezing the seed also helps to break seed dormancy. Pill and Kilian (2000) reported that priming of parsley seed with PEG increased the germination from 74 to between 82-86 per cent, however, it was greatly influenced with temperature, duration and GA application after priming. Sowing rates range from 7-12 kg/ha when grown for the fresh market, higher seed rates of 20-25 kg/ha are used for plants grown for industrial purposes, but 40-60 kg/ha for direct sowing has also been reported. Seed may be sown in double rows, 30-60 cm apart on raised beds, and the seedlings thinned to 10-15 cm apart. Seedlings grown in the nursery can be transplanted when they attain 6 leaves or are 5-8 cm tall.

Interculture

Weeding in parsley, although difficult because of the close spacing, should be done by hoeing, cultivation or mulching. Mulching will produce a cleaner product, since it prevents soil from splashing on the foliage during the rainy season. Both pre-emergence and post-emergence herbicides have been recommended for managing the weed infestation. Fertilizer can be broadcast before planting and N side dressed after germination and after the first leaf harvest, to maintain a continuous production of leaves. Turnip-rooted parsley needs heavy dose of nutrients as compared to leaf types. The usual production period from sowing to last harvest is 4-6 months.

Plant Protection

Parsley is rarely severely attacked by diseases and pests. The most common diseases are *Septoria* leaf spot (*Septoria apiicola*), damping-off (*Pythium* spp.), aster yellows, bacterial soft rot, blight, stem rot (white mould), Texas root rot, heart rot and several viruses. *Septoria* leaf spot is disseminated by seed or rain splash and can be prevented by using clean seed and mulching. Aster yellows can be prevented by controlling leaf hoppers. Heart rot can be prevented by Ca application and careful watering. Important pests of parsley are carrot fly (*Psila rosae*), aphids transmitting carrot mottle dwarf virus, caterpillars, plant bugs and nematodes.

Harvesting and Yield

Harvesting of parsley starts 75 days after sowing or when the plants are about 20 cm tall. In temperate regions, three cuts can be made a year. Parsley is harvested manually by grasping the stalks of the outer leaves, cutting them off with a knife and tying them together to make bunch. New leaves develop continuously from the centre of the plant for later harvests. Single harvesting systems are gaining popularity. Plants are then harvested by slicing off the top of the taproot; this facilitates bunching. Parsley grown for industrial use is harvested mechanically; roots of turnip-rooted parsley are mostly harvested manually by digging. Yields of parsley are generally 150-200 q/ha of fresh herb or 20-30 q/ha of dried herb. There is little difference in yield between single and multiple harvesting systems, but the quality of the product is generally better with frequent harvests. Turnip-rooted parsley may yield up to 250-300 q/ha of fresh roots.

Post Harvest Management

Rapid removal of field heat without excessive drying is essential for retaining green color and freshness of parsley. Parsley can be pre-cooled with ice (package icing or liquid-icing (Cantwell and Reid, 1992) or by vacuum-cooling (Aharoni *et al*., 1989). Forced-air or hydro-cooling are commonly practiced (Joyce *et al*., 1986). Parsley is not chilling sensitive. It should be stored as cold as possible without freezing which occurs at -1.1°C. The recommended conditions for commercial storage of parsley are 0°C with 95 to 100% RH. MAP is effective in extending storage life, but temperature changes and condensation must be avoided. Aharoni *et al.* (1989) reported that non-perforated polyethylene liners delayed yellowing and decay at low temperature. Park *et al.* (1999) achieved 77 days of storage at 0°C or 35 days at 5° with good retention of firmness and vitamin C content, using a 40 µm-thick ceramic film. A pre-harvest application of gibberellic acid may extend storage life (Lers *et al*., 1998).

For the production of parsley flakes, fresh leaves are dried rapidly under controlled conditions of temperature and humidity to retain the essential oil and the bright green leaf colour. For commercial purposes, dried parsley should be free from yellow or discoloured leaves and have a moisture content not exceeding 4.5%. Dried parsley can be packed either in polythene-lined corrugated fibre cases or well-sealed metal containers. They should be stored under cool, dry conditions. Steam distillation of the shoot and the mature fruits yields the commercial parsley herb oil and fruit oil, respectively (Ipor and Oyen, 1999).

Controlled Atmosphere (CA) Storage

Parsley can tolerate 8 to 10% O_2 + 8 to 10% CO_2 (Saltveit, 1997), but it has slight advantage over 0°C. It was observed that 10% O_2 + 11% CO_2 had optimal balance for delaying yellowing in parsley stored at 5°C (Apeland, 1971). Maintaining 10% O_2 + 10% CO_2 in storage (Yamauchi and Watada, 1993) or 10% CO_2 (Lers *et al.*, 1998) delayed yellowing at room temperature. Parsley leaves produce very little ethylene but are very sensitive to ethylene (Joyce *et al.*, 1986; Tsumura *et al.*, 1993). Cantwell and Reid (1993) observed that parsley leaves produced 0.08, 0.44 and 0.80 μL kg^{-1} h^{-1} at 0,10 and 20°C. As little as 0.4 μL L^{-1} is enough to accelerate yellowing if parsley is stored above 0°C (Cantwell and Reid, 1993). Inaba *et al.* (1989) observed that application of ethylene did not stimulate respiration in parsley. Zenoozian (2011) reported that storage of parsley under modified atmosphere packaging caused minimum loss in weight and retained maximum chlorophyll. Parsley had the most molds in perforated packages. However, maximum residual vitamin content was seen in MAP parsley.

Genetic Resources and Improvement

There is little danger of genetic erosion in parsley as it is a popular herb grown almost throughout the world. Several germplasm collections are being maintained, e.g. at the Research Centre for Agrobotany, Tápiószele, Hungary (about 170 accessions, 1990), the N.I. Vavilov Institute of Plant Industry, St. Petersburg, Russian Federation (about 100 accessions, 1990) and the North Central Regional Plant Introduction Station, Ames, United States (176 accessions, 1997). The major objective in parsley breeding is to evolve the variety possessing dark green colour, fineness and leaf form (curliness), re-growing capacity and yield. Hybridization of parsley and celery (*Apium graveolens* L.) has given promising results (Ipor and Oyen, 1999). Pascal celery cv. Golden Spartan was hybridized with parsley which is immune to celery blight (*Septoria apiicola*). Three F_1 hybrids were obtained from over a thousand seedlings grown from the pascal celery parent. The F_1 plants were intermediate for most of the characters. Segregation for petiole color in the F_2 generation demonstrated that a hybrid actually occurred. Segregation for late blight resistance occurred independently from petiole color (Honma, and Lacy, 1980). Inter-generic crosses have been tried between parsley and celery. Madjarova and Bubarova (1978) used three cultivars of celery and two of parsley. The cross parsley 'Lister' x celery 'Pioneer' expressed in a new parsley var. as Festival-68. These crosses had new forms of leaf celery characterized by high levels of vitamin C, carotene, essential oils and aminoacids.

Tissue culture

In a micropropagation protocol study of parsley, Vandemoortele *et al.* (1996) concluded that concentrations of thidiazuron or benzyladenine that were optimal for shoot proliferation, resulted in shoots with a low capability to root. During the rooting treatment, these shoots showed wilting signs. Rooting was increased significantly by using a two-week inductive stage with 2.5 μM naphthaleneacetic acid directly followed by acclimatization. Two proliferation media (5 μM benzyladenine and 0.5 μM naphthaleneacetic acid or 5 μM kinetin and 2.5 μM naphthaleneacetic acid) resulted in moderate proliferation but produced shoots that were easy-to-root. The medium with 5 μM benzyladenine and 0.5

μM naphthaleneacetic acid was optimal. In parsley embryogenic callus was induced by culturing endosperm tissues of germinating seeds on MS basal medium without any exogenous supply of growth substances. No deterioration of the embryogenic potential of this callus strain was observed on this medium for more than half a year through several passages. The endosperm callus consisted of approximately triploid cells, but diploid ones predominated in the embryos originated from this callus (Masuda *et al.*, 1977).

References

Aharoni, N., Reuveni, A. and Dvir, O. (1989. Modified atmospheres in film packages delay senescence and decay of fresh herbs. Acta Hort. 258:255-260.

Apeland, J. (1971). Factors affecting respiration and color during storage of parsley. Acta Hort. 20:43-52.

Athar, N., Spriggs, T.W. and. Liu, P. (1999). The concise New Zealand food composition tables, 4thEd. NZ Inst. Crop & Food Res., Palmerston North, New Zealand.

Benamotz, A. and Fishler, R. (1998). Analysis of carotenoids with emphasis on 9-cis b-carotene in vegetables and fruits commonly consumed in Israel. Food Chem. 62:515-520.

Cantwell, M. I. and Reid. M.S. (1992). Postharvest handling systems: fresh herbs. In: A.A. Kader (ed) Postharvest technology of horticultural crops, 2nd ed., pp. 211-213. Univ. of Calif. Div. Agric. and Nat. Res., Oakland, CA.

Cantwell, M.I. and. Reid, M.S. (1993). Postharvest physiology and handling of fresh culinary herbs. J. Herbs, Spices and Medicinal Plants 1:93-127.

Honma, S. and Lacy, M. L. (1980). Hybridization between Pascal celery and parsley. Euphytica, 29(3):801-805.

Inaba, A., Kubo, Y. and Nakamura. R. (1989). Effect of exogenous ethylene on respiration in fruits and vegetables with special reference to temperature. J. Jap. Soc. Hort. Sci. 58:713-718.

Ipor, I.B. and Oyen, L.P.A., (1999). *Petroselinum crispum* (Miller) Nyman ex A.W. HillIn: de Guzman, C.C. and Siemonsma, J.S. (Eds). Plant Resources of South-East Asia No. 13: Spices. Backhuys Publisher, Leiden, the Netherlands, pp. 172-176.

Joyce, D., Reid, M..S. and Katz. P. (1986). Postharvest handling of fresh culinary herbs. Perishables Handling, Univ. of Calif., Davis CA, 58:1-4.

Lers, A., Jiang, W.B., Lomanies, E. and Aharoni. N. (1998). Gibberellic acid and CO_2 additive effect in retarding postharvest senescence of parsley. J. Food Sci., 63:66-68.

Madjarova, D. and Bubarova, M.G. (1978). New forms obtained by hybridization of *Apium graveolens* and *Petroselinum hortense*. Acta Hortic,73:65-72.

Masuda, K., Koda,Y. and Okazawa, Y. (1977). Callus formation and embryogenesis of endosperm tissues of parsley seed cultured on hormone-free medium. Physiologia Plantarum, 41(2):135-138.

Park, K.W., Kang, H.M. Yang, E.M. and Jung. J.C. (1999). Effects of film package and storage temperature on quality of parsley in modified atmosphere storage. Acta Hort. 483:291-298.

Pill, W.G. and Kilian, E.N. (2000). Germination and emergence of of parsley in response to osmotic and matric seed priming and treatment with gibberellin. Hortscience, 35(5):907-909.

Saltveit, M.E. (1997). A summary of CA and MA requirements and recommendations for harvested vegetables. In: M.E. Saltveit (ed.) CA '97 Proc., Vol. 4: Vegetables and Ornamentals, Univ. of Calif., Davis CA, pp. 98-117.

Tsumura, F., Ohsako, Y. and Haraguchi. Y. (1993). Rapid changes in parsley leaves during storage in controlled or ethylene containing atmosphere. J. Food Sci. 58:619-625.

Vandemoortele, L., Billard, J.P., Boucaud, J. and T. Gaspar, Th. (1996). Micro-propagation of parsley through axillary shoot proliferation. Plant Cell, Tissue and Organ Culture, 44(1): 25-30.

Yamauchi, N. and Watada. A.E. (1993). Pigment changes in parsley leaves during storage in controlled or ethylene-containing atmosphere. J. Food Sci. 58:616-618, 637.

Zenoozian, M.S. (2011). Effect of modified atmosphere packaging on quality changes of fresh parsley, spinach and dill. 2011, 2nd International Conference on Environmental Science and Technology IPCBEE vol. 6 (2011). © (2011) IACSIT Press, Singapore V1, 76-79.

28

FLORENCE FENNEL

Florence fennel is a type of fennel, having fine, feathery foliage and is considered one of the most decorative collections among the exotic vegetables. Plants are generally raised for the succulent, aniseed-flavoured bulb, which develops from the swolled bases of the leaf stalks. Swollen parts are used both raw and cooked. Swollen base of the stem developed from the petiole of the leaves lacks the bitter principles and taste strongly of anise due to high anethole content of the essential oil. Nutritive value of the petiole is given in table (28.1) and chemical composition of essential oil of florence fennel is given in table (28.2).

Table (28.1) : Nutritive value of florence fennel (per 100 g of edible portion)

Constituents	Amount
Water	93 g
Energy	15 K cal
Protein	1.1 g
Fat	0.1 g
Carbohydrates	2.6 g
Calcium	44 mg
Iron	0.8 mg
Magnesium	23.0 mg
Phosphorus	38.0 mg
Vitamin 'A'	100 IU
Vitamin B_1	0.04 mg
Vitamin B_2	0.02 mg
Niacin	0.4 mg
Vitamin C	9.0 mg

Source: Howard *et al.* (1962).

Table (28.2) : Chemical composition of essential oil of florence fennel

Compound	Value
Alfa Pinene	1.65
Camphene	0.08
Sabenene	0.39
Bete Pinene	0.12
Myrecene	0.18
Alfa Phelandrene	0.11
O Cymene	0.46
Limonene	12.53
Eucalyptol	2.05
Gama Terpinene	0.24
Fenchone	7.99
Linalool	0.29
Camphor	0.13
Estragole	11.99
Fenchyl acetate	0.13
Cumic aldehyde	0.18
p-Anisaldehyde	0.40
Trans- Anethole	61.13

Source: Molecules (2011)

Botanical Description

Florence fennel comes under the genus *Foeniculum* which is frequently stated to have three species, *Foeniculum vulgare* (fennel), *F. azoricum* Mill. (Florence fennel, a vegetable) and *F. dulce* Mill. (sweet fennel, also a vegetable). *F. vulgare* has no subspecies although subdivision into subsp. *piperitum* (wild taxa) and subsp. *vulgare* (cultivated taxa) has been proposed, as has division into *F. vulgare* var. azoricum, *F. vulgare* var. dulce and *F. vulgare* var. vulgare. Its chromosome number (2n) is =22.Taxonomic position of the genus *Foeniculum* is given as under:

Taxonomic position of the genus *Foeniculum*

Kingdom	Plantae
Plants Subkingdom	Tracheobionta
Superdivision	Spermatophyta
Division	*Magnoliophyta*
Class	*Magnoliopsida*

Subclass	Rosidae
Order	Apiales
Family	Apiaceae (Umbelliferae)
Genus	*Foeniculum*
Species:	*Foeniculum vulgare* Mill.

Although the Apiaceaes (formerly Umbelliferae) or carrot, family comprise one of the largest assemblages of herbaceous plants, *Foeniculum vulgare* P. Mill., commonly known as fennel, is the only known species of the genus *Foeniculum.* F*oeniculum vulgare* has been cultivated and used by humans for centuries and like many members of the Apiaceae family it is a popular herb and spice. It has a strong aroma that is very similar to anise or licorice and has long been a popular food source in many cultures. Although all parts of the plant are edible, sweet fennel is not currently commercially cultivated, because unlike its cultivated relatives, sweet fennel does not form the bulb-like base that is a popular vegetable. Where naturalized or cultivated in the small garden, however, the leaves are used as an herb or garnish, the seeds make a useful spice, and the young stems are eaten both raw and cooked.

Foeniculum vulgare is the only cultivated species in the genus *Foeniculum*, which is one of approximately 300 genera in the Apiaceae (carrot) family. Apiaceae is the largest of three families that comprise the order of flowering plants called Apiales and the two smaller families are Araliaceae (ginseng) and Pittosporaceae. Apiaceae and Araliaceae are the closest relatives and both are characterized by their wide distribution, ability to self-pollinate or be pollinated by a variety of organisms and by their distinctive umbrella-like arrangement of tiny flowers called umbels . The flowers are usually white or yellow and their group arrangement is an evolutionary trait developed to attract pollinators. Another evolutionary trait developed by many Apiaceae is the production of ethereal oils that repel predators. It is the presence of these aromatic oils that make them popular herbs and spices for humans and their swollen roots make many Apiaceae an important food source. Most Apiaceae are herbaceous and many, including celery, carrots, dill, cumin, parsley and anise, in addition to *Foeniculum vulgare* (fennel) are economically valuable. It is also true, however, that some are highly toxic, such as the poison hemlock that poisoned Socrates.

Pollen records for Apiales can be traced back as far as the Cretaceous and records of Eocene fossil pollen for Apiaceae have also been identified. Because of their longevity, wide distribution and anthropogenic history, Apiaceae were one of the first plant groups to be systematically classified in the end of the 16th century, however, taxonomy of the time was based largely on morphology and it is only now that Apiales are being reclassified according to their genetic traits. Until recently, Apiales were constituents of the sub-class of Magnoliopsida titled Rosidae, but as scientific knowledge has progressed it has become evident that the genetic make-up of Apiales falls more closely into the sub-class Asteracaes and further that they are part of the Eurasterids II clade. As knowledge progresses so does additional classification and it is now contended that the Apiaceae family may be further divided into three subfamilies and twelve tribes (Downie, 2003). As *Foeniculum vulgare* is the smallest genus in the family it is not surprising that it is the

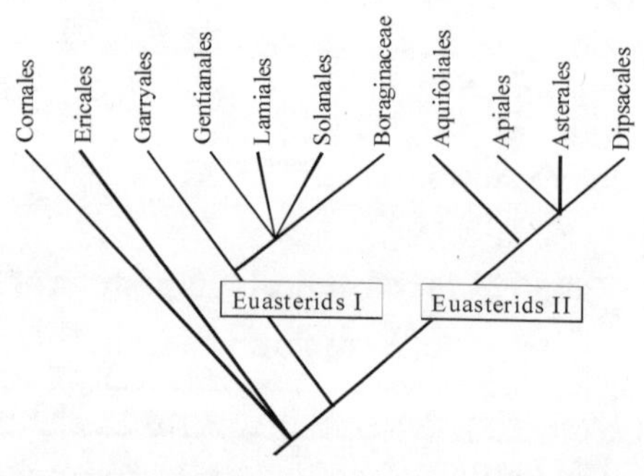

least studied, but as it continues to invade natural areas research is steadily advancing. However, for the most part the goal is to find effective eradication measures and not to build a phylogenetic framework for the species.

The order Apiales, that is part of the Eurasterids II clade, is subdivided into three families: Apiaceae, Araliaceae and Pittosporaceae.

Distribution and Habitat

Foeniculum vulgare is native to southern Europe where it has been cultivated for centuries (Garland, 1989), but is now found throughout the Mediterranean region and temperate climates in the western hemisphere where it has proliferated after repeatedly escaping cultivation. As with most genera in the Apiaceae family, *Foeniculum vulgare* is eurytopic and distribution in temperate regions with dry, warm summers and cool, moderately moist winters is continuous and nearly cosmopolitan. Reproduction easily occurs through root division and seed germination and may take place at any time of the year. Although plants will not flower until they mature between 18 months to 2 years after germination (Bean and Russo, 1988), seed dispersal methods ranging from animal droppings and carriage by waterways, to ejection from agricultural machinery and automobiles are highly effective (Klinger, 2000). While many organisms utilize *Foeniculum vulgare* as a food source, they are not necessary for survival because each plant continuously self-pollinates while flowering and finally produces copious amounts of seeds. Moreover, seeds remain viable for several years and will germinate in full sun without ground cover and little moisture. Additionally, *Foeniculum vulgare* tolerates a wide range of soil types. It is a drought tolerant and grows easily in disturbed areas where it is extremely difficult to eradicate once it becomes established. Distribution is limited in Mediterranean climates only by frost and elevations greater than 1000 feet.

Fennel (*Foeniculum vulgare*) is a perennial herb. It is erect, glaucous green, and grows to heights of up to 2.5 m, with hollow stems. Plants of forence fennel has bright green to the blue green leaves in alternate phyllotaxy, leaves compound with a leaf sheath forming an open cylinder at the base embracing the stem. The leaves grow up to 40 cm long; they are finely dissected, with the ultimate segments filiform (thread-like), about 0.5 mm wide. The flowers are produced in terminal compound umbels 5-15 cm wide, each umbel section having 20-50 tiny yellow flowers on short pedicels. The fruit is a dry seed from 4-10 mm long, half as wide or less, and grooved.

Varieties

Perfection, Sirio, Sweet Florence, Cantino and Zefa Fino are recommended varieties of florence fennel.

Cultural Practices

Florence fennel does well on fertile, drained and light sandy soils, however, it can be grown in all kinds of soils provided with plenty of organic matter and drainage facilities. Crop also needs plenty of moisture throughout growth. It can be grown in plains of northern India during winter season. For raising the crop, seeds can directly be sown in well prepared bed at the distance 30 cm apart from each way. The optimum temperature for germination lies between 20 and 25°C (Thomas, 1996). Further, seeds harvested from secondary umbels germinated better both in the light and dark than those taken from primary umbels. Transplanting can also be done by raising the seedlings. However, care should be taken to avoid the root disturbance during the planting. Seedlings of 3-4 leaf stage are considered ideal for planting. After planting, a light irrigation should be given and bed should be kept weed free. Plants grow fast in suitable conditions and bcome ready 70-80 days after sowing. Once the stem start to swell they can be earthed up halfway with soil. This makes swollen structure whiter and sweeter.

Harvesting

Harvesting of the swollen bulb can be done after 15-20 days earthing up. When the bulbs are formed, they are cut from the ground level, leaving the stump in the ground. Aniseed flavoured bulb used raw or braised.

Micropropagation

Different NAA plus kinetin or BA combinations were tested for callus induction and plant regeneration. The highest frequency of shoot regeneration was found when the auxin and kinetin were used at a 1:1 ratio. Moreover, a prolonged culture period increased shoot formation. Two different protocols were used for somatic embryo induction. Using the first protocol among the different fennel genotypes tested, only Francia Pernod showed embryogenic capacity. In this case, from a primary non-embryogenic callus cultured for 12 months in presence of 2, 4-D, an embryogenic secondary callus was produced. When transferred to the medium without 2, 4-D, this gave embryogenic plants in high frequency. As far as the second embryogenic method is concerned, secondary embryogenic callus developed only in the presence of 2,4-D plus kinetin in Francia Pernod genotype. Thereafter, the replacement of those growth regulators by GA_3 into the medium greatly increased the somatic embryo development, especially in 'Francia Pernod', but also in 'Aboca erbe' callus, a population with a very poor embryogenic capacity. In Francia Pernod, the primary and secondary (embryogenic) calli showed different morphological and histological responses, either when the secondary callus was induced by 2,4-D alone or by 2,4-D plus kinetin (Anzidei *et al.*, 2000). Two different fennel subspecies, *Foeniculum vulgare* Mill subsp. *vulgare* var. *azoricum* and *Foeniculum vulgare* subsp. *piperitum*, have been compared to test their ability to produce somatic embryos. It has been observed that most organs are able to produce callus, but only in response to some treatments, and some whitish tissue swellings with a completely different morphology appear on the surface of calli. These callus masses, transferred to a shaking hormone-free liquid medium, develop a cell culture which produces embryos. Both subspecies produced somatic embryos, even if the subsp. *piperitum* appears more adaptable because wilder (Ruta *et al.*, 2000).

References

Anzidei, M., Bennici, A., Schiff, S., Tani, C. and Mori, B. (2000). Organogenesis and somatic embryogenesis in *Foeniculum vulgare* : histological observations of developing embryogenic callus. Plant cell, Tissue and Organ Culture, 61(1):69-79.

Bean, C. and Russo, M.J. (1988). Element stewardship abstract for *Foeniculum vulgare*. http://tncweeds.ucdavis.edu/esadocs/documnts/foenvul.html

Downie, S. (2003). Apiaceae (Umbelliferae)SubfamilyApioideae.http://www.life.uiuc.edu/downie/apiaceae.html.

Garland, S. (1989). Gran libro de las hierbasy especias. Editorial Blume SA, Barcelona, España.

Howard, F.D., MacGillivary, J.H. and Yamaguchi, M. (1962). Nutrient composition of fresh California grown vegetables. Bull Nr 788, Calif Agric Expt Stn, University of California, Berkeley.

Klinger, R. (2000). *Foeniculum vulgare* Miller. pp. 198-202. In: Bossard, C.C., M.C. Hoshovsky and J.M. Randall. Invasive Plants of California's Wildlands. Berkeley and Los Angeles: University of California Press.

Molecules (2011). 16 p.1368.

Ruta, C., Morone Fortunato, I. and Tagarelli, A. (2000). Somatic embryogenesis in fennel (*Foeniculum vulgare* Mill.) Atti V Giornate Scientifiche S.O.I. 1:185-186.

Thomas, Tudor H. (1996). Responses of florence fennel (*Foeniculum vulgare azoricum*) seeds to light, temperature and gibberellins. Biomedical and Life Sciences, Plant Growth Regulation, 14(2):139-143.

29

CELERY

Celery is grown for its fresh herb. It is a biennial in temperate areas, but cultivated as an annual for vegetable purpose. In the vegetative phase, the plants produce mainly leaves and the stem does not elongate and stalks grows to about 50 cm in height. The leaves and stalk are used fresh or after blanching in soups, sauces, and purees. It can be cooked as vegetable and canned for further use. The leaves of all types of celery can be dried and used to flavour soups. Most of the self blanching cultivars have off-white or yellowish stalks, which can rendered whiter and slightly sweeter by close planting or excluding the direct sunlight. Celery leaves are quite rich in vitamin 'A' and 'C' content. Leaves are also good source of iron. Its herb also contains essential oil with the chemical composition of £ and â-pinene, myrcene, transfarnesene, humulene, limonene, cis-Bcis-B Ocimene, G-terpenene, Trans-B Ocimene, apiol, B-selinine, senkyuonlide and neocnidilide. Besides, choline ascorbate from the celery leaves and enzyme inositol trisphosphate have also been isolated. The characteristic smell of the celery plant is due to the lactone sedanolide (3-butyl-3a, 4, 5, 6-tetrahydrophthalide), sedanenolide (3-butyl-4, 5-dihydrophthalide) and related phthalides (Schippers, 2004). Nutritional composition of the celery leaves and its stalk is given in table (20.1).

Table (20.1).Nutritional composition of celery leaves and stalk (per 100g of edible portion)

Constituents	Amount	
	Leaves	**Stalk**
Moisture	88.0 g	93.5 g
Protein	6.3 g	0.8 g
Minerals	2.1 g	0.9 g
Fibre	1.4 g	1.2 g
Carbohydrates	1.6 g	3.5 g
Energy	37 K cal	8 K cal
Calcium	230 mg	30 mg
Phosphorus	140 mg	38 mg
Iron	6.3 mg	4.8 mg

Source: Gopalan *et al.* (2004)

Celery is specially grown for its seed, which contains a valuable volatile oil used in the perfume industry. In our country, it is mostly grown for seeds which are exported. Year wise quantity exported and value received is mentioned in table (20.2).

Table (20.2).Year wise quantity exported and value received from celery seed

Year	Quantity (tonnes)	Value (Rs. Lakh)
2005-06	4165	1500.64
2006-07	4294	1563.36
2007-08	2900	1325.00
2008-09	3650	2333.00
2009-10	5000	2662.50
2010-11 (E)	3750	2585.90

Source: DGCI&S, Calcutta/ Shipping Bill/ Exporters' Return

Origin and Distribution

Celery is native to the Mediterranean and the Middle East and is found in the wild state in much of the north temperate old world. In our country, celery is cultivated mainly in the states of Punjab (Jallandhar, Gudaspur and Amritsar districts), Haryana and western Uttar Pradesh (Ladhwa and Saharanpur districts) over an area of about 5000 ha. About 90% of the total produce comes from Punjab. Year wise area and production of celery is given in table (20.3).

Table (20.3). Year wise area and production of celery

Year	Area (ha)	Production (metric tonnes)
2002-03	2800	3000
2003-04	2800	3000
2004-05	1197	4577
2005-06	1154	1492
2006-07	1799	2350
2007-08	3158	4239
2008-09	4177	5329

Source: Spices Board of India

Botanical Description

Celery is a member of the Apiaceae (Umbelliferae) family. This family comprises unusually aromatic plants with hollow stems, including the cumin, parsley, carrot, parsnip, dill, caraway, fennel, and other relatives. The Apiaceae family is a large family with about 300 to 400 genera and between 2500 and 3000 species. They are distributed throughout a wide variety of habitats, principally in the north temperate regions of the world and rarely in tropical regions. The Apiaceae family is closely related to the Araliaceae family and both the families have minor differentiable features. The plants of Apiaceae family are herbaceous, biennials or perennial, although some grow tough stems and there are a few woody tree-like or shrubby species in tropical regions. Leaves are arranged in alternate and usually compound sometimes simple, often fern-like, or feather-divided; the leaves widen at the base into a sheath that clasps the stem.

The stems are hollow between the leaf-joints, often furrowed. The flowers are very uniform. The most obvious distinctive feature of the family is the inflorescence, the flowers grow in clusters, almost always concentrated in flat-topped simple or compound umbels; the rays of the primary umbel giving rise to a secondary umbel with the flower-bearing pedicels. The flowers are actinomorphic, have 5 petals, usually uneven, and 5 stamens; the flowers usually bisexual but are at times functionally pistillate or staminate. Functionally staminate flowers have a pistil but have no ovules capable of being fertilized. Functionally pistillate flowers have stamens, but their anthers do not produce viable pollen. They are very often white, sometimes cream, yellow, or pink. The outer flowers of the umbels are the first to open. Individual flowers are small; by themselves, they would not be readily apparent to pollinators. The inflorescence, however, is highly visible, and it is this that attracts the pollinators. The ovary of the flowers is inferior. Most members of the family are promiscuous. They are generally self-compatible and entomophilous. The fruits are dry schizocarps which split at maturity into two seeds; the seeds are often conspicuously ribbed, and sometime winged. Some part of the plant generally possesses a strong aroma. Taxonomic position of the genus *Apium* is given as under:

Taxonomic position of the genus *Apium*

Kingdom	Plantae
Subkingdom	Tracheobionta
Super Division	Spermatophyta
Division	Magnoliophyta
Class	Magnoliopsida
Subclass	Rosidae
Order	Apiales
Family	Apiaceae
Genus	*Apium* L.
Species	*Apium graveolens* L.
Variety	*Apium graveolens* L. var. dulce (Mill.) DC.

Morphological Description

Biennial, erect, glabrous herb up to 100 cm tall, with a fusiform to tuber-like fleshy taproot; stem strongly grooved. Leaves alternate, lower ones pinnate, long-petioled, upper ones tri -fid; stipules absent, but distinct sheath present; leaflets deltoid-rhomboid, 2–5 cm × 1.5–3 cm, often deeply 3-lobed, cuneate at base, glossy. Inflorescence a sessile or short-peduncled umbel, terminal or opposite the leaves; primary rays 4–15, rather unequal; involucres and involucels absent; umbellules 6–25-flowered. Flowers bisexual, penta merous; pedicel 1–5 mm long; calyx teeth obsolete; petals free, ovate to orbicular, c. 0.5 mm long, with inflexed apex, greenish white; stamens free, alternating with petals; ovary inferior, 2-celled, styles 2, divergent. Fruit a broadly ovoid to globose schizocarp, splitting into two 1-seeded parts up to 1.5 mm long, distinctly ribbed. Seedling with epigeal germination; hypocotyl 1–2 cm long, epicotyl absent; cotyledons stalked, blade ovate-oblong, up to 6 mm long, herbaceous. Its chromosome number (2n) is = 22 (Schippers, 2004).

Cytogenetics

The celery genome consists of 11 large chromosomes, nine submetacentric, one metacentric and one telocentric (Murata and Orton, 1984). The genome size of celery is fairly large compared to carrot, approximately 3.5 pg per 2C nucleus or 3x109 bp per haploid nucleus (Quiros, 1998). Cytologically, a single pair of chromosomes in the genome organizes a nucleolus, containing the 18S-25S rRNA gene family.

It is formed by a tandem repeat unit of 9.3 Kbp, which is invariable in all three *A. graveolens* cultivated types. There is, however, variation for unit size and restriction sites in the *Apium* wild species. In addition to the 18S-25S genes, very few repetitive coding sequences have been observed after RFLP mapping based on a celery cDNA library clones (Huestis *et al.*, 1993). The proportion of single-copy coding sequences in the celery has been estimated to range between 59 per cent and 78 per cent. It indicates that celery is a true diploid species whose genome has not suffered extensive duplications for coding genes. The level of polymorphism in *A. graveolens* is relatively low. cDNA clones tested in celery and celeriac for five restriction enzymes generated only 23% polymorphic loci. A similar level of polymorphism was detected for RAPD markers (Yang and Quiros, 1993). The most likely mechanism for RFLPs in celery seems to be due to base deletion and insertion more than to base substitution.

Varieties

Celery breeding has been done exclusively at the diploid level. Polyploidy and aneuploidy have not been exploited in celery breeding and genetics. Most celery varieties are open pollinated. Existing commercial celery hybrids are few and are not as popular as the existing open pollinated varieties. Basically, there are two types of celery varieties, self-blanching or yellow, and green or Pascal celery. In North America, generally green celery varieties are commercially grown, whereas in Europe and in many other parts of the world, mostly self-blanching varieties are produced.

During, 1887 the French varieties 'Paris Golden Self-Blanching' and 'Pascal Celery' a green celery line selected from the former, were introduced to North America under the names of 'White Plume' for the self-blanching type and 'Giant Pascal' for the green type. These two are the important progenitors of our modern varieties, with little genetic contribution from accessions of other origins (Quiros, 1993).

Celery Triumph: It matures in 85-90 days, has dark green colour attractive foliage and medium long petioles. It has good standing ability.

Ajmer Cel-1 : The variety has been developed through selection from Punjab Local at NRCSS, Ajmer, This variety is suitable for cultivation in semi-arid regions under irrigated conditions. It produces an average yield of 8.0 q/ha, however, under good climatic and soil conditions and with scientific cultivation practices celery can give the grain yield upto 12q/ha. Seeds of celery contain 2-2.5% essential oil which may be costlier in international market (Lal *et al.*, 2013).

Celery varieties grown in the United States belong either to the golden (or yellow class), or to the green class. Commercial celery production in California and Florida is of the green class. The common green cultivar is Utah-52 70 R, which may be harvested after 26 weeks, Golden Self Blanch is another variety which may be harvested after 14 weeks.

Climate and Soil

Celery is a cool season crop northern and central India, including the hills having cold and dry climate are best suitable for its cultivation. It is cultivated in the foothills of north western Himalayas and the outlying hills of Punjab, Himachal Pradesh, and Uttar Pradesh. A combination of 12-15°C and 22-25°C either day-night or night-day gives 80 per cent germination. At a temperature of 18-21°C the plants maintain vigorous growth. It can be grown in all types of soil except saline, alkaline and water logging. However, plants thrive well in loamy soils rich in organic matter and retentive of soil moisture. Clay and silt soils produce excellent crops when they are well manured and irrigated.

Nursery Raising and Transplanting

Celery is propagated through seed. Seeds are very small and approximately 3000 seeds count per gram. Application of FYM/ organic manures maintains the soil moisture as well as green lustrous colour of celery plants. For a hectare planting, 100 g seeds are required. Germination of celery seed is characteristically slow and variable. In the nursery, seeds are sown on flat beds and grown for 5-6 weeks then transplanted in the field. October is the best month for raising the seedlings in the northern plains of India. Fully developed seedlings are transplanted at the spacing of 20x15 cm.

Manures and Fertilizers

For succulent 'string free' quality crop of celery, large amounts of nutrients are applied. One hectare of crop removes about 330 kg N, 80 kg P_2O_5 and 70 kg K_2O. Depending upon the fertility of the soil, 150 kg N, 80 kg P_2O_5 and 60 kg K_2O should be applied for a hectare field.

Irrigation

Celery is a shallow rooted crop with most of the roots within the first 60 cm of soil. For succulent and tender stalks high soil moisture is necessary. Water must be supplied regularly at frequent intervals and more than adequate moisture is needed in last period before harvest as the most rapid growth occurs at this time. About 75 cm of water are required to grow a crop of celery to maturity.

Blanching

Blanching may be done to make the petioles crisp, improve flavour and keep the plant tender. It may be done either by wrapping the paper or earthing up the soil around the stem. When plants are 30 cm tall, blanching is done by paper and is completed in 10-14 days interval.

Harvesting and Yield

Petioles are cut by using the knife. Care should be taken for proper harvesting at right stage of maturity because as time goes on, there will be increases in size of stalks and overly delayed harvesting aggravates defects, such as pithiness, yellow leaves and seed stalks. High quality celery consists of stalks which are well formed, have thick petioles, and are compact, have minimal petiole twisting, and have a light green and fresh appearance. In addition to these, quality indices are stalk and midrib length, freedom from defects such as blackheart, pithy petioles, seed stalks, cracks or splits, and freedom from insect damage and decay. The average yield of foliage varies from 400-500 quintals per hectare.

Controlled Atmosphere Storage

At optimum conditions, celery should have good quality after storage up to 5 to 7 weeks. Commonly, celery is rapidly pre-cooled and then stored at 0 to 2°C. If storage is intended to be less than one month, storing celery at 5°C is not recommended for more than 2 weeks in order to maintain good visual and sensory quality. Some continued growth of inner stalks will occur post-harvest at temperatures >0°C (Suslow and Cantwell, 1998). Controlled or modified atmospheres offer moderate benefit to celery. Delayed senescence and decay development have been observed at 2-4% O_2 and 3-5% CO_2. Injury from low O_2 (<2%) or elevated CO_2 (<10%) will induce off-odors, off-flavors, and internal leaf browning. CA for mixed storage or long distance transport of celery and lettuce has some commercial application. Elevated CO_2 levels delay leaf yellowing and decay but could not be used in mixed loads with lettuce (lettuce does not tolerate CO_2 enriched atmosphere).

Major Diseases

Celery is subjected to a number of diseases caused by fungi, bacteria, and viruses. Amongst them early and late blight diseases are most destructive. Some of the major diseases attacking the crop are briefly described as under:

1. Early blight

It is seed borne fungal disease caused by *Cercospora apii*. It first appears mainly on older leaves as scattered small circular brownish yellow spots with concentric rings, which enlarge rapidly having a light brown central area. It's development ceases when temperature goes below 40°C. The disease appears first on the leaves as small circular, yellowish brown spots. These spots enlarge rapidly under favourable conditions and in a few days they have a light brown central area. Gradually turning to dark brown surrounded by a band of yellow. It attacks leaves in all stages of growth but is most prevalent on older leaves.

Control Measures

1. Prior to sowing dip the seeds in 1:300 formaldehyde solutions.
2. Spraying or dusting the seedlings with copper fungicide.
3. Crop rotation and field sanitation should also be followed.

2. Late blight

It is also a fungal disease caused by *Septoria apiigraveolentis*. Its symptoms appear as small circular water-soaked area on tips and margins of older leaves in January having black fruiting bodies. Later on, these areas turn black and all plant parts situated above the ground are affected. It develops rapidly in cool moist weather, especially in late sowing.

Control measures

1. Hot water treatment of seed at 48 to 49^0 C avoids the occurrence of disease.
2. Spraying with Dithane Z.78 is suggested.

3. Damping off

This is caused by *Pythium* spp. toppling over of infected seedlings at any time after emergence is generally observed. Infection usually occurs at or below the ground level. Infected tissue becomes soft and water soaked as the disease advances. The stem becomes constructed at the base and ultimately collapses even within a day.

Control Measures

1. Seed treatment with *Trichoderma* @ 4-5 g /kg seed.
2. Spray copper fungicides in the nursery beds.

4. *Fusarium* yellow

It is caused by *Fusarium oxysporum*. Symptoms are stunted growth of the plant, vascular discolouration, rotting of crowns, petioles, and root.

Control Measures

1. Follow long term crop rotation.
2. Grow resistant varieties.
3. Soil drenching with 0.2 per cent Dithane M-45 is also beneficial to manage the disease.

5. Bacterial diseases

Among the bacterial infections, bacterial soft rot caused by *Erwinia carotovora* and bacterial leaf spot caused by *Pseudomonas apii* are known to infect the crop. In bacterial leaf spot, symptoms on plant are small circular reddish spots, with pale yellow borders which develop rapidly in hot humid weather.

Control Measures

1. Apply copper hydroxide in the field.

6. *Phoma* root rot

It is caused by fungus *Phoma apicola*. It attacks the plants in the seed bed where it causes shunting, yellowing of the outer leaves and rotting of the roots. The disease usually starts at the base of the stem and the tissues turn bluish green but gradually become dark brown. The fungus causes rot while in storage also.

Control Measures

1. Follow long term crop rotations, deep ploughing to bury the fruiting bodies and sterilization of seed bed.

7. Brown rot

This disease is caused by a fungus *Cephalosporium apii*. The main symptoms of brown rot are irregular tan to brown shallow lesions on any of the above ground parts of the plant. The lesions often are so numerous that they coalesce to form a brown streak extending the full length of the petiole. Some distortion of the mature petiole may occur.

Control Measures

1. Grow resistant varieties like Utah 52-70, Utah-15, Utah-16 and Tall Ford Hook.
2. Apply Mancozeb @0.2 per cent or Copper oxychloride @ 0.2 per cent as foliar spray.

8. Viral diseases

The viral infections which are observed in celery include celery mosaic causing leaf mottling and ultimately leaf spots. The other virus is celery mosaic showing early symptoms as mottling and greyish colour of the inner leaves. Leaflet turns fern-like and growth becomes stunted.

Control Measures

1. Grow resistant varieties.
2. Manage vector population to avoid the spread of disease.

Major Pests

Leaf minor, carrot weevil, green celery worm, thrips, leaf hopper, celery caterpillar, carrot rust fly, aphid and slug are some of the pests which attack the crop. A brief description of important pests of celery is given as under:

1. Carrot rust fly (*Psila rosae*)

The larva or maggot attacks celery eating off the small roots. The adult is small, shining, dark green fly and the larva is a slender, straw coloured maggot about 3/10 inches long.

Control Measures

1. Apply Phorate granules @ 1g/m^2 area in soil.
2. Spray Quinalphos @ 0.5 per cent on foliage.

2. Tarnished plant bug

This is a sucking bug. The main injury to celery results from the feeding of the adult bugs, which puncture the stalk especially at the joints and cause a blackening of the tissue. The injury is known by celery growers as black joint. The adult is brownish but 1/5 inches long mottled with shade of red and yellow brown spots. The insect hibernates in leaves and other trashes. They emerge in the spring and the female, inserts her eggs in the green tissues of celery plants. The larvae cause serious danger to the heart of the plants.

Control Measures

1. Natural enemies may help to control tarnished plant bug populations.

Pollination Management and Crossing Technique

Celery flowers are very small, which precludes the removal of individual anthers for emasculation. Furthermore, the different developmental stages of the flowers in the umbels make it difficult to control pollination. The standard hybridization technique in celery consists in selecting flower buds of the same size, eliminating older and younger flowers. The umbelets are covered with glycine paper bags for 5 to 10 days, when the stigmas become receptive. At this time, pollen or umbelets shedding pollen from selected male parents are rubbed onto the receptive stigmas of the female parent. An improvement of this technique consist of washing away the anthers from open, unreceptive flowers with a water stream, covering them with the paper bags and pollinating them a few days later when stigmas become receptive. The advantage of this technique is the lower rate of accidental selfing caused by the failure of the anthers to abscise before stigma receptivity (Ochoa *et al.*, 1986). In order to induce seed stalk development during the vegetative phase, the plants require a period of vernalization, from 6 to 10 weeks at 5 to 8°C when they are at least 4 weeks old (Honma,1959). Due to a wide range of response to the cold treatment, it is often difficult to synchronize crosses since plants will flower at different times. However, pollen can be stored for 6 to 8 months at -10°C in silica gel or calcium chloride, and with a viability decline of approximately 20 to 40%, thus providing flexibility to perform crosses most of the time. For selfing, the plant or selected umbels are caged in cloth bags. These are shaken several times during the day to promote pollen release. House flies (*Musca domestica*) can also be introduced weekly into the bags for pollination.

Genetic Resources and Improvement

Important germplasm collections are maintained at the Research and Plant Breeding Institute for Vegetables in Olomouc, Czech Republic, the National Bureau of Plant Genetic Resources, New Delhi, India, the Vavilov Institute of Plant Industry, Petersburg, Russia, the Institute of Horticultural Research, Wellesbourne, United Kingdom, and at the Northeast Regional Plant Introduction Station, Geneva, New York, United States. The main breeding objectives for celery are to improve disease resistance and tolerance of high temperatures, for stalk celery the tenderness of the petioles and the self-blanching character (Schippers, 2004).

Interspecific hybridization efforts have been made in celery because of the unavailability of wild species. Ochoa and Quiros (1989) succeeded in hybridizing celery to *A. chilense* and *A. panul*. The hybridization between these three species can be achieved in both directions regardless of which species is used as female. The celery x *A. panul*, and celery x *A. chilense* F_1 hybrids are very vigorous but display a pollen fertility range from 0 to 20%. These two wild species are important as sources for late blight (*Septoria apicola* f.s. apii) resistance. The sterility seems to be due to chromosomal inversions. The sterility of the hybrids precludes obtaining F_2 progeny, although backcross progeny to celery are readily obtained. After two backcross, the petiole characteristics of celery is partially recovered. On the other hand, *A. chilense* x *A. panul* F_1 hybrids were fully fertile, and did not show any evidence of chromosomal rearrangements. This indicates that the two wild species have the same chromosomal constitution and are closely related to each other. Quiros (1994) reported successful hybridization between celery and *A. prostratum*, which is resistant to leaf miner (*Lyriomiza trifolii*) (Trumble and Quiros, 1988), beet army worm *Spodoptera exigua* (Diawara *et al.* 1992) and Western yellow mosaic, a viral disease. The F_1 hybrids are quite vigorous, but have reduced pollen fertility approximately 25 per cent. There is no evidence of meiotic abnormalities or chromosomal behavior in the hybrids. *A. nodiflorum*, another species resistant to *Septoria* and leaf miners has also been hybridized to celery.

Crossing relationships between celery and three wild species

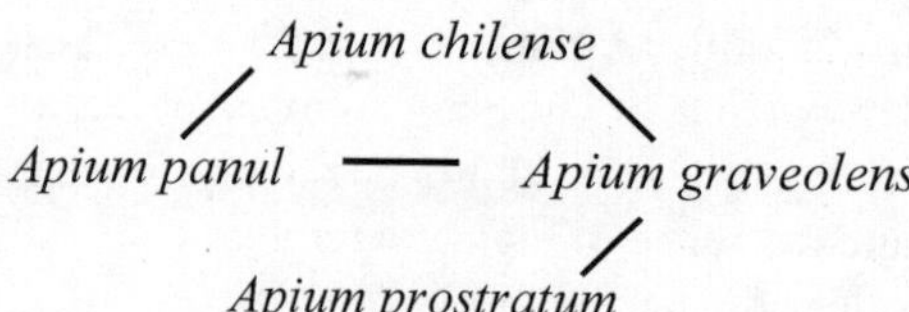

Source : Quiros (1998)

Inheritance of most important traits

The inheritance of some phenotypic traits has been determined in celery. These include mainly morphological characters and several disease resistance genes. These traits have economic importance and a description of their inheritance is summarized below:

Hollow petioles: This character segregates as a dominant, monogenic trait (Emsweller 1933; Townsend *et al.,* 1946). Quiros (1993) proposed *Ho* as the symbol for this gene. Hollow petioles are widespread in celeriacs and in some smallage accessions. Solid stems most likely were selected and fixed for the domestication of celery stalks.

Flowering behavior: Annual habit is a partially dominant, monogenic trait determined by the gene *Hb*. This locus forms a linkage group with two isozyme loci and gene *A*, coding for anthocyanin pigmentation in the plant (Quiros *et al*., 1987).

Bouwkamp and Honma (1970) reported that early bolting was dominant over slow bolting and determined by a single gene denominated *Vr*. It is possible that *Hb* and *Vr* are allelic. In contrast to the simple genetic determination of annual *vs* biennial habit, resistance to bolting in biennial types is a complex trait, most likely polygenic and affected by environment.

***Fusarium* resistance:** Resistance to *Fusarium* yellows is partially dominant and determined by two loci (Orton *et al.,* 1984). The resistance based on these two genes has a quantitative nature, where the dominant allele Fu_1 from celeriac contributes the largest effect. The dominant allele Fu_2 of the second locus, is found in tolerant celery varieties. This allele has a small contribution to resistance (Quiros, 1987). Based on a series of F_2 and backcross progenies the following genotypes are postulated for disease reaction.

Table (4): Resistance attributes

Resistant	Tolerant	Susceptible
Fu1Fu1Fu2Fu2	*Fu1fu1fu2fu2*	*fu1fu1Fu2fu2*
Fu1Fu1Fu2fu2	*fu1fu1Fu2Fu2*	*fu1fu1fu2fu2*
Fu1Fu1fu2fu2		
Fu1fu1Fu2Fu2		
Fu1fu1Fu2fu2		

Source: Quiros (1998)

In order to maximize resistance, when possible the breeder should aim to fix both loci in homozygous dominant condition.

Virus resistance: Celery mosaic virus resistance has been found in a feral accession of *A. graveolens*. Resistance is monogenic and recessive (*cmv*). Molecular markers have been identified allowing marker assisted selection.

Male-sterility: Single genetic male sterile have been reported in celery. It is recessive and determined by a single gene named *ms-1* (Quiros *et al*., 1986). It was found as an spontaneous mutant in a weedy Iranian accession. Therefore, substantial breeding activity will be necessary to transfer it into a useful genetic background. Male sterility is due to tapetal degeneration in the anthers. Nectar production in the male sterile mutants is not impaired; therefore their flowers still attract pollinators (Quiros *et al*., 1986). Cytoplasmic male sterility (*CMS*) celery is useful for hybrid seed production. Asymmetric protop last fusion of fertile celery and petaloid *CMS* carrot was carried out and a total twelve somatic hybrids were obtained and plants regenerated on MS_0 medium. AFLP analysis indicated that all of them were somewhat asymmetrically fused. STS marker analysis suggested that petaloid *CMS* were successfully introduced into 11 regenerated plants (Fang *et al*., 2009).

Stem color: Anthocyanin pigmentation is determined by single dominant gene designated *A* (Townsend *et al*., 1946, Arus and Orton, 1984, Quiros *et al*., 1987). This gene has been found to be tightly linked, approximately 2cM, to the isozyme locus *Aco-1*, coding for the enzyme aconitase (Huestis *et* *al*., 1993). Yellow celery (*y*) is recessive to green and it is determined by a single gene (Townsend *et al*., 1946). This trait often has been used as a marker for hybrid identification.

Leaf shape: Bouwkamp and Honma (1970) reported that deeply toothed leaf is recessive and determine by the gene dt.

Pest Resistance

Apium graveolens L. (celery, var 'Conquistador'), *A. chilense*, 23 backcross accessions of *A. graveolens* × *A. chilense*, and two relatives were examined for resistance to a major pest *Liriomyza trifolii* (Burgess) and of these accessions, several appeared to have excellent leaf miner resistance. Compared to all plants except the wild relatives, these plants had fewer feeding punctures and allowed significantly less survival during development. Resistance was caused by the carcinogenic and mutagenic linear furanocoumarins, the amounts of psoralen, bergapten and xanthotoxin present in both the leaves and petioles of each accession were quantified. Results indicated that *L. trifolii* resistance was not due to these undesirable compounds, but some of the accessions were eliminated from further immediate consideration by unacceptably high concentrations (Trumble *et al.*,1998).

Tissue Culture

The formation of globular, heart-shaped and torpedo-shaped embryoids was initiated in a cell suspension culture of differentiating celery callus, however, no cell differentiation was observed in the globular and heart-shaped stages but cell differentiation and tissue organization were present at the torpedo stage. The structure of the plantlet was simpler than that of the seedling but this may have been due to the conditions in the culture flask. Both plantlet and seedling leaf showed comparable levels of organization. It indicates that biochemical development in the embryoid may be related to the onset of cell differentiation (Abta and Collin, 1978). Celery cell suspension immobilized in Ca-alginate and maintained in a medium containing 0.5 mg l"1 2,4-dichlorophenoxyacetic acid (2,4-D) or 3,5-dichlorophenoxyacetic acid (3,5-D) as the auxin sources were compared with freely suspended cells in the same media. In a 2,4-D-containing medium the immobilized cells showed a reduced dry weight and a more uniform respiration rate, whereas the uptake of sucrose, phosphate and ammonia nitrogen was similar in both immobilized and freely suspended cells. In a 3,5-D medium, no increase in dry weight occurred in the immobilized cells, but a small increase in respiration rate showed that the cells were still viable. Uptake of sucrose, phosphate and ammonia nitrogen was reduced in the immobilized cells (Watts and Collin, 1985).

The relationship between greening and production of flavour compounds was investigated in blanched and green young (10-week) and mature (20-week) celery plants, and in colourless and green celery cell suspensions. In the young plants the concentrations of limonene and other terpenoids increased in the petiole and leaf-extract as the level of chlorophyll increased. In the cell cultures, greening was induced by transferring cells to a medium where 2,4-dichlorophenoxyacetic acid (2, 4-D) was replaced by 3,5-dichlorophenoxyacetic acid (3, 5-D). During the first subculture, phthalides were produced in the growth phase, but were present in trace amounts only in the second and third subculture. Limonene, however, increased after the first subculture and during the growth phase of the third subculture was comparable to the levels in the young celery plant. Greening and aggregation of the cells had also increased by the third subculture in the 3, 5-D-containing medium. It was suggested that phthalide production was stimulated by the transfer from a 2, 4-D- to a 3, 5-D-containing medium, but as the level of aggregation and greening increased in the culture, phthalide production was reduced and limonene production increased. Further, it was showed that, despite the greening and aggregation, there was no chloroplast formation or cell differentiation in the cultures incubated with 3, 5-D (Watts *et al*., 1985).

References

Abta, S. and Collin , H. A. (1978). Cell differentiation in embryoids and plantlets of celery tissue cultures. New Phytologist, 80 (3):517-521.

Arus, P. and Orton. T.J. (1984). Inheritance patterns and linkage relationships of eight genes in celery. J. Hered., 75: 11-14.

Bouwkamp, J.C. and Honma, S. (1970). Vernalization response in celery. J. of Hered. 61:115.

Diawara, M.M. *et al*. (1992). Resistance to *Spodoptera exigua* in *Apium prostratum.* Entomol. Exp. Appl. 64:125-133.

Emsweller, S.L. (1933). An hereditary pithiness in celery. Amer. Soc. Hort. Sci. Proc. 29:480-485.

Fang, T., Huo-lin, S. and Shuai, W. (2009). Preliminary study of asymmetric protoplast fusion between celery (*Apium graveolens* L.) and CMS carrot (*Daucus carota* L.) Acta Horticulturae SINICA, V 36(8):1169-1176.

Gopalan, G., Rama Sastri, B.V. and Balasubramanian, S.C. (2004). Nutritive Value of Indian Foods. National Institute of Nutrition, I.C.M.R., Hyderabad, India.

Honma, S. (1959). A method for evaluating resistance to bolting in celery. Proc. ASHS, 74:506 13.

Huestis, G. (1992). Genetic mapping in celery (*Apium graveolens* L.) and its wild allies based on restriction fragment length polymorphisms. Ph. D. diss. Univ. of California, Davis.

Huestis, G., MacGrath, M. and Quiros. C.F. (1993). Development of genetic markers in celery based on restriction fragment length polymorphisms. Theor. Appl. Genet. 85:889-896.

Lal, G., Singh, B., Mehta, R.S., Meena, R.S. and Meena, N.K. (2013). High-tech Seed Spices Production, National Research Centre on Seed Spices, Ajmer, 203-206 , India.

Murata, M. and Orton, T. (1984). G-banding like differentiation in meiotic prometa phase chromosomes of celery, J. Hered., 75:225-28.

Ochoa, O. and Quiros, C.F. (1989). *Apium* wild species: novel sources for resistance to late blight in celery. Plant Breed, 102:317-321.

Ochoa, O. *et al*. (1986). Techniques for water emasculation and cut seeds talk pollination in celery. Hort Sci., 21:1455-56.

Orton, T.J. and Arus, P. (1984). Out crossing in celery. Euphytica , 33:471-480.

Orton, T.J. *et al*. (1984). Studies on the inheritance of resistance to *Fusarium oxysporum* f. sp. *apii* in celery Plant Disease , 68:574-578.

Quiros *et al*. (1987). Use of stem proteins and isozymes for the identification of celery varieties. Plant Cell Repts. 6:114-17.

Quiros, C.F. (1994). Celery. In: Genetic Improvement of Vegetable Crops, Kalloo,G. and Bergh, B.O. (Eds.). Pergamon Press.

Quiros, C. F. *et al*. (1986). Cytological and genetical studies of male sterile celery. Euphytica, 35:867-75.

Quiros, C. F. *et al*. (1987). Inheritance of annul habit in celery: cosegregation with isozyme and anthocyanin markers. TAG, 74:203-208.

Quiros, C.F. (1993). Celery breeding program at the Deptt. of Vegetable Crops, UCD. Hort Sci., 28:(cover).

Quiros, C.F. (1998). Celery note on the crop of celery Deptt. of Vegetable Crops,UCD.

Quiros, C.F., Douches, D. and D'Antonio. V. (1987). Inheritance of annual habit in celery: Cosegregation with isozyme and anthocyanin markers. Theor. Appl. Genet., 74:203-208.

Schippers, R. R. (2004). *Apium graveolens* L. Record from Protabase. Grubben, G.J.H. & Denton, O.A. (Eds.). PROTA (Plant Resources of Tropical Africa) Wageningen, Netherlands.

Suslow, T. and Cantwell, M. (1998). Celery: recommendations for maintaining postharvest quality Department of Plant Sciences, University of California, Davis.

Townsend, G. R., Emerson, R.A. and Newhall, A. G. (1946). Resistance to *Cercospora apii* Fres. in celery (*Apium graveolens* var. dulce). Phytopathology, 36:980-982.

Trumble, J.T. and Quiros, C.F. (1988). Antixenotic and antibiotic resistance in *Apium* species to *Lyriomiza trifolii* J. Econ. Entomol. 81:602-607.

Trumble, J.T. Diawara, M.M. and Quiros, C.F. (1998). Breeding resistance in *Aapium graveolens* to *Liriomyza trifolii:* antibiosis and linear Furanocoumarin content. Acta Hort. (ISHS), 513:29-38.

Watts, M. J., Galpin , I. J. and Collin, H. A. (1985). The Effect of greening on flavour production in celery tissue cultures . New Phytologist, 100(1): 45-56.

Watts, M.J. and Collin, H.A. (1985). Growth and nutrient uptake by immobilized tissue culture cells of celery (*Apium graveolens*). Plant Science, 42(1):67-72.

Yang, X.F. and Quiros, C.F. (1993). Identification and classification of celery cultivars with RAPD markers. Theor. Appl. Genet. 86:205-212.

Yang, X.F. and Quiros, C.F. (1995). Construction of a genetic linkage map in celery using DNA-based markers. Genome, 38:36-44.

30

CELERIAC

Celeriac is also known as celery root, turnip-rooted celery or knob celery. It is a kind of celery, grown as a root vegetable for its large and bulbous hypocotyl rather than for its stem and leaves. The swollen hypocotyl is typically used when it is about 10-12 cm in diameter; about the size of a large potato. Unlike other root vegetables, which store a large amount of starch, celery root is only about 5-6% starch by weight. Celeriac may be used raw or cooked. It has a tough, furrowed, outer surface which is usually sliced off before use because it is too rough to peel. Celeriac has a celery flavor and the volatile compounds of celeric are γ-terpinene and α-pinene, however, E)-3-butylidene-4, 5-dihydrophthalide (or (E)-ligustilide) and cis,syn-3-butyl-3a, 4, 5, 7a-tetrahydrophthalide (or cnidilide) could not be detected in celeriac (MacLeod and Ames, 1989). It is often used as a flavouring in soups and stews. Nutritional composition of celeriac is given in table (30.1).

Table (30.1). Nutritional composition of celeriac (per 100 g of edible portion)

Constituents	Amount
Calories	42.0 kcal
Protein	1.50 g
Fat	0.30 g
Carbohydrate	9.20 g
Fiber	1.80 g
Sugar	1.60 g
Calcium	43.00 mg
Iron	0.70 mg
Magnesium	20.00 mg
Phosphorus	115.00 mg
Potassium	300.0 mg
Sodium	100.0 mg
Zinc	0.33 mg
Copper	0.07 mg
Manganese	0.16 mg
Selenium	0.70 mg
Vitamin C	8.00 mg
Thiamin	0.05 mg
Riboflavin	0.06 mg
Niacin	0.70 mg
Vitamin B_6	0.17 μg
Folate	8.00 μg
Food - Folate	8.00 μg
Folate - DFE	8.00 μg
Vitamin B_{12}	0.00 μg
Vitamin A	0.00 I.U.
Retinol	0.00 μg
Vitamin E	0.36 mg
Vitamin K	41.00 μg
Fat - Saturated	0.08 g
Fat - Monounsaturated	0.06 g
Fat - Polyunsaturated	0.15 g
Cholesterol	0.00 mg

Botanical Description

Celeriac is member of family Apiaceae and belongs to the genus *Apium.* A herbaceous plant with a swollen turnip-like structure which consists of stem, swollen hypocotyl and root. When the plants become reproductive, the stem which is solid and ribbed elongates and may reach a height of 1 m.Taxonomic position of the genus is given as under:

Taxonomic position of the genus *Apium*

Kingdom	Plantae
Subkingdom	Tracheobionta
Superdivision	Spermatophyta
Division	Magnoliophyta
Class	Magnoliopsida
Subclass	Rosidae
Order	Apiales
Family	Apiaceae
Genus	*Apium* L.
Species	*Apium graveolens* L.

Variety	*Apium graveolens* L. var. rapaceum (Mill.) Gaudin

Origin and Distribution

Celeriac is grown world-wide, especially in Europe, Southwest Asia, Siberia, North Africa and North America. Cultivated celeriac originates from wild celery (var. graveolens), which is found in the whole Mediterranean area and in North Europe on saliferous soils. Celeriac was already used in early history by the Egyptians, Greeks and Romans, mainly for religious purposes. Its use as a vegetable started during middle age. Celeriac first achieved importance as a vegetable plant in the middle age. During the process of improvement, the root, which comprises one third each of main root, hypocotyl and shoot has been increased in size.

Varieties

There are a number of cultivars available, especially in Europe. Among them are 'Prinz', 'Diamant', 'Ibis', and 'Kojak', which all received Royal Horticultural Society Award of Garden Merit designation in the trial in 2000. Other important varieties are Alabaster (120 days) and Giant Prague (120 days).

Agrotechniques

Celeriac germinates best at temperatures between 10-19°C Celeriac does well in cool weather conditions. The seeds are very slow to germinate. Seeds should be sown 2.5-3.5 cm deep in rows 60 to 75 cm apart, and when the seedlings are large enough to handle, thin them to 15 to 20 cm apart. The crop of celeriac can be grown through raising nursery and transplanting the seedlings in well prepared fields. Celeriac tolerates light shade and prefers rich soil that is high in organic matter, well able to hold moisture but with good drainage. It needs constant moisture and does well in wet locations. It is a heavy feeder and needs plenty of fertilizer to keep it growing quickly. At the time of field preparation apply 10-15 t/ha FYM and at the time of planting NPK may be given @150:80:80 kg/ha. Apply one third nitrogen and full dose of phosphorus and potassium at the time of planting and give rest nitrogen dose in form of top dressing at 30 and 45 DAT. Frequent watering is important; celeriac, like celery, is shallow-rooted, and a lack of soil moisture can reduce its growth. Celeriac cannot compete with weeds. During weeding attention should be paid not to disturb the shallow roots. Celeriac has no serious pest and disease problems.

Flowering

Booij and Meurs (1995) observed that the requirements for flowering and bolting were different in celeriac. For stalk elongation a longer period of low temperatures is required than for flowering. However, the low temperature needed for stalk elongation could be replaced by a long photoperiod.

Harvesting and Storage

Time from planting to harvest is 110 to 120 days from seed. Harvesting should be done when the swollen root is12-15 cm wide. Celeriac increases in flavor after the first frost, but should be harvested before the first hard freeze. Keep the roots in the refrigerator up to one week, or store them in a cold, moist place for two to three months. They will keep in the ground in areas where freezing weather is not a problem. Celeriac normally keeps well and should last three to four months if stored between 0°C and 5°C.

Genetic Resources and Improvement

Important breeding aims are higher root yields, firmness of the root, good taste, shoot stability, resistance to fungal diseases and reduced tendency to become hollow and discolour. Smooth surface and small leaf and root connections should reduce the cleaning work. Genetic relationships among elite celeriac varieties and celeriac accessions conserved in genebanks have not been fully utilized. Genetic relationships were estimated in 34 elite celeriac varieties bred in Europe and 28 celeriac accessions conserved at the German genebank. Two varieties of celery, two varieties of leaf celery and three genebank accessions of wild *Apium* species were additionally analysed. Fifteen EcoRI/MseI-based AFLP primer combinations were used. Average GS estimate in elite germplasm (GS=0.90) was higher than in exotic germplasm (GS=0.80). An AMOVA (analysis of molecular variance) revealed that a high proportion of variation was due to variation within elite celeriac varieties and genebank accessions. Cluster analyses revealed clear genetic groupings of celeriac germplasm, which was supported by morphological traits. Elite, moderately bred and exotic varieties formed distinct clusters, indicating that only a part of the available genetic diversity in celeriac germplasm has been exploited in breeding. Distinct *Apium* species might be useful for the introgression of new genes into cultivated celeriac material. Broadening of celeriac collections in genebanks and detection of new genetic resources are vital for improvements in celeriac breeding (Jasmina *et al.*, 2004).

References

Booij, R. and Meurs, E. J. J. (1995). Effect of photoperiod on flower stalk elongation in celeriac (*Apium graveolens* L. var. rapaceum (Mill.) DC.). Scientia Horticulturae, 63(3-4):143-154.

Jasmina Muminovi, Albrecht E. Melchinger and Thomas Lübberstedt (2004). Prospects for celeriac (*Apium graveolens* var. rapaceum) improvement by using genetic resources of *Apium*, as determined by AFLP markers and morphological characterization. Plant Genetic Resources: Characterization and Utilization, 2: 189-198.

MacLeod, G. and Ames, J.M. (1989). Volatile components of celery and celeriac. Phytochemistry, 28(7):1817-1824.

31

SWISS CHARD

Swiss chard derived from green spinach beet, is a non- tuberous variety of *Beta vulgaris* used as spinach. It is a vegetable with two uses. The petiole and main hypertrophied leaf midribs are eaten after boiling. A folacin content of about 297m g/100 g fresh wt has been reported (Gami and Chen, 1985). Although lower values of folacin *i.e.*, 11 mg/100 g were reported earlier (Olsen *et al.*, 1947). Vitamin K, or phylloquinone content in Swiss chard varies according to the geographical location; maturation also influences vitamin K, content in the leaves (Ferland and Sadowski, 1992). Analysis of the leaves is given in table (31.1). A protein concentrate prepared from the leaves consists of pure protein, 91.3 per cent with arginine, 6.4; histidine 1.3; and lysine, 5.6 per cent as essential amino acid . The leaves also contain 0.06-0.13 ppm cobalt. The leaf stalks and leaves contain three flavoring constituents, 3-alkyl-2-methoxypyrazines where alkyl radical is isopropyl, isobutyl and sec butyl.

Recently five flavonoids from Swiss chard leaves have been isolated and identified as vitexin 2"-xylloside and its 6" malunated derivative, kaemferol 3-gentiobioside, isorhamnetin 3-gentiobioside and iso rhamnetin 3-vicianoside. 6" malonyl. 2" - xylostyl vitexin is a new naturally occurring flavonoid. Flavonoid C-glycoside acylated with a dicarboxylic acid was first time reported in nature (Gil *et al.*, 1998). Flavonoid content in fresh leaves (green cultivar) varies in the range of 2.4-3.0 mg/g fresh weight and yellow cultivars contain flavones C-glycosides.

Table (31.1): Nutritive value of Swiss chard leaves (per 100g of edible portion)

Constituents	Amount
Moisture	91.0 g
Protein	2.4 g
Fat	0.3 g
Carbohydrates	4.6.
Energy	25 kcal
Calcium	88 mg
Iron	3.2 mg
Sodium	147 mg
Potassium	550 mg
Vitamin A	6500 IU
Thiamin	0.06 mg
Riboflavin	0.17 mg
Niacin	0.5 mg
Vitamin C	32 mg

Origin and Distribution

Red chard was first introduced by Aristotle about 350 BC. Other early Greeks and Romans wrote about chard. Mention of chard appears in writing of the Chinese as early as the 7th century. It is a favorite crop of the people of Switzerland. Swiss settlers introduced into the United States in 1806 (Splittstoesser, 1990). Chard is not commonly grown in India but is recommended for growing in the hills and plains of North India during winter season.

Botanical Description

Swiss chard is member of family Chenopodiaceae and its genus is *Beta*. Chromosome number of Swiss chard (2n) is 18. Genus *Beta* is glabrous succulent herbs with swollen roots containing sugars, commonly called beets found in Europe and the Mediterranean region. The cultivated beets include sugar beet, garden beet, leaf beet, chard or Swiss chard and mangles.

Taxonomic position of the genus *Beta* L. is given as under:

Kingdom	Plantae
Subkingdom	Tracheobionta
Superdivision	Spermatophyta
Division	Magnoliophyta
Class	Magnoliopsida
Subclass	Caryophyllidae
Order	Caryophyllales
Family	Chenopodiaceae
Genus	*Beta* L.
Species	*Beta vulgaris* L.

In general, cultivated beets have been place into groups : (i) The *Cicla* group (*B. vulgaris* var. cicla L.) including the beets grown as leafy vegetables, viz., leaf beet or spinach beet and chard or Swiss chard, and (ii) the crassa group (*B. vulgaris* var. crassa (Alef.) T. Helm.) including the types grown for their roots, used as a vegetable (garden beet or beet root), forage (mangels) or as a source of sugar (sugar beet). Some taxonomic raise spinach beet and chard to specific levels under *B. orientalis* (Roth.) Moq. and *B. cicla* L. respectively, and placing the rest of the beets under *B. vulgaris*. While a group of botanists retains chard under *B. cicla* L. but place mangels under a seprate species. *B. maritima* L. with a ssp. *orientalis* (Roth.) Buren. Comprising the spinach beet; the table and sugar beets are treated under *B. vulgaris* L. others reduce the leafy beets to *B. vulgaris* F. rapo Dumm. still other workers classify them under *B. vulgaris* into ssp. variously named ssp. *vulgaris* or ssp. *esculentae* (Salisb.) Gurk. with swollen roots and ssp. *maritima* (L.) Archangeli or ssp. *cicla* (L.) Moq. with less swollen roots.

Ford and Styles (1986) have given detailed account of classification of cultivated and wild beets. They studied the factors affecting inter-specific variation and have proposed a systematic scheme. It is considered to retain classification of *Beta vulgaris* to the level of ssp. *vulgaris* (beet root, mangel, fodder beet and sugar beets), *cicla* (chard and spinach beet), *maritima* and *macrocarpa* (the wild forms equivalent to limited *B. maritima* and *B. macrocarpa*).

Plants of Swiss chard are annual or biennial, about 70 cm high, leaves much developed, blades up to 38 cm long and 25 cm wide or larger, sometimes ruffled and puckered, often highly coloured and in some cases with petioles and midribs very thick.

Climate and Soil

Swiss chard needs a cool climate for its proper growth. In the plains of North India, it is grown as winter crop where as in low and mid hills of Himachal Pradesh, it is generally grown in October and September, respectively. In the high hills, sowing starts from March and continues up to June. Swiss chard does well in loams and is sensitive to soil acidity. However, plant is tolerant to slightly saline soils.

Seed Rate and Methods of Sowing

Approximately 1g seed counts 60 seeds and about 4.0 kg seeds are required for a hectare crop. Field are prepared by 3-4 ploughing and removing the weed segments and other undesired materials. Sowing is done in rows made at 30 cm apart and seeds are placed 2.5 cm deep. Weak seedling are rouged out to maintain the proper spacing of 10 cm between the plants.

Manures and Fertilizers

Swiss chard is a foliage plants and needs good nutrient balance for its succulent growth of the leaves. In Himachal Pradesh, application of 20 tonnes FYM and 400 kg Calcium Ammonium Nitrate per hectare have been recommend by Shukla *et al.*(1992). Nitrogen should be given in splits, first at the time of planting and second after a month of its germination. Yalcin and Topcuoglu (1994) observed that plant DM increased with increasing rates of N and P. although N was more effective. As the N application rate increased, plant oxalic acid, nitrate, total N, Ca and Mg concentration increased and P and K concentrations decreased. As P application increased, oxalic acid, total N, Ca and Mg concentrations decreased and P concentration increased.

Metal toxicity

Swiss chard is sensitive to Pb absorption. Chisholm (1972) reported that higher Pb and As concentrations in the surface soil (0-15 cm) application raised the levels of both elements in Swiss chard. The Canadian residue tolerance for Chard is 2 ppm. In a study on heavy metal content in Swiss chard with sludge application on Zinc and Cadmium availability Kuo *et al.* (1985) reported that both inclusions of pH amorphous Fe oxide content along with total metal content had the best prediction of metal availability. Their findings suggested that soil high in the oxide content in conjunction with soil liming may give the best control of metal availability. Cochran *et al.* (1983) reported that application of volcanic ash up to 400 kg/ha to the Swiss chard crop had no adverse effect on growth of the crop and it increased Zn and decreased the Cd concentration. Swiss chard grown selanate laden soil accumulated higher concentrations of Se than grown in selanite laden soil. While studing the effect of Cl and Cd uptake by Swiss chard in nutrient solutions, Smolders and Mclaughlin (1996) reported that either $CdClN^{-2h}$ species can be taken up by plant roots or that Cl enhances uptake of Cd^{2+}through enhanced diffusion of the uncomplexed metal to uptake sites.

Chromium accumulation was observed in Swiss chard var. Ford Hook Giant with the application of tannery waste containing higher concentration of Chromium. Plant growth inhibition was noted at the 10 and 15 per cent rates of tannery waste soil incorporation (Gru-binger *et al.*, 1994).

Irrigation and Inter-culture

For succulent growth of leaves, adequate moisture content in soil is necessary. Removal of weeds at early stage of plant growth is essential. For chemical weed control, application of pyrazon @ 2.4-2.8 kg/ha as pre emergence is effective (Mitidieri and Bianchini, 1973).

Harvesting, Yield and Storage

Large leaves and stems are harvested when they are 50-60 days old. Outer leaves are harvested first before they become full grown and fibrous. It takes about 12 weeks time from sowing to harvesting. Average yield per plant is 350-400 g and a hectare crop yields about 150-200 quintal green leaf. Chard is highly perishable and should be refrigerated as close 32^0F as possible and desirable relative humidity is 90-95 per cent. Vitamins and quality are retained when wilting is prevented. Modified Atmosphere Packaging (MAP) (7 per cent O_2 and 10 per cent CO_2 had no effect on total flavonoid content after 8 days of storage, although it increased flavonoid extraction during cooking in boiling water. In contrast, vitamin C content decreased, especially in MAP stored Swiss chard, to reach levels below 50 per cent of the initial content after 8 days of cold storage (Gil *et al.*, 1998).

Plant Protection Measures

1. Radicle Necrosis: It is caused by *Phoma betae* Frank. Susceptible cultivars show low germination and radicle necrosis of seedlings. This was first time reported by Wolcan and Parello (1989) in Agrentina.

2. Cucumber mosaic virus: This is the first report of natural infection of Swiss Chard by CMV (Krajacic and Juretic, 1993). The identification was base on best plant reactions, type of virus particles, serology and viral replicative ds RNA analysis.

Other important diseases are bacterial soft rot caused by *Erwinia carotovora* and *Cercospora* leaf spot caused by *Cercospora beticala*.

Important varieties

1. Lucullus: It is a leading yellowish green type variety, has very long crumpled, dark green leaves with greenish-white leaf stems.

2. Giant Perpetual: It has broad and light green leaves.

3. Ford Hook Giant: It has heavy crincled and dark green leaves.

4. Dark Green White Ribbed: It has broad prominent white ribs and is famous for its crispness.

5. Rhubarb Chard: Plant with crumpled leaves featuring dark red to white red veins.

6. Burpee's Rhubarb Chard: It has stalk like rhubarb and dark green with heavily crumpled leaves.

Seed Production

Seeds are produced from plants subjected to a cold period which develops leaf stock formation. The plants selected for seed production should produce seed in their second season and bolting in the first season should be avoided. Crop of Swiss chard should be well isolated from other ssp. of *B. vulgaris*, by a distance of at least 1000 m (Tindal, 1983).

Genetic Resources and Improvement

A collection of local populations and bred varieties of Swiss chard (*B. vulgaris* ssp. *provulgaris* var. flavescens) and spinach beet (var. cicla) has been established at Changins in Switzerland in 1980 for the conservation and utilization of genetic resources (Badoux, 1995). The morphological observations indicate that wide genetic diversity between and within population. The yield capacity of the local populations of Swiss chard is not related to the colour and shape of the leaves. Bred varieties, however, are more uniform and high yielder.

Gene flow and introgression from cultivated plants may have important consequences for the conservation of wild plant populations. Cultivated beets (sugar beet, red beet and Swiss chard, *Beta vulgaris* ssp. *vulgaris*) are of particular concern because they are cross-compatible with the wild taxon, sea beet (*B. maritima*). Cultivated beet seed production areas are sometimes adjacent to sea beet population; the number of flowering individuals in the former typically out number those in the populations of the latter. In such situations, gene flow from cultivated beets has the potential to alter the genetic composition of nearby wild populations (Bartsch *et al.*, 1999).

References

Badoux, S. (1995). Study of variation within a collection of Swiss chard and spinach beet. Revue. Suisse. De-Viticulture. D Arboriculture et ad', Horticulture, 27(6):333-337.

Bartsch, D., Lehnen, M., Clegg, J., Pohl, O. M., Schuphan, I. and Ellstrand, N. C. (1999). Impact of gene flow from cultivated beet on genetic diversity of wild sea beet populations. Molecular Ecology, 8(10):1733-1741.

Bayuelos, G. S. and Meek, D. W. (1989). Selenium accumulation in selected vegetables. J. Pl. Nutri. 12(10):1255-1272.

Chisholm, D. (1972). Lead, arsenic and copper contents of crop grown on lead arsenate treated and untreated soils. Canadian J. Pl. Sci. 52(4):583-588.

Cochran, V. L., Bezdicek, D. F., Elliott, L. F. and Papendick, R. I. (1983). The effect of Mount St. Helens. volcamic ash on plant growth and mineral uptake. J. Environ. Quality, 12(3):415-415.

Ferland, G. and Sadowski, J. A. (1992). Vitamin K_1 (phylloquinone) content of green vegetables: effect of plant maturation and geographical growth location. J. Agril Food Chem. 40(10):1874-1877.

Ford, L. B. V. and Styles, B. T. (1986). Inraspecific variation in wild and cultivated beets and its effects upon intraspecific classificatoin. Systematics Assoc. 29.

Gami, D.B. and Chem, T.S. (1985). Kinetic of folacin destruction in Swiss chard during storage. J. Food Sci. 50:447.

Gil, M. I., Ferreres, F. and Tomas Barberan, F. A. (1998). Effect of modified atmosphere packaging on the flovonoids and vitamin C content of minimally processed Swiss Chard (*Beta vulgaris* ssp.*cycla*). J. Agril. Food Chem. 46(5):2007-2012.

Grubinger, V. P., Gutenmann, W. H., Doss, G. J., Rutzke, M. and Lisk, D. J. (1994). Chromium in Swiss chard grown on soil amended with tannery meal fertilizer. Chemosphere, 28(4):717-720.

Krajacic, M. and Juretic, N. (1993). Natural infection of Swiss chard by cucumber mosaic virus in Croatia. Petria, 3(2):93-98.

Kuo, S., Jellum, E. J. and Baker, A. S. (1985). Effect of soil type, liming and sludge application an Zinc and Cadmium availability to Swiss Chard. Soil Sci. 139(2):122-130.

Mitidieri, A. and Bianchini, P. R. (1973). Evaluation of herbicides for weed control in spinach, spinach beet and beet root. Idia, 301:54-68.

Olsen, O. E., Burris, R. H. and Elvehjem, C. A. (1947). A preliminary report of the folic acid content of certain foods. J. Amer. Diet. Assoc, 23:200.

Shukla, Y.R., Jain, Y.C.and Sharma, O.P. (1992). Cultural Practices for Vegetable Production in Himachal Pradesh. Directorate of Extension Education. Dr. Y.S.Parmar University of Horticulture and Forestry, Solon, H. P.

Smolders, E. and Mc Laughlin (1996). Effect of Cl and Cd uptake by Swiss chard in nutrient solutios. Pl. Soil, 179(1):57-64.

Splittstoesser,W.E. (1990). Vegetable Growing Hand book, Springer, pp.362.

Tindal, H.(1983). Vegetables in the Tropics, Mac Millan, London, United Kingdom.

Wolcan, S.M. and Parello, A.E. (1989). Susceptibility of different organs of spinach beet and fodder beet to various isolates of *Phoma betae* frank. Revista de la facultad de Agronomia La Plata 65(1 & 2):13-18.

Yalcin, S.R. and Tocuoglu, B. (1994). Effect of nitrogen and phosphorus on accumulation of oxalic acid and nitrate and some nutrient contents in chard (*Beta vulgaris* var. cicla). Ankara Univ. Ziraat Fakultesi Yilligi, 44(1 & 2):217-228.

32

KALE

Kale is a minor temperate leafy *Brassica* vegetable. It was introduced in India in 19th century. Its cultivation on commercial scale is very rare in India. Kale is considered one of the most nutritious vegetables and it contains higher ∝ carotene, beta carotene, ∝-tocopherol, gamma tocopherol and ascorbic acid content than broccoli, Brussels sprouts, cabbage and cauliflower (Kurilich *et al.*, 1999). Muller (1997) recorded 34.8 mg carotenoids per 100 g edible portion of kale. The major glucosinolates present in kale are 2-propenyl, 3-mehylsulfinyl propyl and indlol-3-ylemethyl which account of 35, 25 and 29 per cent of the total glucosinalate content, respectively (Rosa, 1997). Free glutamine (17-52%) is the dominant free amino compound in kale (Eppendorfer and Bille, 1996). It also possesses comparatively high quantity of polycyclin aromatic compound (Speer *et al.*, 1990). Kale is a good source of vitamin K (phyllquinone) and possesses significant antioxidant activity against hydroxyl radicals (Cao *et al.*, 1996). Kale's risk-lowering benefits for cancer have recently been extended to at least five different types of cancer. These types include cancer of the bladder, breast, colon, ovary, and prostate. Isothiocyanates made from glucosinolates in kale play a primary role in achieving these risk-lowering benefits. Nutritive value of kale is given in table (32.1).

Table (32.1): Nutritive value of kale leaves (per 100 g of edible portion)

Constituent	Amount
Moisture	83.0 g
Protein	9.0 g
Fat	0.8 g
Carbohydrates	9.0 g
Energy	53 kcal
Calcium	249 mg
Phosphorus	93 mg
Sodium	75 mg
Potassium	378 mg
Vitamin A	10,00 I.U
Thiamin	0.16 mg
Riboflavin	0.26 mg
Niacin	2.10 mg
Ascorbic acid	186.00 mg

Source: Watt and Merrill (1963)

Origin and Distribution

Kale is one of the oldest forms of cabbage, originating in the Eastern Mediterranean or in Asia Minor. Kale was known to the ancient Greek and several cultivars were described by Cato around 200 B.C. (Decoteau, 2000). Travelers and immigrants through the ages have introduced this leafy vegetable to many parts of the world and it was grown in the United States in the 17th century. In India, it is grown on a limited scale in Jammu and Kashmir, Himachal Pradesh, Utterakhand , Punjab, Haryana and in Nilgiri hills in Tamil Nadu. In recent past, cultivation of kale has picked up around metropolitan cities due to increasing demand in hotels and tourist restaurants.

Botanical Description

Kale belongs to the genus *Brassica* and species *oleracea* and botanical variety *acephala*. Taxonomic position of the genus *Brassica* L. is given as under:

Kingdom	Plantae
Subkingdom	Tracheobionta
Superdivision	Spermatophyta
Division	Magnoliophyta
Class	Magnoliopsida
Subclass	Dilleniidae
Order	Capparales

Family	Brassicaceae
Genus	*Brassica* L.
Species	*Brassica oleracea* L.

Chromosome number of kale (2n) is 18. There are different types of kales classified as sub-varieties of *Brassica oleracea* var. *acephala* most of which bear a rosette of leaves at the top of the stem. These are curly kales possessing finely curled foliage (sub var. *laciniata*). Smooth leafed kales (sub var. *plana*), thousand headed kales with sprouted axillary buds forming loose leafy rosette (sub var. *millecapitata*), tree kales having more plant height up to 2 m (sub- var. *palmifolia*) and narrow stem kale with long swollen stem (sub var. *modullosa*) (Sane *et al.*, 2000). It is the curly kale which is grown for edible leafy green. The top leafy rosette or individual leaves are harvested for vegetable. The other types of kales are used as fodder for livestock. Some forms of kales found with variably coloured foliage are used for ornamental purpose. 'Karam Sag' of Kashmir regarded as kale is closely related to Chinese kale *Brassica alboglabra* (Snogerup, 1980). Chinese kale is a cultigens native to Southern and Eastern China, mainly grown for its young stem, leaves and young inflorescences which are consumed cooked or fried, sometimes raw also(He-Chengkun *et al.*,1994). Chinese kale which is widely cultivated in China bearing both white and yellow flowers is a variety of *Brassica oleraca* namely var. chinensis. This was contrary to current classification as *B. alboglabra*. Smooth leafed Siberian kale (Hanover salad) is not commonly grown in the United States and is classified as *Brassica napus* var. *pabularia* (Decoteau, 2000). There is close resemblance between kale and collard, however, some distinguishing features is given in table (32.2)

Climate and Soil

Kale is mostly cultivated in temperate climate but it can be grown in regions having cool winters. It is the hardiest crop and can withstand temperature as low as -10 to -15^{0}C. Some varieties are less hardier which are grown in tropical regions too (Chadha, 2000).

Kale grows well in different types of soils, however, proper drainage and moisture supply is essential. Soil rich in organic matters promotes growth and quality of produce. Soil rich in mineral contents imparts favorably higher mineral composition of leaf also, however, Ca concentration in laminae and K in petioles showed less de-

Table (32.2): The distinguisting characters between kale and collards

Kale	Collards
1. Growth open type.	1. Growth habit upright.
2. Terminal portion ending in rosette of leaves ie-non heading intermediate between heading and non heading cabbage i.e. transitional type of cabbage.	2. Terminal portion ending in a loose head of cabbage like leaves.
3. Plant dwarf (20-30 cm)to10 tall (180 cm in thousand headed kale).	3. Plant intermediate in size (90-100 cm) between dwarf and tall kales.
4. Leaves curled, smooth.	4. Leaves smooth.
5. Varieties highly variable with different size, height, colour and form of leaves.	5. Varieties less variable.
6. The stem though coarse but neither branched nor markedly thickened.	6. The stem unbranched thick and stiff.
7. Winter hardy and plants not killed even in the severe winter.	7. Besides being winter hardy, collards tolerate more heat than cabbage.
8. Requires low temperature for chilling.	8. Requires low temperature for chilling.

Source: Sane *et al.* (2000)

pendable on soil characteristics (Almeida and Rosa, 1996). Soil should not contain high amount of thallium because kale has close relationship between its uptake and in the degree of Tl solubility in the soil (Crossmran, 1994).

Agrotechniques

Land preparation: The field is prepared properly by giving 3-4 ploughing like cauliflower or cabbage and is divided into plots of convenient size to ensure proper irrigation as well as drainage.

Manures and Fertilizers

Well rotten cow dung manure or FYM @ 10-12 t/ha should be added 4 to 6 weeks before transplanting. A dose of 150 kg N and 75 kg each of P and K should be applied. The N is given in equal split doses at planting time, 30-40 days after planting and 15-20 days prior to first harvesting. Fillipo and Neri (1998) reported that N increased total and leaf biomass up to 150 and 50 kg N/ha, respectively, causing a linear decrease of leaf thickness and dry mater content, an improvement in leaf colour and an increase in leaf fibre compared with controls. Highest yield of kale was obtained with the application of N @ 112 kg/ha (Dangler and Wood, 1993). Petiole and more particularly lamina thiocynate concentration decreased as N application rate was increased (Chweya, 1990). In nutrition of kale, sulphur is very important element. Stuiver *et al.* (1997) reported that after 1 week of S deprivation in hidroponically grown kale, the plants started to show symptoms of S deficiency, resulting in yellowing, negligible shoot growth and increase in shoot dry matter content. Shoots were more rapidly affected by S deprivation than roots. The appearance of S deficiency symptoms was preceded by a dramatic decrease in sulfate and thiol content in both shoots and roots. S deprivation had a decisive impact on the uptake and assimilation of nitrate. Nitrate uptake was strongly reduced in sulphur deficient plants and symptoms was accompanied by an accumulation of nitrate and free aminoacids and a loss of soluble proteins.

Nursery raising and Transplaning

Crop of kale raised by transplanting the seedlings in the main field. About 350-400 g of seed is required for a hectare. In the plains of North India, seeds are sown in month September- October and in Himachal Pradesh, Jammu and Kashmir and Nilgiri Hills, ideal time of sowing is August-September. Seeds are treated with *Trichoderma* @ 5-10 g/kg seed to avoid the problem of nursery diseases.

When seedlings attain 4-6 leaf stage and they are 25-30 days old, are planted in field at a spacing of 45 cm×30 cm. Plant spacing vary with the soil type, cultivars and fertility. Plant densities of 200 and 40 plants/m^2 have been found effective for mechanical harvesting or top picking for the fresh market, respectively (Fillipo and Neri, 1998). Paul (1991) reported that plant DW and leaf and cotyledon area per plant decreased with increasing sowing depth and increased with increasing seed size. Ne Smith (1998) reported that individual plant leafy dry mass decreased linearly with increased plant population. The maximum yields per ha were obtained with the 15 cm within row spacing (Dangler and Wood, 1993).

Irrigation and interculture

Kale does well in water supplied soil. Fast growing crop is less fibrous and more tender than the slow growing. First irrigation soon after transplanting is essential for proper establishment of newly transplanted seedlings. Thompson and Doerge (1995) reported that excessive irrigation (SWT soil water tension) <5-6 k Pa resulted in reduced yields and N uptake and high rates of un-utilized fertilizer N. In early stage of plant growth, weed infestation should be avoided by giving a hand weeding 15-20 days after transplanting. In chemical weed control, Metolachlor @ 1.5 kg/ha or lower as pre-emergence can be used safely for weed control in direct seeded kale, however, application in susceptible cultivars should be avoided. (Harrison *et al.*, 1998).

Harvesting and Storage

The lower leaves of kale can be individually harvested when they are small and tender. In rosette type dwarfing cultivars, whole leaves are harvested. Its harvesting starts from November and continues till January end. Average yield of kale varies between 150 and 250 q/ha. There is close relationship between water content in leaves before harvest and post harvest senescence in kale. Harvesting the leaves at 13.00 h when they showed marked deficit, resulted in a one day postharvest delay in senescence compared with leaves harvested at 6.00 h. The leaves harvested at 13.00 h recovered their turgor during storage. They also showed 2 fold and 3 fold higher initial contents of TSS and starch, respectively, than did the leaves harvested at 6.00 h. and this was probably associated with the greater post-harvest longevity. (Amarante and Puschmann, 1993).

Shelf life of kale is limited by moisture loss, yellowing and microbial activity. Beaulieu *et al.* (1997) reported total chlorophyll content decrease by 11, 16 and 18 per cent after 1, 2 and 3 days, respectively, at 4^{0}C whereas loss was 45 per cent after 3 days at 20^{0}C. Effect of heat treatment on post harvest quality of kale was studied by Wang *et al.* (2000) and they observed that kale leaves treated at 45^{0}C for 30 minutes and stored at 15^{0}C resulted in higher post harvest quality, delayed yellowing and less decline of sugar and organic acids. Kale leaves can be stored for 21 d or longer in strayrofoam containers under refrigerated conditions (Magee, 1987).

Plant Protection

The major serious disease is damping off in nursery, which can be controlled by treating seeds as well as nursery beds with treament of *Trichoderma*. Downy mildew caused by *Peronospora parasitica* is another serious disease during wet and cool season. The major pests of kale are flea beetle, diamond back moth, borers and cabbage loopers. Spraying of neem based insecticides and adoption of IPM approaches are suggested to avoid the insecticide residue in edible green leaves.

Seed Production

The crop of kale continues making rosette of leaves at top of the stem till late winter. With the rise of temperature, plants start bolting and thus there is no dormant stage while transitioning from vegetative to reproductive phase. Hence the 'replant method' is not successful and the seed is produced by *in- situ* method, allowing the plant to grow, over-winter, flower and produce seed at the same place (Sane *et al.*, 2000). In Katrain's condition of Himachal Pradesh, bolting starts in March and flowering occurs in April. The bolting takes place in response to low temperature during winter. Flowering and seed-setting takes place satisfactorily between 20 and 30^{0}C. Bees (*Apis mellifera*) is the major pollinizer and other pollinizers are bumble bees, *Osmia sp. Halictus sp.*, *Megachile sp.* and *Ceratina* sp. (Pinzauti and Maganani, 1994).

The effect of different pollination techniques with and without emasculation and delayed pollination on seed setting were investigated after self pollination and after intraspecific and inter-specific crosses of kale. After controlled self pollination, the best seed set occurred in buds 3-10 for the intra-specific crosses, the youngest flower and the oldest bud produced the largest number of developed ovules, but bud pollination was productive to bud 8. The yields from thee two pollination types were best when the female parent was not emasculated (Brown *et al.*, 1990).

Genetic Resources and Improvement

Small germplasm collection of Chinese kale are maintained at various institutions in South-East Asia such as national gene banks, universities and seed companies (Sangwansupyakorn, 1993). Monteiro and William (1989) collected the 23 accession of 9 land races from different geographical origins and reported resistence to *Peronospora parasitica* in all the accessions. Type a resistance to *Fusarium oxysporum* F. sp. *conglutinams* race 1 was present in most of the land races. Resistance to *Plasmodiphora brassicae* race 6 was found in 1 accession of the Portuguese tree kale. High resistance to *Leptosphaeria maculans* and *Albugo candida* was not detected, although several accession showed 20-30 per cent of plants with intermediate expressions of resistance. All Portuguese kale accessions were susceptible to black rot (*Xanthomonas campestris* pv. *campestris*).

The major breeding objective in kales are evolving varieties possessing high yield, quality attributes, resistance to biotic sresses, less bolting and with less anti-nutritional factors. Selection of kale genotypes having high glucosinolate and glucobrassicin content may given priority owing to their health promoting properties. In the kale most predominant glucosinolates are sinigrin (10.4 micro/mol/g DW) and glucobrassicin 1.2 μ mo/g DW (Kushad *et al*., 1999). Besides these, selection for low S-methyl cysteine sulphoxide content in kale is also priority. Gowers and Show (1999) obtained a wide range of kale accessions including lines derived from crosses between kale and rape. Leaf samples from mature plants were frozen and freeze dried, then assessed for S. methylcysteine sulfoxide content. On a DM basis MCSO contents were 1.5-2 per cent for standard kale cultivars, 0.5-1% for most kohlrabi and 0.64-1.7 per cent for crosses between kale and kohlrabi. Most of the calcium in kale leaf is present in form of calcium oxalate and as such is unlikely to be available to the body. It is desirable that the nutritional value of kale may be improved by reducing the water soluble oxalate content (Kasaye and Kelbessa, 1995).

Kale contains thicyonate which is considered to be a major goitrogen. In process of four generations of half sub-family selection for high and low SCN content in young leaves resulted in sub-populations wih an almost 2 difference in their means when bulks of sub-populationswere crossed. SPN contents were 104.5 and 58.5 mg/100 g DM, respectively. Cultivar 'Merlin' had the lowest (63.5) and 'Protector' the highest content (89.6 mg/100 g DM) (Bradshaw and Griffiths, 1990).

Black rot caused by the bacterium *Xanthomonas campestris* pv. *campestris* is one of the most serious diseases of *B. oleracea*. Since source of resistance to the disease within *B. oleracea* are insufficient and control means are limited, the development of resistant breeding line is extremely desirable. Certain line of *B. napus* contain very high resistance controlled by a dominant gene but crossing the two species sexually is very difficult. Therefore, somatic hybrids were produced by protoplast fusion between rapid cycling *B. oleracea* (CRGC 3-1) and a *B. napus* line (PI 1999 47) highly resistant to *X. campestris pv. campestris*. Hybrid identity was confirmed by morphological studies, flow cytoplasmic estimation of nuclear DNA content and analysis of random amplified polymorphic DNA (RAPD). Inoculation with the pathogen identified four somatic hybrids with high resistance. The resistant hybrid plants were fertile and set seed when selfed or crossed reciprocally to the bridge line 15. Direct crossed to *B. oleracea* were unsuccessful, but embryo rescue facilitated the production of a first backcross generation. The BC_1 plants were resistant to the pathogen. Progeny from the crosses to line 15 were all susceptible. Embryo rescue techniques were not obligatory for the development of a second backcross generation, and several resistant BC_2 plants were obtained (Hansen and Earle, 1995).

Studies were carried out on specific characteristics in the inheritance of resistance to pathotype 16/11/31 of *Plasmodiphora brassicae* in compatible lines of kale in the system of diallel crossings and in F_2, BC_{11} and BC_{21} progenies. Variations between the lines with respect to common combining ability were mainly determined by recessive gene. In the line MSK 1-2, it was controlled by two. When producing lines of *B. brassicae* resistant kale, use of incompatible lines MSK 2-1, MSK 1-2 and Chghk 1-1 was recommended (Monakhos and Ushanov, 1998). In evaluation against resistance to the green peach aphid (*Myzus persicae*), Leite *et al.* (1996) reported Crista- de- Galinha to be the most resistant, followed by Joenes and Raxo whereas Manteigo was the most susceptible. Developing late bolting cultivar is also priority. Furnham and Garrett (1996) reported that significant genetic control of the long standing (delayed bolting) phenotype is present in kale. In ploidy breeding, germinating seeds of 4 diploid lines were treated with colchicines (0.05%). Higher chloroplast number per stomata cell were recorded in the auto tetraploids obtained. The proportion of autotetraploids produced varied the line used. Meiotic behaviour was similar to that found in the diploids, good pairing was observed and pollen fertility was high at 60-70 per cent (Chevre *et al.*, 1989).

Wild kale possesses desirable gene which can be introduced in Chinese cabbage by developing the somatic hybrids. Yamagishi *et al.* (1988) reported that somatic hybrids between Chinese cabbage and wild kale had a leaf morphology intermediate between that of the parents but did not form a head. Pollen and seed fertility following selfing were normal, seeds were obtained by back crossing to Chinese cabbage but not to kale. Anther culture was successful in 6 of the 18 genotypes studied and the highest yield was 17 embryos per 100 anthers plated. Two stages of anthers development were identified as being responsive to the anther culture. The first and most responsive was that corresponding to the late uninucleate stage and second to the late binucleate stage. These stage correspond with the on set of mitotic events in microspores. Low pollen viability was noted which declined to zero after 9 days of anther culture. The initial viability level, however, was not clearly related to androgenic ability (Kieffer *et al.*, 1993).

References

Almeida, D. and Rosa, E. (1996). Protein and mineral concentration of Portuguese kale (*Brassica oleracea* var. acephala) related to soil composition. Acta Horticultrae, 407:269-276.

Amarante, C.V.T. and Puschmann, R. (1993) Relationship between harvest time and senescence in kale leaves. Revista Brassileira de fisiologia vegetal, 5(1):25-29.

Beaulieu, J.C., Oliveira, F.A.R., Fernandes-Delgado T., Fonseca, S.C. and Brecht, J.K. (1997). Fresh cut kale: quality assessment of Portuguese store-supplied product for development of a MAP system, p.145. In Proceedings of the 7th International Controlled Atmosphere Research Conference, Gorny, J.R. (ed.), vol. 5, Davis, California, USA.

Bradshaw, J.E. and Griffiths, D.W. (1990). Selection for high and low thiocynate ion content in kale (*Brassica oleracea* var. acephala). Annals of Applied Biology, 116(1):157-167.

Brown, A.P., Brown, J. and Dyer, A.F. (1990). Optimal pollination conditions for seed set after a self pollination, and intra-specific and inter-specific cross of narrow stem kale (*Brassica oleracea* L. var. acephala). Euphytica, 51(3):207-214.

Cao, G., Sofic, E. and Prior, R.L. (1996). Antioxidant capacity of tea and common vegetables. Journal of Agricultural and Food Chemistry, 44(11):3426-3431.

Chadha, K.L. (2000). Hand Book of Horticulture. Published by Indian Council of Agricultural Research, New Delhi.

Chevre, A.M., Eber, F., Thomas, G. and Bahon, F. (1989). Cytological studies of tetraploid kale (*Brassica oleracea* L. spp. acephala) obtained from diploid lines after colchicines treatment. Agronomie, 9(5):521-525.

Chweya, J.A. (1990) Effect of nitrogen on thiocyanate content of *Brassica oleracea* var. *acephala* leaves. Plant and Soil ,124(2):261-263.

Crossmann, G. (1994). Transfer and accumulation of thallium in vegetable and fruit species. Kongressband 1994, Jena. Vortrage Zum Ganeralthema des 106, VDLUFA-Kongresses Vom 19. 24-9. 1994 in Jena : Alter nativen in der flachennutzung, der Erzeugung und vewertung Landwirt Schaftlicher Produkte. Nachwa Chsende Rohstoffe, Extensivierung, Stillegung, 513-516.

Dangler, J.M. and Wood, C.W. (1993). Nitrogen rate, cultivar and within row spacing affect collard yield and leaf nutrient concentration. Hort.Science, 28(7):701-703.

Decoteau, D.R. (2000). Vegetable Crops. Prentice Hall, Upper saddle River. N.J.07458

Eppendorfer, W.H. and Bille, S.W. (1996). Free and total amino acid composition of edible parts of beans, kale, spinach, cauliflower and potatoes as influences by nitrogen fertilization and phosphorus and potassium deficiency. Journal of the Science of Food and Agriculture, 71(4):449-458.

Ferland, G. and Sadowski, J. A. (1992). Vitamin K (phylloquinone) content of green vegetables: effect of plant maturation and geographical growth location. Journal of Agricultural and Food Chemistry, 40(10):1874-1877.

Filippo D'Antuono, L. and Neri, R., (1998). Characterisation and potential new uses of palm tree kale (*Brassica Oleracea* L., ssp. Acephala DC, Var. Sabcllica L.). Acta Hort. (ISHS) 459:97-104.

Furnham, M.W. and Garrett, J.T. (1996). Importance of collard and kale genotypes for winter production in South-Eastern United states. Hort.Science, 31(7):1210-1214.

Gowers, S. and Show, M.L. (1999). Selection for low S. methylcysteine sulphoxide content in kale. Cruciferae Newsletter, 21:89-90.

Hansen, L.N. and Earle, E.D. (1995). Transfer of resistance to *Xanthomonas campestris* pv campestris into *Brassica oleracea* L. by protoplast fusion. Theoretical and Applied Genetics, 91(8):1293-1300.

Harrison, H.P., Farnham, M.W. and Peterson, J.K. (1998). Differential response of collard and kale cultivars to pre-emergence application of metolachlor. Crop Protection, 17(4):293-297.

He-Chengkun, Dong, G. and Meng, L.Y. (1994). On the taxonomy of kale. Advances in Horticultue, 155-159.

Kasaye, J. and kelbessa, U. (1995). Forms and contents of oxalate and calcium in some vegetables in Ethiopia. Ethiopian Journal of Health Development, 9(1):13-18.

Kiefter, M., Fuller, M.P., Chauvin, J.E. and Schlesser, A. (1993). Anther culture of kale (*Brassica oleracea* L. convar. acephala DC.) Plant Cell Tissue and Organ Culture. 33(3): 303-313.

Kurilich, A.C., Tsau, G..J., Brown, A., Howard, L. Klein, B.P., Jeffery, E.H., Kushad, M., Walling, M.A. and Juvik, M.A. (1999). Carotene, tocopherol and ascorbate contents in sub-species of *Brassica oleracea.* Journal of Agricultural and Food Chemistry, 47(4):1576-1581.

Kushad, M.M., Brown, A.F., Kurilich, A.C., Juvik, J.A., Klein, B.K., Wallig, M.A.and Jeffery, E.H. (1999). Variation of glucosinolates in vegetable subspecies of *Brassica oleracea.* J. Agric. Food Chem., 47: 1541-1548.

Leite, G.L.D., Picanco, M., Bastos, C.S., Araujo, J.M. de, Azevedo, A.A. and De-Araujo, J.M. (1996). Resistance to kale clones to the green peach aphid. Horticultural Brasileira, 14(2):178-181.

Magee, C. (1987). Leafy vegetables storage in an ice-walled container. Paper American Society of Agricultural Engineers, 87:6001.

Monakhos, G.F. and Ushanov, A.A. (1998). Investigation of resistance to club root (*Plasmodiphora brassicae* Wor.) in lines of kale (*Brassica oleracea* spp. *acephala*). Izvestiya Timirya Zevskoi- Sel-Skokhozyaistvennoi Akademii, 2:106-114.

Monteiro, A.A. and Williams, P.H. (1989). The exploration of genetic resources of Portuguese cabbage and kale for resistance to several *Brassica* diseases. Euphytica, 4(2):215-225.

Muller, H. (1997). Determination of the carotenoid content in selected vegetables and fruit by HPLC and photodiode array detection. Zeitschrift fur Labensmittel Untersuchung und forschung, 204(2):88-94.

Ne Smith, D.S. (1998). Effect of plant population on yields of once over harvest collards (*Brassica oleracea* L. var. *acephala* group). Hort.Science, 33(1):36-38.

Paul, N.K. (1991). Effect of sowing depth and seed size on emergence and seedling growth of *Brassica* species. Bangladesh Journal of Scientific and Industrial Research, 26(1-4):208-212.

Pinzauti, M. and Magnani, G. (1994). The effect of pollination by insects on the seed production in kale (*Brassica oleracea* L. var. Aceplala DC.). Apicta, 29(4):75-82.

Rosa, E.A.S. (1997). Glucosinolates from flower buds of Portuguese *Brassica* crops. Phytochmistry, 44(8):1415-1419.

Sagwansupyakorn, C. (1993). *Brassica oleracea* L. cv. group Chinese Kale. In : PROSEA. Plant Resources of South-East Asia 8 Vegetables. Pudoc Scientific Publishers, Wageningen.

Sane, P.V., Sharma, S.C. and Verma T.S. (2000). Producing Seeds of Biennial Vegetables in Temperate Regions Published by Indian Council of Agricultural Research , New Delhi.

Snogerup, S. (1980.). The wild forms of the *Brassica oleracea* group (2n=18) and their possible relations to the cultivated ones. 121–132 In: Tsunoda, S. *et al.*, *Brassica* crop and wild allies, Biology and Breeding, 121–132.

Speer, K., Nortsmann, P., Steeg, E., Kuhn, T. and Thontag, A. (1990). Analysis of polycyclic compounds in vegetable samples. Zeitschrift fur Lebensmittel Unter Suchung and Forschung, 191(6):442-448.

Stuiver, C.E.E., Kok, K.J. de, Westerman, S. and De Kok, L.J., (1997). Sulphur deficiency in *Brassica oleracea* L. development, biochemical characterization and sulphur/nitrogen interactions. Russian Journal of Plant Physiology, 44(4):505-513.

Thompson, T.L. and Doerge, T.A. (1995). Nitrogen and water rates for surface trickle irrigate of collard, mustard and spinach. Hort. Science, 30(7):1382-1387.

Wang, C.Y., Herregods, M. Nicolai, B., Jager, A. de and Roy, S.K. (2000). Effect of heat treatment on post harvest quality of kale, collard and Brussels sprouts. Proceedings of the XXV International Horticultural Congress. Part 8. Quality of Horticultural Products, Storage and Processing new outlooks on post harvest biology and technology, potentiality of processsing of underutilized fruits of the tropics, Brussels, Belgium, 2-7 August, 1998. Acta Harticulturae, 2000, No. 518, 71-78.

Watt, B.K.and Merill, A.L. (1963). Composition of Food, Raw, Processed, Prepared. USDA Agricultural Handbook,8.

Yamagishi, H., Yoshikawa, H. and Yui, S. (1988). Somatic hybrids between Chinese cabbage (*Brassica campestris* L.) and wild kale (*B. oleracea* L.). Bull. Nat. Res. Inst. Veg. Ornam. Plants and Tea.A2 : 209-216. (In Japanese with English summary).

33

COLLARD

Collard is a form of kale resembling non-heading cabbage. The edible portion is numerous tender leaves of the top of the stalk. Collard is an important green vegetable crop in the United States. Collards are said to be extremely nutritious owing to low in calorie value and containing substantial amounts of vitamin A and C. Fung *et al.* (1978) analysed all 14 essential elements in collard except Cr, Co and Mo. However, retention of Cl, Cu, Mg, Mn, Ni, P and Zn in frozen collard ranged from 82-97 per cent. The retention of Fe, K and Na was 74 , 86 and 73 per cent, respectively. Frozen collard contains generally more Ca than raw ones. Feeding mice on diets enriched with dried collard resulted in a significant decrease in the number of pulmonary metastases after the mice were injected intravenously tumour cells and it was concluded that collard is benificial in prevention of canccr.

In collard, indole-3 carboxaldehyde and traces of indole-3- acetonitrile are mainly responsible for anti-mutagenic factors. It was observed that when collards were cooked in a pressure cooker there was considerable loss in content of indole-3-acetonitrile and certainn other compounds especially after 30 minutes (Wall *et al.*, 1988). Schwartz *et al.* (1984) observed that magneshium absorption from collard is comparatively high as compared to other leafy vegetable. Nutritive value of collard is given in table (33.1).

Table (33.1) Nutritive value of collard (per 100 g of edible portion)

Constitutents	Amount
Moisture	85.0 g
Protein	4.8 g
Fat	0.8 g
Carbohydrates	7.5g
Energy	45 kcal
Calcium	250 mg
Phosphorus	82 g
Iron	1.5 mg
Potassium	450 mg
Vitamin A	9300 IU
Thiamine	0.16 mg
Riboflavin	0.31 mg
Niacin	1.7 mg
Ascorbic acid	152 mg

Origin and Distribution

The collard is closely related to kale and is one of the most primitive member of the cole crops. It resembles to the wild form of cabbage of ancient times. Domestication of the collard goes back to antiquity, well head of the Christian era. The ancient Greeks and Romans were known to have used collard along with kale. The collard was introduced in to the British Isles in the 4th century BC. Mention of collard (called Colewarts) in American literature goes back to 1669, where the plant was described as growing in Virginia. There is also reference of the collard appearing in Haiti in 1565. Collard is not much of commercial importance in India. Its cultivation is restricted to hilly tracts of Himachal Pradesh and Jammu &Kashmir by amateur kitchen gardeners. Collard is a favorite vegetable in the Southern United Sates, where they are traditionally boiled with salt pork or hog jowl. In Unite States, Georgia produces almost 1/3rd of the total, followed by Florida, Southern Carolina, Virginia and Alabama. Other southern states also produce but their share are minimal.

Botanical Description

Collard is a member of family Brassicaceae and it is botanically classified as *Brassica oleracea* L. var. acephala. However, collard is botanically placed under *B.*

oleracea L. var. virdis. Taxonomic position of the genus *Brassica* L. is given as under:

Kingdom	Plantae
Subkingdom	Tracheobionta
Superdivision	Spermatophyta
Division	Magnoliophyta
Class	Magnoliopsida
Subclass	Dilleniidae
Order	Capparales
Family	Brassicaceae
Genus	*Brassica* L.
Species	*Brassica oleracea* L.

Chromosome number of collard (2n) is=18. In some of the literatures, same scientific name *B. oleracea* L. var. acephala has been mentioned for both kale and collard, however, they differ in appearance but hav more resemblance to wild cabbage than crinkled leaf blade. The collard has the same colour and texture as cabbage, but the veining is obovate or shield shaped with irregular toothed margins. Like cabbage, the older leaves become more evenly rounded in shape with finely toothed or serrate margins, often with two lobes at the base.

Climate and Soil

Collard is tolerant to low temperature and its plant can withstand 10^0C temperature. Low temperature favors sugar and flavor in plants. Collard can be raised in different kinds of soils, however, organic rich loam soil is preferred for early and enhanced yield. Ample provision of drainage is necessary. Soil having pH value 6.5 is considered ideal, however, it can successfully be raised in soil with pH range of 5.5 to 6.0. Soil solarization of collard field by covering 50 cm thick clear polythene resulted in 91 per cent reduction of weeds mainly *Cyperus rotundus*, *Digitaria citilaris*, *Eichnochloa crusgali*, *Richardia scabra* and *Amaranthus retroflexus*. Solarization also enhanced collard yield, root branches, fresh weight and thermo-tolerant fungi and bacteria in the rhizosphere (Stevens *et al.*, 1990).

Nursery Raising and Planting Time

Crop of collard is raised as like cabbage by raising the seedlings in nursery bed and thereafter transplanting seedling in the main field. As per the growing habit, 500-700 g seeds are needed for dwarf, 400-500 g seeds for medium and 350-400 g seeds per hectare for tall cultivars. Guertal *et al.* (1997) concluded that uniformly prepared waste material are suitable for production of collard seedlings. They also observed that N content of seedlings was higher for first 16 days. In the plains of Northern India, seeds are sown in month of October- November, while in hills, August-September are considered ideal months for transplanting.

Manures and Fertilizers

Before transplanting, FY M @ 20-25 t/ha should be incorporated. Besides it, 200 kg N, 80 kg P_20_5 and 100 kg K_20/ ha may be given for good yield and lustrous plant growth. One forth dose of nitrogen should be given at the time planting and remaining dose of nitrogen may be given at the interval of 20-25 days in form of top dressing. Split application of N stimulates yield and growth of collard. A maximum yield of 222-250 lb/acre was obtained with applicationof N @ 120 lb/acre (Guertal *et al.*, 1997). Guertal and Edward (1996) reported that collard yield increased with increasing rates of N with a maximum yield of 163 kg N/ha. Hochmuth (1994) observed that petiole sap nitrate N and K concentrations in collard correlated highly with whole leaf N and K concentrations and decreased through the season. Dangler and Wood (1993) found that leaf N concentration increased linearly with increasing plant population and increasing rate of N enhanced the leaf N and P concentrations. The glutathione content of collard young and old leaves was significantly high in sludge applied soil. This may be due to stress factor imparted by the sludge amendments. Putresline, spermidine and spermine content of collards also increased positively by induction due to cadmium accumulation or to general metal stress. There is inverse relationship between soil pH and Cd. Peer *et al.* (1980) observed elevated level of Cd in collard field with low soil pH and/or elevated soil Cd.

Transplanting

Three to four week old seedlings are transplanted in duly prepared field. In general, 60×45 cm spacing is provided for dwarf and medium tall cultivars while tall cultivars are spaced at 90×60 cm. Ne Smith (1980) suggested that population of 5-7 plants/m^2 for once over harvest of whole plant bunch collard optimize yield and crop quality, especially under favourable cropping systems. Dufault *et al.* (1992) obtained good yield with 36 cm row spacing and 20 cm between transplant. Dangler and Wood (1993) obtained maximum yield with the 15 cm within row spacing.

Irrigation and interculture

Collard is shallows rooted (45-60 cm) plant. Hence, light irrigation at frequent interval is necessary for optimum growth and tender leaf. Initially field should be kept free from the weed infestation. Pathak *et al.* (1982) reported that Methane Sodium reduced the weed population and increased the growth and yield of collard.

Harvesting and Yield

Leaves are used for vegetable purpose and when plants attain a height of 15-25 cm, they are harvested from the ground level. Leaves of collard can be stored for 10-14 days at 6^0C and 90-95 per cent relative humidity. Wilting during storage should be prevented to avoid the loss of vitamin content and quality.

Important Disease

1) *Fusarium* yellow : It is caused by fungus. Leaves of infected plants become yellow .

Control:

i) Adopt long term crop rotation.

ii) Grow resistant variety.

2) Black rot: It is bacterial disease caused by *Xanthomonas campestris*. Infected plants show the symptoms of 'V' shaped structure on leaf margin. Growth spots progress towards centre and veins become black.

Control:

i) Select disease free seeds.

ii) Crop residue should be brunt and adopt long term crop rotation.

iii) Prior to sowing seeds should be treated in hot water (48.9^0C) for 30 minutes.

3) Alternaria leaf spot.: Symptoms of this disease appear as round and dark spots on leaves.

Control:

i) Grow disease resistant varieties.

ii) Adopt log term crop rotation.

4) Downy mildew: This is caused by fungus *Peronospora parasitica*. Small irregular spots appear on the upper surface of leaves of infected plants. Thereafter white and violet coloured mold develops.

Control:

i) Select disease resistant varieties.

ii) Seed treatment prior to sowing.

Seed Production

Chilling treatment is essential for production of collard seeds. Hence, its seeds are produced in areas where seeds of other member of cole groups like cabbage and Brussels sprouts are produced. Replanting is not required and *in -situ* condition collard seeds are produced. In collards, bolting starts in month of March and flowering occurs in April in condition of Himachal Pradesh. The inflorescence in collards develop directly from the rosette of leaves at the terminal end of the stem. Staking should be given to prevent lodging of plant. A temperature range of 20-30^0C is considered optimum for flowering and seed setting. GA_3 (100 and 250 ppm) promotes flowering in collard (Kahangi and Waithaka, 1981). Pollen sterility and low seed setting may due to either very low or high temperature. Low population of bees also considered one of the main reasons of poor seed setting.

Crop of collards should be harvested when about 70 per cent pods have changed to yellow but before they are dry. The method of harvesting, curing, threshing are the same as of kale. The collard seed yield varies from 500-600 kg/ha.

Important varieties

1) Georgia: It produces plants ranging from 75-90 cm height. They have large blue green and crumpled leaves. Although no head is produced, the plant yields greens under adverse conditions of heat and poor soil. Ist harvesting may be taken 80 days after transplanting.

2) Vates: It is a promising variety. The plant holds its colour well even in cold weather. This variety has relatively good resistance to bolting in early spring. The plants are dwarf 50 to 60 cm, erect, spreading and relatively compact. The yield of large, thick, dark green leaves is good. From seed to harvest it requires about 75 days.

3) Morris Heading: Plants of this variety attain 75-85 cm height. The leaves are broad, wavy, medium green with short stems. This variety produces a moderately tight head something like cabbage. The plant is slow bolting and thus good for winter planting.

4) Lousisiana Sweet: This variety was developed by Lusiana State University Experiment Station. It is similar to Georgia variety, but has a larger leaf area and less stem. The leaves are thick and tender. From seed to harvest, it requires about 75 days.

Genetic Resources and Improvement

Yugoslav Plant Gene Bank initiated *Brassica* collection in 1987 and collection of 15 accession of collards, 11 accessions of intermediate collards were made. Collection was mainly made from regions of Croatia and Bosnia Hercegovina (Zldovec, 1997). To assess genetic variation and relationships among collards and other entries of *Brassica*, evaluate intrapopultion variation of OP collard lines and determine the potential of collard land races to provide new *Brassica oleracea* genes, a collection of collard germplsm including 13 cultivars or breeding lines and 5 land races was evaluated using RAPD DNA markers. Two hundred and nine RAPD bands were scored from 18 oligonucleotide decamer primers. When collards and other *B. oleracea* were compared of these 147 (70%) were polymorphic and 29 were specific to collards. Collards entries were most closely related to cabbage (similarity index = 0.83) and Brussels sprouts entries (index = 0.80). Analysis of individuals of an OP cultivars and land race indicated that interpolation genetic variance accounts for as much as that observed between population. RAPD indicated collard land races as unique DNA markers. The systematic collection of collard land races showed enhance diversity of the *B. oleracea* germplasm pool and provide gene for future crop improvement (Farnham, 1996).

Bagget and Kean (1989) while studying the inheritance of annual flowering in *Brassica oleraceea* reported that annual habit was dominat over biennial and was controlled by several major genes with a strong effect of modifiers from both annual and biennial parent. Times of heading of annual plant in F_2 progenies appeared to be controlled by quantitative mainly additive factors. This suggests that the biennial parent especially collards considered dominant genes for late maturity.

References

Baggett, J. R. and Kean, D. (1989). Inheritance of annual flowering in *Brassica oleracea*. Hort.Sci., 24(4):662-664.

Dangler, J. M. and Wood, C. W. (1993). Nitrogen rate, cultivar and within row spacing affect collard yield and leaf nutrient concentration. Hort.Science, 28(7):701-703.

Dufault, R. J., Baat, K. D., Decoteu, D., Carrett, J. T., Cranberry, D., Mc Laurin, W., Wagata, R., Perry, K. B. and Sanders, D. (1992). Scheduling collard planting dates regionally to lengthen the production period. Hort. Technology, 2(1):64-66.

Farnham, M. W. (1996). Genetic variation among and within United States collard cultivars and landraces determined by Randemly Amplified Polymorphic DNA markers. J. American Soc. Hort. Sci., 121(3):374-379.

Fung, A.C., Lopez, A. and Cooler, F.W. (1978). Essential elements in fresh and frozen spinach and collards. Journal of Food Science,43(3):897-899.

Guertal, E. A. and Edward, J. H. (1996). Organic mulch and nitrogen affect spring and fall collard yields. Hort.Science, 31(5):823-826.

Guertal, E. A., Snten, E van and Van, Snten E.(1997). Nitrogen rate and timing effects on collard yield and plant nitrogen concentration. J. Production Agril. 10(3):438-441.

Hochmuth, G. J. (1994). Efficiency ranges for nitrate - nitrogen and potassium for vegetable petiole sap quick tests. Hort. Technology, 4(3):218-222.

Kahangi, E. M. and Waithaka, K. (1981). Flowering of cabbage and kale in Kenya as influenced by altitude and GA application. Journal of Horticultural Science, 56(3):185-17.

Ne Smith, D. S. (1998). Effect of plant population on yields of once over harvest collards (*Brassica oleracea* L. var. acephala Group). Hort Science, 33(1):36-38.

Pathak, S. C., Young J. R. and Summer, D. R. (1982). Effect of methane sodium applied through over head irrigation system on weed control and yield of vegetables. Proc. Second Nat. Symp. on Chemigation, 23-27.

Peer. J. R., Sekhon, H. S., Week, J.Jr. and Stephens, B. R. (1980). Heavy metals in garden soil and vegetables in Washington, DC. Trace Substances in Environmental Health. XIV., 516-521.

Schwartz. R., Spencer, A. and Wolsh, J.J. (1984). Magnesium absorption in human subjects from leafy vegetables, intrinsically labeled with stable 26 Mg. American J. Clinical Nutrition, 39(4):571-576.

Stevens, C., Khan, V. A., Okoronkwol, T., Tang, A. H., Wilson, M. A. Lu, J. and Brown, J. E. (1990). Soil solarization and Dacthal influence on weeds growth and root microflora of collards. Hort.Science, 25(10):1260-1262.

Wall, M. L., Taylor, H., Perara, P. and Wani, M. C. (1988). Indoles in edible members of Cruciferae. Journal of Natural Products, 51(1):129-135.

Zldovec. V. (1997). HBBG status of the nation *Brassica* collection in Croatia. Sjemenarstvo, 14(1 & 2):117-118.

34

WATERCRESS

Watercress is a perennial herb, grown for its young shoots and leaves which are eaten raw or salads and is used as garnishing for meat dishes. Plant contains two aromatic compounds (i) 8-methylthiootanonitrile and (ii) 9-methyl thiononanonitrile which impart mustardy flavour, when used as raw (Macleod and Macleod, 1977). A tender aerial part of watercress is used as a cooked vegetable. Besides rich in calcium, iron, carotene, riboflavin and ascorbic acid content, watercress is rich source of vitamin E (α-tocopherol) which ranges around 75 mg/100 g F.W. Further, tocopherol content does not affect with processing and storage, however, long term storage of frozen vegetables reduces it (Horbowicz, 1989). Watercress has several medicinal values and is used as a stimulant and laxative. It yields a pungent volatile oil containing phenylethyl isothio-cynate (PEITC) which is of interest due to its claimed anti carcinogenic activity (Palaniswamy *et al.*, 1997). The nutritive value of watercress is given in table (34.1).

Table (34.1) : Nutritive value of watercress
(per 100 g of edible portion)

Constituent	Value
Moisture	89.2 g
Protein	2.9 g
Fat	0.2 g
Mineral	2.2 g
Fibres	0.6 g
Carbohydrates	4.9 g
Energy	33 kcal
Calcium	290 mg
Phosphorus	140 mg
Iron	4.6 mg
Carotene	2803 µg
Thiamin	0.12 mg
Riboflavin	0.38 mg
Niacin	0.8 mg
Vitamin C	13.0 mg

Origin and Distribution

Watercress is supposed to have got its origin in eastern Mediterranean and adjoining areas of Asia, and possibly Ethiopia. It is now widely distributed in South and Central Europe, Western Asia and many parts of the tropics (Tindall, 1983). Watercress is commercially cultivated in many western countries. The UK. is the largest producer of watercress with about 100 ha under cultivation, followed by France, the Benelux countries, Florida, Hawaii and Portugal. In India, it is grown as a leafy vegetable in Punjab, Odisha and West Bengal at very limited scale. In Sikkim, it is much common green vegetable and is brought in bundles to the market where sale goes year round but during winter. It is more in demand as a substitute for regular vegetable items.

Botanical Description

The genus *Nasturtium* thought to be closely related and are placed together in the tribe *Arabideae* DC. (Al-Shehbaz, 1988). *Nasturtium* which is often reduced to a synonymy of *Rorippa*, is recognized as a distinct genus with five species in the world (Al-Shehbaz and Price, 1998). However, molecular analysis do not support the incorporation of *Nasturtium* within *Rorippa* (Franzke *et al.*, 1998). The fruits of *Nasturtium officinale* are shorter and wider compared with *N. microphyllum, N. officinale* has coarsely reticulate seeds with less than 60 areolae per side in contrast to the moderately to minuetly reticulate seeds with less than 130 areolae per side found in *N. microphyllum.* Moreover, *N. officinale* is a tetraploid species (2n=4x=32) while *N. microphyllum* is octoploid (2n=8x=64). There are commonly 3 cultivated species (i) *R. heterophylla* (ii) *R. nasturtium aquaticum;* and (iii) *R. schlechteri. R.*

heterophylla is distributed in Japan and Myanmar while *R. schlechteri* is endemic to Papua New Guinea.Taxonomic position of the genus *Nasturtium* is given as under:

Taxonomic position of the genus *Nasturtium*

Kingdom	Plantae
Subkingdom	Tracheobionta
Superdivision	Spermatophyta
Division	Magnoliophyta
Class	Magnoliopsida
Subclass	Dilleniidae
Order	Capparales
Family	Brassicaceae
Genus	*Nasturtium* R. Br.
Species	*Nasturtium officinale* W.T. Aiton - watercres

Description of these three species under the genus *Nasturtium* has been given by Rahmansyah (1993), described as under :

1. ***R. nasturtium aquaticum :*** Perennial, much branched, glabrous, aquatic herb, upto 1 m tall. Stem prosprate to ascending; juicy, hollow, usually rooting at the lower nodes. Leaves petiolate, auriculate, pinnate, oblong in outline, up to 10 cm long; lateral leaflets in 2-9 pairs, sessile, narrowly ovate, elliptical or nearly orbicular, entire to fairly dentate; terminal leaflet similar but larger. Racemes ca 10 cm long; flowers with white petals; pedicels curved, ca 1 cm long. Fruit broadly linear, 10-18 mm × 1.5 - 2.5 mm, more than 1.5 mm thick, often curved and torulose, 7-12 times as long as broad, with seeds arranged in two distinct raws. Seed shiny, red brown, covered by a distinct reticulum.

2. *R. heterophylla :* Annual herb, 5-50 cm tall. Stem erect or ascending, one or more from the base, along the stem; lower ones narrowly obovate in outline, 1.5-8.0 cm long with 1-6 cm long petioles, undivided or lyrate - pinnati partite with 1-3 pairs of small lateral lobes and a much larger terminal lobe; upper leaves rather few, undivided, narrowly ovate. Racemes 2-10 cm long; flowers with reduced or lacking petals, sepals green; pedicels straight, 2-8 mm long. Fruit linear, 14-25×0.7-1.3 mm, straight, less than 1.5 mm thick, with seed in one row only. Seed red brown, minutely foreolate.

3.*R. schlechteri:* Annual to short lived perennial,15-50 cm tall, stem, solitory, ercet, hardly branched. Leaves indistincly petiolate, auriculate in outline,1.5-15 cm × 1-3.5cm, lyrato pinnate with 1-3 pairs of oblong lateral lobes, 0.2-1.5 cm × 0.1-0.6 cm, and a much larger elliptical -ovate terminal lobe. Racemes up to 15 cm long; flowers with yellowish petals; pedicels curved, 3-10 mm long. Fruit inflated, semiglobose, 5-9 mm-3-6 mm, more than 1.5 mm thick. Seed red brown, minutely foreolate.

Climate and Soil

Watercress grows abundantly in region of high alitutude. Plant thrives well in running water and marshy lands. It is a long day plant and does not flower under short day photo period conditions. Palaniswamy *et al.* (1997) reported that application of enhanced photosynthetic photon flux (PPF) before harvest increased phenethyl isothiocynate in watercress under 8 h photoperiod. Watercress grows in almost all types of soil having good water holding capacity and is rich in humus; a pH range of 6.5-7.5 would appear to be optimum (Tindall, 1983).

Propagation

Watercress can be raised either by seeds or cuttings.

Raising the Seedlings : Seeds from a temperate source can be sown in shaded seed beds and seedlings are planted when they attain 6-8 cm height. Dormancy in water cress seed has been observed and early ripening seed is said to be more dormant than late ripening and further, seeds become progressively less dormant following harvesting. Germination of seeds at 20^0C in the dark can be induced by treatment with a combination of GA_4, GA_7 and BA (Biddington and Ling, 1983). Biddington and Ling (1983) reported that dark coloured seeds of water cress germinated readily than pale ones when the seeds were freshly harvested but with old seeds germination was generally greater with paler seeds than darker ones.

By cuttings : Cuttings strike root readily and give early yield as compared to seeded plants. Cuttings 10-15 cm long are planted in well prepared bed at the spacing of 15-20 cm apart. Beds should be flooded with water soon after planting the cuttings.

Nutritional requirement

Watercress is grown as outdoor as well as under controlled conditions. Under soilless culture, supply of N @ 105 mg/litre and bench flooding ten times very day gave sufficient yield and quality with plant nitrate content below the European Commission Limit (Michalsky *et al.*, 1997). Triper (1984) obtained optimum yield of greenhouse cultivated water cress on Grodan rock with 3 NHL as 33 of the total N and NO_3 as 67%. Further, he reported that an anion/cation ratio was best for realizing maximum yield. As watercress plant has both root systems basal and adventitious. Cumbus and Robinson (1977) reported that greater proportion of P and K were absorbed by adventitious than by the basal root systems, although the latter was more massive. Nutritional studies made on plants grown in cross bed simulation tanks showed that critical levels of nutrient in leaf tissue were 5.7% N, 0.5% P and 2.8% K (Robinson and Cumbus, 1977). Total translocation of Fe by intact plant of water cress was inhibited by increasing levels of P, Zn and Mn in the nutrient medium. The high P and Zn concentrations enhanced Fe absorption into the roots but increased Fe retention in the roots and stem; high Mn reduced initial Fe uptake into the roots (Cumbus *et al.*, 1977).

Irrigation

Flowing water induces the vegetative growth of the plant. MacHugh and Nishimuto (1980) observed that overhead sprinkling for 12-15 minute period/day increased the quality of watercress grown in the warm season and increased yield from 1.25 kg/m^2 to 1.96 kg/m^2. However, sprinkling had no effect on yield, quality or leaf temperature grown in cool season.

Harvesting and Yield

Harvesting is done by removing the upper top portion of the plant. Crop of watercress raised by cutting becomes ready for harvest within 4-6 weeks of planting. It is a very high yielding crop and 150-200 quintals yield of green leaves can be obtained from one hectare.

Storage and Packaging

Watercress is very perishable and cannot be kept for longer period under ambient conditions. During storage chlorophyll breakdown is closely linked to lipid oxidation, while proteolysis seems to proceed independently of these two associated processes (Philosoph *et al.*, 1994). Chlorine treatment is effective in Fr reducing the microbial population and improving the keeping quality. Treatment with 100 ppm chlorine effectively reduced the microbial load of the produce without significantly quality losses. However, effectiveness of chlorine treatment was limited to short term storage (Park and Dongsun, 1995). Packaging in non-perforated PE (NPPE) lined cartons resulted in a marked retardation of both yellowing and decay. The best results for watercress were obtained by packaging in sealed PE (SPE) lined cartons. However sealed film packaging is applicable only if the temperature during transit and storage is well controlled (Ahroni *et al.*, 1989).

Major Diseases

Crook root disease : This is fungal disease caused by *Spongospora subterranea* f. sp. *nasturtii*. It can be controlled by adding 0.3-0.5 g Zn/ml to the inlet water supply to the crop (Tomlinson and Hunt, 1987).

Watercress yellow spot virus

It is transmitted by zoospores of the crook root fungus, *Spongospora subterranea* f. sp. nasturtii. Adoption of WYSV tolerant lines is the only solution to check the spread of disease. Walsh (1992) reported that watercress lines originating in the U.K. are most susceptible to both crook root and WYSV whereas Japanese lines are less susceptible.

Turnip mosaic virus (TUMV) : Turnip mosaic virus is not seed transmitted and can be controlled by raising of plants from seed (Tomlinson and Hunt, 1987).

Major Pests

Diamond back moth (*Plutella xylostella*), aphids (*Aphis gossypii, Mazus persicae*), Hyadaphis crysimi (*Lipaphis erysimi*) and *Aphis nasturtii* are major pests of watercress. For controlling the infestation of these pests, integrated approach is most judicious. Nakahara *et al.* (1985) reported that *P. xylostella* was effectively and economically controlled through the implementation of an intermitted overhead sprinkler system operating between 8.00 and 22.0 h and weekly release of *Braconid apanteles plutelle*. They further, reported that different species of aphids were effectively controlled by established predators and timely application of diazinon.

Cultivars

European Seed Catalogues list watercress, but there are no officially registered cultivars. Recently a variety Vicred has been evolved by Crowe (1998). Its detailed characters are given as under :

Vicred : It is a perennial, succulent, temperate watercress variety. It was selected from a spontaneous mutation in *N. officinale* on the basis of colour, vigour and foliage density. The young leaves are green with the grey margins (RHS 137C) and the adult leaves are purple grey (RHS 200A) through the interveinal area with green (RHS 137C).

Genetic Resources and Improvement

In the U.K. Horticultural Development Council has funded the Vegetable Gene Bank Horticulture Research International, Wellesbourne, UK, to collect watercress germplasm. Over 200 accessions have been collected from around the world (Pink, 1993). Stevens (1983) has suggested following objectives for watercress breeding in addition to resistance diseases and pests; dark green, glossy, round leaves of medium size which are produced consistently through out the year; the ability to tiller and produce high-density swords, particularly in winter, frost hardiness; late flowering and ample seed set. Shoots should ideally be around 10 cm long and completely free of roots; a pronounced but not overstrong; mustard flavour is also desirable. A uniform growth is also desirable attribute for mechanized harvesting. Since watercress is self compatible it is necessary to emasculate flower of female parent plant prior to crossing. Female flowers also need to be enclosed in cellophane bags to prevent accidental pollination (Pink, 1993). Johnson (1968) suggested that progeny testing might be most appropriate method of improving the crop. During the late 1970s and early 1980s the simple method of mass selection was used to improve the crop (Royle, 1982). The two species of genus *Nasturtium*, *N. officinale* and *N. microphyllum* make a hybrid state, *Nasturtium* x sterile, with intermediate morphological and chromosomal features (Naqinezhad, 2006).

***In vitro* culture**

Nodal explants of watercress taken from 4 uniforms, pot grown clones were culture *in vitro* on growth regulator free MS medium. The plantlets produced were grown on 25, 50 or 100% strengthened MS media supplemented with 1, 2 or 4% sucrose. The highest number of nodes (40.9%) occurred on 25% MS with 2% sucrose, but some shoots (14%) become vitrified. Two node propagules produced more nodes than shoot tip or single node propagule because both auxillary buds developed shoots, although the number of nodes/shoot was lower than for other types of propagule and there was not difference in final dry weights. Propagules derived from the distal part of the shoot produced new shoot with significantly more nodes (37.5) than those derived from the mid-dle or proximal (20.5) sections. *In vivo*, 99% of the two node propagules were successfully established by rooting directly in a 50:50 mixture of vermiculite and compost, compared with 79% in 100% vermiculite (Wainwright and Marsh, 1986). Callus was initiated on MS medium supplemented with auxin and cytokinin in the dark at 25^{0}C. The regeneration media contained no auxin, and the callus was cultured at 25^{0}C with a 16 h photoperiod. The cytokinin thidiazuron was better than benzyla-denine for shoot regeneration and auxin was regulated in the initiation medium. NAA simulated shoot production from callus on initiation medium while with 2, 4-D shoot production was at its highest frequency at 24-32 weeks (Gilby and Wain Wright, 1989).

References

Ahroni, N., Reuveni, A. and Dvir, O.(1989). Modified atmosphere in film packages delay senescence and decay of fresh herbs. Acta Horticulturae, 258: 225-262.

Al Shehbaz I.A. (1988) .The genera of *Arabideae* (Cruciferae; Brassicaceae) in the south eastern United States. J. Arnold Arbor. 69: 85-166.

Al Shehbaz, I.A. and Price, R.A. (1998). Delimitation of the genus *Nasturtium* Brassicaceae) Novon, 8: 124-126.

Biddington, N.L. and Ling, B. (1983). The germination of watercress (*Rorippa nasturtium aquaticum*) seeds I. The effect of age, storage, temperature, light and hormones on germination. Journal of Horticultural Science, 58(3): 417-426.

Crowe, F.D. (1998). Variety: Vicred, Application No. 97/171. Plant Varieties Journal, 11(2):50.

Cumbus, I.P. and Robinson, I.W. (1977). The function of root system in mineral nutrition of watercress (*Rorippa nasturtium aquaticum* L.). Plant and Soil, 47(2):395-406.

Cumbus, I.P., Hornsey, D.J. and Robinson L.W. (1977). The influence of phosphorus, zinc and manganese on absorption and translocation of iron in watercress. Plant and Soil, 48(3): 651-660.

Franzke, A., Pollman, K., Bleeker, W, Kohrt, R. and Hurka, H. (1998). Molecular systematics of *Cardamine* and allied genera (Brassicaceae): ITS and non-coding chloroplast DNA. Folia Geobotanica, 33 (3) 225-240.

Gilby, A.C. and Wain Wright, H. (1989). Use of tissue culture in improvement of watercress (*Rorippa nasturtium aquaticum* L. Hayek). Acta Horticulturae, 244: 105-113.

Horbowicz, M. (1989). Vegetables as a source of vitamin E. Biuletyn Warzywniczy Suppl. II. 205-209.

Johnson, A.G. (1968). Watercress, Rep. natn. Veg. Res. Stn. for, 1967, 25, 1968.

Mac Hugh, J.J. Jr. and Nishimuto, R.K. (1980). Effect of overhead sprinkler irrigation on watercress yield, quality and leaf temperature. HortScience, 15(6): 80-82.

Macleod, A.J. and Macleod, G. (1977). Synthesis and natural occurrence of 8. Methylthiootanonitrile and 9-methylthiononanonitrile. Phytochemistry, 16(7): 907-909.

Michalsky, F., Hanke, A. and Schnifler, W.H. (1997). Yield and quality in ebb and flow irrigation. Gemüse (München), 33(10): 575-578.

Nakahara, L.M., Mac Hugh, J.J, Jr., Otsuka, C.K., Funasaki, G.Y. and Lai, P.Y.(1985). Integrated control of diamond back moth and other insect pests using an overhead sprinkler system an insecticide and biological control agent on a watercress farm in Hawaii. In diamond back moth management proceedings of the I[st] International Workshop, Tainan, Taiwan, 11-15, March, 1985.

Naqinezhad, A. (2006). A short note on the genus *Nasturtium* (Cruciferae), and a new hybrid state from this genus for Iran. Iran. Journ. Bot. 12(1): 75-77.

Palaniswamy, U., Macavoy, R., and Bible, B. (1997). Supplemental light before harvest increases phenethyl isothiocynate in watercress under 8 hours photo period. Hort Science, 32(2): 222-232.

Park, Woopo and Dongsun, Lee (1995). Effect of chlorine treatment on cut watercress and onion. Journal of Food Quality, 18(5): 415-424.

Philosoph, Hadas, S., Meir, S., Akiri, B. and Kanner, J. (1994). Oxidative defense systems in leaves of 3 edible herb species in relation to their senescence rates. Journal of Agricultural and Food Chemistry, 42 (1): 2376-2381.

Pink, D.A.C. (1993). Watercress. In Genetic Improvment of Vegetable Crops. G. Kalloo and B.O. Bergh (eds.). Pergamon Press, Oxford, New York, Seoul, Tokyo, pp. 579-583.

Rahmansyah, M. (1993). *Rorippa*. In: PROSEA Plant Resources of South-East Asia, No. 8, Vegetables. Pudoc Scientific Publisher, Wageningen, the Netherlands.

Robinson, L.W. and Cumbus, I.P. (1977). Determination of critical levels of nutrients in watercress (*Rorippa nasturtium aquaticum* (L.) Hayek) grown in different concentration of N, P and K. Journal of Horticultural Science, 52(3): 383-390.

Royle, D. (1982). More cress strains sorted out. The Grower, 97:5.

Stevens, C.P. (1983). Watercress ADAS/ MAFF Reference Book No. 136, Grower Books, London, 54.

Thompson, H.C. and Kelly, W.C. (1957). Vegetable Crops, MC Graw Hill, New York.

Tindall, H. (1983). Vegetables in the Tropics. MacMillan, London, United Kingdom.

Tomlinson, J.A. and Hunt, J. (1987). Studies on watercress chlorotic leaf spot virus and on the control of the fungus vector (*Spongospora subterranea*) with zinc. Annals of Applied Biology, 110(1):75-88.

Triper, A. (1984). Growing watercress on rockwool. Revue Horticole Suisse, 57(3): 73-79.

Wainwright, H. and Marsh, J. (1986). The micropropagation of watercress (*Rorippa nasturtium aquaticum* L.). Journal of Horticultural Science, 61(2): 251-256.

Walsh, J.A., (1992). The epidemiology and control of watercress yellow spot virus. In: Recent Advances in Vegetable Virus Research, 7th conference ISHS Vegetable Virus Working Group, Athens, Greece, July 12-16, 1992.

35

WINTER CRESS

Winter cress is also called upland cress. Its leaves possess strong flavour similar to water cress (*Rorippa nasturtium-aquaticum*) and are used as a water cress substitute, either raw or cooked (Brickell, 1998). Fatty acid composition of winter cress seed was dominated by erucic (50%) .The major glucosinolates found were gluconasturtiin (106 µmol g^{-1}) (Andersson *et al.*, 1999). Barillari *et al.* (2001) reported that winter cress is a source of 2-phenylethyl glucosinolate which is precursor of cancer chemo-preventivephenylethyl isothiocyanate. . Nutritive value of winter cress is given in table (35.1).

Table (35.1): Nutritive value of winter cress (per 100 g of edible portion)

Constituents	Amount
Water	90 per cent
Calories	30 kcal
Carbohydrate	6.0 g
Protein	3.0 g
Fat	1.0 g
Fibre	1.0 g
Vitamin A	9300 IU
Vitamin C	70.0 mg
Vitamin B_1	0.1 mg
Vitamin B_2	0.3 mg
Niacin	1.0 mg
Calcium	80.0 mg
Phosphorus	80.0 mg
Potassium	610 .0mg
Sodium	10 .0mg
Iron	1.0 mg

Source : Rubatzky and Yamaguchi (1997)

Origin and Distribution

Winter cress is mainly cultivated in France and Belgium, however, its occurrence in wild state has been reported from western and south-western Europe. It has also been introduced to Africa, America and Asia. In Malaysia, it is cultivated in limited extent (Siemonsma and Piluek, 1993).

Botanical Description

Winter cress *Barbarea verna* Syn. *Barbarea praecox* (Smith) R.Br. is a plant of family Brassicaceae. It is a hardy biennial but is grown as an annual or winter perennial. Plant height upto 75 cm, leaves pinnately lobed with 4 or more pairs of lobes and a larger terminal lobe. Its related species is *Berna vulgaris* which has been introduced in East Java and Indonesia. The genus *Barbarea* includes about 97 species. Taxonomic position of the genus *Barbarea* is given as under:

Taxonomic position of the genus *Barbarea*

Domain	Eukaryota
Kingdom	Plantae
Subkingdom	Viridaeplantae
Phylum	Tracheophyta
Subphylum	Euphyllophytina
Infraphylum	Radiatopses
Class	Magnoliopsida
Subclass	Dilleniidae
Superorder	Violanae
Order	Capparales
sub order	Capparineae
Family	Brassicaceae
Genus:	*Barbarea*
Specific epithet	*verna* (P. Mill.) Aschers

Important species under the genus *Barbarea:*

B. abyssinica · *B. affinis* · *B. alpicola* · *B. alpina* · *B. altaica* · *B. americana* · *B. apoiensis* · *B. arcuata* · *B. arisanense* · *B. arisanensis* · *B. augustana* · *B. aurea* · *B. auriculata* · *B. australis* (Native Wintercress) · *B. balcana* · *B. barbarea* · *B. barbareae* · *B. bosniaca* · *B. brachycarpa* ·

B. bracteosa · *B. brevicaulis* · *B. brevistyla* · *B. ceretana* · *B. cilicica* · *B. ciliifera* · *B. cochlearifolia* · *B. conferta* · *B. croatica* · *B. elata* · *B. erysimoides* · *B. gradlii* · *B. grandiflora* · *B. grayi* · *B. hedgeana* · *B. heterophylla* · *B. hirsuta* · *B. hondoensis* · *B. hongii* · *B. humilis* · *B. iberica* · *B. integrifolia* · *B. intermedia* (Medium Flowered Winter Cress) · *B. kayserii* · *B. ketzkhovelii* · *B. lepuznica* · *B. linnaei* · *B. lippizensis* · *B. longirostris* · *B. longisiliqua* · *B. lutea* · *B. lyrata* · *B. macrocarpa* · *B. macrophylla* · *B. minor* · *B. morrisonensis* · *B. oligosperma* · *B. orthocera* · *B. orthoceras* (American Winter Cress) · *B. orthoceras dolichocarpa* · *B. palustris* · *B. parviflora* · *B. patens* · *B. patula* · *B. perennis* · *B. pinnata* · *B. planisiliqua* · *B. plantaginea* · *B. platycarpa* · *B. praecox* · *B. prostrata* · *B. pyrenaica* · *B. rivularis* · *B. rupestris* · *B. rupicola* · *B. schulzeana* · *B. sibirica* · *B. sicula* · *B. silvestris* · *B. stolonifera* · *B. stricta* · *B. sylvestris* · *B. taiwaniana* · *B. taurica* · *B. trichopoda* · *B. verna* (American Winter Cress) · *B. vicina* · *B. vulgaris* (Garden Yellow Rocket) · *B. vulgaris arcuata* · *B. vulgaris* 'Flore Pleno'

Morphological Description of Plant

The leaves are very small, almost square shaped, with slight notching. The plant is low and spreading, with 15-20 cm long leaf stems. It resembles watercress in flavor and structure, but is easier to grow. Although it is similar to *Barbarea vulgaris* (Yellow Rocket), it is not the same plant. It will reseed itself profusely if allowed. It is an edible vegetable and is treated mainly as a biennial, so it grows best over the course of two years. Normally plants attain an height of 10.0 cm. This plant tends to bloom in early summer, followed by first harvests in late spring.

Agro techniques

Winter cress is very hardy plant and it remains green during winter. Plant thrives best in the shade in heat it runs to seed rapidly, becoming course and very hot tasting. Winter cress grows best in fairly rich soil and tolerates moist soils (Joy Larkcom, 1981). Plants of land cress can be raised through direct seedling in the main field or by raising the seedling and thereafter planting in the main fields. Seeds are planted 50-60 mm deep in row 30 cm apart. After germination when plants are established they are thinned to stand 10 cm apart. Winter cress may be grown in unheated greenhouses in winter. If few plants are left to run seed in spring they seed freely and seedlings may be transplanted. Plant is almost free from attack of major pests and diseases. At early stage of growth sometimes flea beetle infests the crop.Winter cress is fast growing plant. After 6-7 weeks of sowing, plants become ready for harvest. Only soft and tender leaves are harvested for consuming either raw as a salad or cooked.

References

Andersson, A. A. M., Merker, A., Nilsson, P., Sørensen, H. and Åman, P. (1999). Chemical composition of the potential new oilseed crops *Barbarea vulgaris*, *Barbarea verna* and *Lepidium campestre*. Journal of the Science of Food and Agriculture, 79: 179-186.

Barillari, J. D., Rollin, P. and Iori, R. (2001). *Barbarea verna* as a source of 2-phenylethyl glucosinolate, precursor of cancer chemopreventive phenylethyl isothiocyanate. Fitoterapia,72(7):760-764.

Brickell, C. (1998). Encyclopedia, Gardening. The Royal Horticultural Society, London, New York.

Joy Larkcom (1981). The vegetable garden displayed. The Royal Horticultural Society, 80, Vincent Square, London.

Rubatzky, V.E. and Yamaguchi, M. (1997). World Vegetables IInd Edition. Chapman &Hall.

Siemonsma, J. S. and Piluek, K. (1993). Plant Resources of South-East Asia. No. 8. Vegetables. Pudoc Scientific Publishers, Wageningen, the Netherlands.

36

RHUBARB

Rhubarb is grown primarily for its fleshy stalks, known as petioles. The use of rhubarb stems as food is a relatively new, first recorded in 17th century in England. Commonly it is stewed with sugar or used in pies and desserts. Rhubarb can be dehydrated and infused with fruit juice. It is rich in phyto-nutrients such as dietary fiber, poly-phenolic anti-oxidants, minerals, and vitamins. The stalks are rich in folates, riboflavin, niacin, vitamin B_6 (pyridoxine), thiamin, and pantothenic acid. It also contains poly phenolic flavonoid compounds like ß-carotene, zeaxanthin, and lutein. Rhubarb stalks are also rich in vitamin K. Rhubarb leaves are poisonous and they contain anthroquinone and of free oxalic acid. Nutritional composition of rhubarb is given in table (36.1).

Table (36.1). Nutritional composition of rhubarb (per 100 g of edible portion)

Composition	Value
Energy	21 Kcal
Carbohydrates	4.54 g
Protein	0.90 g
Total Fat	0.20 g
Cholesterol	0 mg
Dietary Fiber	1.8 g
Folates	7 µg
Niacin	0.300 mg
Pantothenic acid	0.085 mg
Pyridoxine	0.024 mg
Riboflavin	0.030 mg
Thiamin	0.020 mg
Vitamin A	102 IU
Vitamin C	8 mg
Vitamin E	0.27 mg
Vitamin K	29.3 µg
Sodium	4 mg
Potassium	288 mg
Calcium	86 mg
Copper	0.021 mg
Iron	0.22 mg
Magnesium	12 mg
Manganese	0.196 mg
Phosphorus	14 mg
Selenium	1.1 µg
Zinc	0.10 mg
Carotene-	0 µg
Carotene-ß	61 µg
Crypto-xanthin-ß	0 µg
Lutein-zeaxanthin	170 µg

Origin and Distribution

Rhubarb has been cultivation in China for medical purposes, and the earliest record of its use is back to about 2700 B.C. Instead of *R. officinale* or *R. undulatum*, *R. palmatum* was mainly used as a medicinal rhubarb (Turner, 1938). The use of rhubarb root eventually reached Europe *via* the Arabs and in the beginning of the Christian era, the drug was used in Greek and Roman medicine. The drug had to be imported from Asia until European rhubarb cultivation started in the 16th century. The first *Rheum* species grown in Europe was probably *R. rhaponticum* (Turner, 1938). In the middle of the 17th century, rhubarb was grown for medical purposes in England and Germany and eventually, the cultivation spread to the Scandinavian countries. However, the imported roots from Asia yielded a drug with much higher quality and European rhubarb cultivation reduced (Hintze, 1951). In the 18th century, it was found that rhubarb had edible stalks and *R. rhaponticum, R. undulatum* and *R. hybridum* were grown as a vegetable crop in England (Turner, 1938), and soon after, also in the Scandinavian countries (Hintze, 1951). At an early stage, the growers in England started to develop culinary rhubarbs and some of the cultivars are still grown e.g. 'Victoria' and 'Prince Albert'. In the breeding of new cultivars, *R. hybridum*, which has an unknown origin, was often used as a parent.

Botanical Description

Rhubarb *Rheum* × *cultorum* (*Rheum rhabarbarum* L.) belongs to the family Polygonaceae. Taxonomic position of the genus *Rheum* is given as under:

Taxonomic position of the genus *Rheum*

Kingdom	Plantae
Subkingdom	Tracheobionta
Superdivision	Spermatophyta
Division	Magnoliophyta
Class	Magnoliopsida
Subclass	Caryophyllidae
Order	Polygonales
Family	Polygonaceae
Genus	*Rheum* L.
Species	*Rheum rhabarbarum* L.

Under the genus *Rheum* about 50 species have been described (Flora of China, 1996) and most of the species are native to the northern and central regions of Asia (Turner, 1938). Chromosome numbers differ between taxa within the genus, the most common being 2n=22 (diploids), 2n=44 (tetraploids) and 2n=66 (hexaploids). Culinary rhubarb cultivars are tetraploids, and exhibit a continuous variation in morphology between *R. rhaponticum* (2n=22) on the one hand, and *R. rhabarbarum* (2n=44, though some cases of 2n=22 have been reported) (Englund, 1983).

Description of Plant : A large, erect, perennial herb, up to 2.5 m tall; rhizome and roots thick, branched, almost fleshy, bright yellow. Leaves partly in a radical rosette, partly spirally arranged on an erect stem; leaves of rosette borne on a thick, subcylindrical petiole, up to 30 cm long, blade orbicular, 50—90 cm x 50—70 cm, base cordate, deeply palmately lobed, 3—5 veined, lobes ovate-oblong or lanceolate, apex acute, entire, dentate or pinnatifid, upper surface normally rough; ocrea large, membranaceous. Inflorescence a loose, hairy panicle with racemiform branches, foliated to the tip. Flowers bisexual, fascicled; tepals in 2 whorls of 3, mostly red, sometimes pink or whitish, little or not accrescent after anthesis; stamens 9; ovary trigynous, styles 3, recurved, stigmas thick. Fruit a trigonous nutlet, 10 mm x 8 mm, brown, much longer than perianth, 3-winged, scarious (Schmelzer and Horsten, 2001)

Cytotaxonomy

The center of primary origin of the *Rheum* species appears to be some where in the vicinity of the Himalaya Mountains, possibly extending into Central China. Emodin and chrysophanic acid were found to occur in the rhizomes and roots of *R. officinale* Baill., *R. palmatum* L., *R. palmatum* var. tanguticum, and Indian and American grown *R. emodi* Wallich. Rhaponticin did not occur in any of the aforementioned except American-grown *R. emodi* Wallich. Rhaponticin was found in the rhizomes and roots of *R. rhaponticum* L., *R. undulatum* L., *R. compactum* L. and in the American grown *R. emodi* Wallich. *R. franzenbachii* Muent. was found to have only a trace of emodin in the rhizome and roots, but chrysophanic acid and rhaponticin were found to be present. The diploid numbers of chromosomes were found to be 22 for *R. franzenbachii* Muent., *R. emodi* Wallich and *R. undulatum* L. and 44 for *R. tataricum* L., *R. compactum* L., *R. altaicum* A. Los. and *R. australe* D. Don. *R. emodi* Wallich shows much larger meiotic chromosomes in comparison with *R. franzenbachii* Muent., but there is no apparent difference between these species in the size of mitotic chromosomes. The size difference is limited to meiosis. The comparative morphology of the somatic chromosomes showed that *R. palmatum* L. and *R. franzenbachii* Muent. has each one pair of chromosomes with satellites and *R. emodi*

Wallich and the 2n *R. undulatum* L. each has two pairs of chromosomes with satellites. The tetraploid *R. undulatum* L. has 6 chromosomes with satellites which suggests the existence of two genomes of 11 chromosomes in the genus of *Rheum* and that this tetraploid form arose by amphidiploidy (Chin and Youngken, 1947).

Varieties

Red stalk types:

Crimson: It produces brightly colored red stalks with the unique characteristic of being red throughout under normal temperature and moisture conditions. Other red varieties are Valentine and Cherry Red.

Speckled types (pink)

Victoria produces large stalks of excellent quality, long, round with smooth ribs. It develops pink speckling on a light green stalk with the pink color being more intense at the bottom of the stalk, fading to a solid green near the top. It is commonly used for forcing. Strawberry is very similar to Victoria, and may be the same variety. MacDonald is another pink type that produces well. German Wine is similar to Victoria but slightly more vigorous and more intense in color, typically with a darker pink speckling on a green stem.

Green types: Riverside Giant, a cold-hardy, vigorous producer with large diameter, long, green stalks.

Climate and Soil

Crop of rhubarb thrives well in temperate conditions, however, in the tropics its cultivation is possible up to 1000 m altitude. The optimum temperature for growth is 15-20°C. A well drained but moisture holding soil with a pH of 5.5 to 7.0 is ideal for cultivation of rhubarb. The light structured soils produce an earlier crop but require more irrigation. Deep plowing is needed for field preparation and it should be free of perennial weeds.

Propagation

Rhubarb is commnly propagated by vegetative means. Seed propagation is not advocated owing to the variability in the progenies. Healthy, vigorous 3 to 4-year old crowns are divided to obtain two or more buds per seed-piece. Five or six year old crowns will yield 8-10 good quality pieces. The crown pieces are planted 10-15 cm deep, 60-75 cm apart in rows. For mechanised harvesting, a spacing of 45 cm apart in rows 120 cm apart is followed. This spacing would require about 7200 plants per acre. Rhubarb requires a dormancy period of temperatures below 40°F to break dormancy and stimulate the production of leaf petioles. When temperatures begin to exceed 45-50°F, crown buds begin to develop. Early growth may be enhanced ten days to two weeks by the use of white plastic mulch. Application of gibberellic acid has also been found effective in enhancing the growth.

Manures and Fertilizers

Application of FYM @ 20-25t/ha is advocated at least 2 months before of planting. Rhubarb is heavy feeder crop and for good harvest, apply NPK @ 150:80:100 kg/ha. Same dose may be repeated in next season also. Among the micronutrients, application of boron gives beneficial influence on plant growth and quality produce. Nitrogen fertilizer applications should be split into 3-4 top dressings. Nitrogen rates may be reduced in the first two years with manure applications.

Irrigation

Adequate soil moisture should be maintained throughout the crop growth. Soil type does not affect the amount of total water needed, however, it does influence frequency of water application. Light textured soils need more frequent water applications, but less water applied per application.

Harvesting, Yield and Storage

Yields of rhubarb depend on the number of pickings, and the age and condition of the field. A yield of 20-25t/ha is most common at the first harvest. Fields may be harvested a second time with yields generally reduced by 50 per cent from the first harvest. In general, green varieties are higher yielder. A well-maintained field may remain productive for 15 or more years. For processing, both ends of the petiole are trimmed so that no leaf tissue remains. For fresh market a small amount (1cm) of leaf tissue is usually left attached to the petiole and the basal end is not trimmed. Now a days mechanical harevesting is also common. While harvesting, stalks should not be pulled during the first year of growth. Stalk color is best after the field is 2 to 3 years old. Plants should not be over-pulled at any time, as a certain amount of foliage is required for the development of the present crop as well as next year's crop. Harvested produce may be stored at 32°F and 95 to 100% relative humidity. Fresh rhubarb stalks in good condition can be stored 2 to 4 weeks at 32°F and high relative humidity. Rhubarb can be hydro-cooled or air-cooled, and the temperature of the stalks should reach 32° or 33°F within 1 day of harvest. The topped bunches or loose stalks should be packed in crates, and the crates should be stacked to allow ample air circulation. Moisture loss in storage is reduced with packed in crates lined with perforated polyethylene film.

Genetic Resources and Improvement

A germplasm collection of *Rheum* species is maintained in the Department of Plant Breedding, Swedish University of agricultural Sciences, Uppsala. A good scope of Rhubarb improvement exists as different species hybridize easily and promising selection can propagated vegetatively (Huibers -Govaert, 1993). Thick, streight , red petioles free from stringiness, devoid of strong acid taste, resistance to biotic and abiotic stresses and less oxalate content are major priotties of rhubarb improvement (Libert, 1987) Fifteen morphological characters were evaluated to differentiate rhubarb cultivars . The mean oBix observed was 3.8, with a range from 2.2 to 6.1. Flesh colour and basal skin colour were poorly correlated (R2 = 0.462); overall skin colour was more red at the base than in the middle of the petiole. Rhubarb character categories, in particular petiole number and petiole base thickness, need to be modified to better anticipate the range of expected values, and thereby contribute improved reproducibility and reliability to separate cultivars based on morp ological characters (Alberto and Joseph, 2010).

***In vitro* propagation**

In vitro culture of shoot-tips of rhubarb on the MS medium containing 1 mg/l of 6 benzylaminopurine and 1 mg/l of 3-indole-butyric acid, induced the development of axillary buds. Rooting was easily attained in 4-5 days by omitting the cytokinin from the basal medium. The temperature was 25 ± 1° C during the 16-h photoperiod and 20 ± 1° C during the 8-h dark period. The rooted plantlets were transplanted to soil medium with 70% success (Joseph and Marie-Christine, 1979).

References

Alberto, P. and Joseph, C. K. (2010). Morphologic variation in the USDA/ARS rhubarb germplasm collection. Plant Genetic Resources, 8:35-41.

Chin, T. C. and Youngken, H. W. (1947). The cytotaxonomy of *Rheum.* American Journal of Botany, 34 (8):401-407.

Englund, R. (1983). Odlad rabarber och taxonomiska problem inom sliiktet *Rheum*, speciellt sektionen *Rhapontica.* Report, Dept. of Systematic Botany, Uppsala University, Sweden (in Swedish).

Flora of China (1996). http:// flora.huh.harvard.edu/china/0 Flora of China Project (information taken on 2001-07-08

Hintze, S. (1951). Rabarber. In: Svensk viixtforadling Del I1 Tradgirdsviixtema Skogsvaxtema. Natur och Kultur, Stockholm, pp. 389-391 (in Swedish).

Huibers-Govaert, I.I.M. (1993). *Rheum* x cultorum Thorsrud & Reisaeter. In : PROSEA, Plant Resources of South East Asia, 8, Vegetables, Siemonsma, J.S. and Piluek, K. Eds. Pudoc Scientific Publishers, Wageningen, pp.237-239.

Joseph, R. and Marie Christine, C. (1979). Rapid clonal propagation of rhubarb by *in vitro* culture of shoot-tips. Scientia Horticulturae, 11(3): 241-246.

Libert, B. (1987). Breeding a low-oxalate rhubarb (*Rheum* sp. L.). Journal of Horticultural Science , 62:523-529.

Schmelzer, G.H. and Horsten, S.F.A.J. (2001). *Rheum palmatum* L.[Internet] Record from Proseabase. van Valkenburg, J.L.C.H. and Bunyapraphatsara, N. (eds.). PROSEA (Plant Resources of South-East Asia) Foundation, Bogor, Indonesia.

Turner, D.M. (1938). The economic rhubarbs: a historical survey of their cultivation in Britain. Journal of the Royal Horticultural Society, 63:355-370.

37

ASPARAGUS

Asparagusis a perennial vegetable and itsyoung shoots are eaten. It is a low-calorie source of folate and potassium and very low in sodium. It is also a good source of vitamin $B_{6,}$ calcium, magnesium and zinc, and a rich source of dietary fiber, protein, vitamin A, vitamin C, vitamin E, vitamin K, thiamin, riboflavin, rutin, niacin, iron, phosphorus, copper, manganese and selenium. The amino acid asparagine has derived its name from asparagus. The shoots are prepared and served in a number of ways. Its stalks are high in folate and folate is considered critical for pregnant mothers, since it protects against neural tube defects in babies. Besides shoots of asparagus, leaves also have immense therapeutic use. Kim *et al.* (2009) reported that the amino acid and inorganic mineral contents were much higher in the leaves than the shoots. In addition, treatment of $HepG_2$ human hepatoma cells with the leaf extract suppressed more than 70% of the intensity of hydrogen peroxide (1 mM)-stimulated DCF fluorescence, a marker of reactive oxygen species (ROS). Cellular toxicities induced by treatment with hydrogen peroxide, ethanol, or tetrachloride carbon (CC^{14}) were also significantly alleviated in response to treatment with the extracts of leaves and shoots. Additionally, the activities of 2 key enzymes that metabolize ethanol, alcohol dehydrogenase and aldehyde dehydrogenase, were unregulated by more than 2 fold in response to treatment with the leaf and shoot extracts. Nutritional composition of the green and white spears of asparagus is given in table (37.1).

Table (37.1) : Nutritional composition of the green and white spears of asparagus (per 100 g of edible portion)

Constituents	Green spear	White spear
Water	92.2 per cent	93.6 per cent
Energy	22 kcal	20 kcal
Carbohydrates	3.8 g	2.9 g
Protein	2.60 g	1.90 g
Fats	0.21 g	0.14 g
Fibres	0.77 g	0.84 g
Ash	0.79 g	0.62 g
Vitamin 'A'	950 IU	50 IU
Vitamin 'C'	33 mg	21 mg
Thiamine	0.20 mg	0.11 mg
Riboflavin	0.14 mg	0.12 mg
Niacin	2.0 mg	1.0 mg
Calcium	22 mg	21 mg
Phosphorus	67 mg	46 mg
Potassium	271 mg	207 mg
Sodium	2 mg	4 mg
Magnesium	18 mg	20 mg
Iron	0.8 mg	1.0 mg

Area and Production

Worldwide, production of asparagus is around 77,12596MT (FAO, 2010). China ranks first by producing 69,69,357MT and it shares 88.98 per cent in total asparagus production of the world. Peru occupies second position and its total production during 2010 was 3,35209MT. Germany and Mexico are 3rd and 4th rankers by producing 92,404 and 74,660MT, respectively. The United States imports both green fresh asparagus and white fresh asparagus from Peru. While both types are imported and marketed in the United States, the color requirements of the current U.S. grading standards only provide for the grading of green asparagus (USDA 2005). White asparagus is very popular in Germany, there it is known as 'spargel.' Its production, however, is only enough to meet 61 per cent of its consumption demands (Spence, 2006). Worldwide asparagus production figure is given in table (37.2).

Origin and Distribution

Asparagus has been used from early times as a vegetable and medicine, owing to its delicate flavour and diuretic properties. There is a recipe for cooking asparagus in the oldest surviving book of recipes, Apicius's third century AD De re coquinaria, Book III. It was cultivated by the ancient Egyptians, Greeks and Romans, who ate it fresh when in season and dried the vegetable for use in winter. Asparagus is pictured on an Egyptian frieze dating to 3000 B.C. France's Louis XIV had special greenhouses built for growing it. It lost its popularity in the Middle Ages but returned to favour in the seventeenth century (Vaughan *et al.*, 1997).

Climate and Soil

Asparagus is a cool season crop. It performs better in the hills than in the plains. Its optimum temperature requirement is 16-24°C. The ideal daily temperature is 25-30°C and the night temperature 15-20°C. Its plant can slightly tolerate frost. Asparagus grows well in fertile and well drained loam soil. The optimum pH requirement of soil is 6.5-7.5. It has high tolerance to salinity and mild tolerance to acidic and alkali conditions.

Botanical Description

Asparagus is a genus of plants in the family Liliaceae.There are up to 300 species in *Asparagus*, all from the Old World. They have been introduced in many countries in both hemispheres and throughout temperate and tropical regions. Many species from Africa are now included in the genera *Protasparagus* and *Myrsiphyllum.* Members of Asparagus range from herbs to somewhat woody climbers. Most species have flattened stems (phylloclades), that serve the function of leaves. Three species (*Asparagus officinalis*, *Asparagus schoberioides*, and *Asparagus cochinchinensis*) are dioecious species. The others may or may not be hermaphroditic. The best known member of the genus is the vegetable asparagus (*Asparagus officinalis*). Other species of Asparagus are grown as ornamental plants. Some species such as *Asparagus setaceus* have branches that resemble 'ferns' hence they are often called 'Asparagus fern'. They are often used for foliage display, and as houseplants. Commonly grown ornamental species are *Asparagus plumosus*, *Asparagus densiflorus* and *Asparagus sprengeri.*

Table (37.2) : Worldwide asparagus production during 2010

Sl.No.	Country	Production(MT)	Per cent Share
1.	China	69,69,357	88.98
2.	Peru	3,35209	4.28
3.	Germany	92,404	1.18
4.	Mexico	74,660	0.95
5.	Thailand	63,108	0.81
6.	Spain	50,400	0.64
7.	Italy	43,973	0.56
8.	USA	36,240	0.46
9.	Japan	29,700	0.38
10.	France	17545	0.22
	Total	**77,12596**	

Source FAO (2010)

Taxonomic position of genus *Asparagus*

Kingdom	Plantae
Subkingdom	Tracheobionta
Superdivision	Spermatophyta
Division	Magnoliophyta
Class	Liliopsida
Subclass	Liliidae
Order	Liliales
Family	Liliaceae
Genus	*Asparagus* L.
Species	*Asparagus officinalis* L.

Chromosome number of asparagus (*Asparagus officinalis*) 2n is =20. Some other important species under the genus *Asparagus* are:

- *Asparagus aethiopicus*
- *Asparagus africanus*
- *Asparagus asparagoides*
- *Asparagus cochinchinensis*
- *Asparagus declinatus*
- *Asparagus densiflorus*
- *Asparagus falcatus*
- *Asparagus macowanii*
- *Asparagus officinalis*
- *Asparagus officinalis officinalis*
- *Asparagus officinalis prostratus*
- *Asparagus plumosus*
- *Asparagus racemosus*
- *Asparagus scandens*
- *Asparagus setaceus*
- *Asparagus sprengeri*
- *Asparagus umbellatus*
- *Asparagus virgatus*

Morphological Description of Plant

Asparagus is an herbaceous perennial plant growing to 100-150 cm tall, with stout wire like stems with much-branched feathery foliage. The leaves are cladodes (modified stems) in the axils of scale leaves; 6-32 mm long and 1 mm broad, and clustered 4-15 together. The root system is adventitious and the root type is fasciculated. The flowers are bell-shaped, greenish-white to yellowish, 4.5-6.5 mm long, with six tepals partially fused together at the base; they are produced singly or in clusters of 2-3 in the junctions of the branch lets. It is usually dioecious, with male and female flowers on separate plants, but sometimes androemonocious plants occur. The fruit is a small red berry 6-10 mm diameter. Seed-black.

Varieties: Important varieties and hybrids are listed as under:

Open Pollinated

1. **Connovers Colossal:** It is a traditional English open pollinated variety (originated in the USA in about 1885). Reselected strains yield good sized spears.
2. **Crimson Pacific:** Aspara, New Zealand. Mid season asparagus variety producing purple spears.
3. **Mary Washington :** USA. Open pollinated variety of asparagus, it is widely grown in the USA. Many reselections are available in the USA, adapted to specific areas.
4. **Purple Pacific:** Aspara, New Zealand. Moderate yield of purple spears with a sweet flavour. Good for white asparagus production. It is a mid season cultivar.
5. **Purple Passion:** It is an American selection of Violetto d'Albenga. A tetraploid asparagus giving a low yield of large spears with a high sugar content.
6. **Violetto d'Albenga:** Italy. Traditional tetraploid asparagus variety with large purple spears, but very low yield and disease prone.
7. **Perfection:** An exotic variety has been recom mended by IARI, New Delhi. It is an early variety with large green succulent spear, uniform maturity and high productivity.

Hybrids

1. **Apollo - C.A.S.T., USA.** Mid season variety with heavy stems but only average yields.

2. **Ariane - Spargelhochzucht Gast, Germany.** An early cropping asparagus variety.
3. **Atlas - C.A.S.T., USA:** An early mid season asparagus variety with heavy spears and good disease resistance.
4. **Dariana - Planasa, France:** A mid season asparagus variety with good yields of heavy spears.
5. **De Paoli:** It is popular in USA.
6. **Ginger:** A new asparagus variety giving good yields of thick, bright green, anthocyanin free spears.
7. **Grande - C.A.S.T., USA:** Fairly late variety with heavy stems. Good for warmer areas.
8. **Pacific 2000 Aspara, New Zealand:** A diploid asparagus giving good yields of heavy spears with sweet flavour. Mid season.
9. **Stewarts Purple - New Zealand:** A very sweet purple asparagus variety with large spears. Mid season.

Hybrids (androecious)

1. **Backlim - Limseeds, Holland:** A late variety useful for extending the season. Recommended for glasshouse production of both white and green asparagus with thick spears of good size and excellent flavour. Compact foliage.
2. **Eros - Italy:** Mid season variety which will grow in heavier soils, medium yield of large spears. Suitable for both green and white asparagus production.
3. **Gijnlim - Limseeds, Holland:** Early and very high yielding with spears of medium size and excellent quality. A standard variety for European commercial asparagus growers.
4. **Grolim - Limseeds, Holland:** An early variety producing a good yield of very heavy spears. Recommended for white asparagus production, and for warm areas, but sensitive to high water table.
5. **Millennium - Guelph, Canada:** Late maturing and useful for extending the asparagus season into July in the UK. Spears start to emerge two to three weeks later than other varieties, often escaping late frosts. Although well adapted to cooler areas, the excellent spear quality is maintained in hot weather.
6. **Jersey Giant - Rutgers, USA:** Moderate yield of medium to large spears. Tends to produce rather short spears at temperatures in excess of 70° F.
7. **Jersey Knight - Rutgers, USA:** A later maturing asparagus with medium to large spears. Spears tend to fern out when rather short at temperatures in excess of 70° F.
8. **Lucullus :** A late maturing variety with medium sized spears.
9. **Marte, Istituto Sperimentale, Italy:** Mid season, good yield potential. Suitable for both green and white asparagus production.
10. **Mondeo - Spargelhochzucht Gast, Germany:** A new high yielding, disease resistant variety with a long harvest season. Good flavour, and suitable for both green and white asparagus production.
11. **Thielim - Limseeds, Holland:** Mid season variety with good yields of thick, excellent quality spears. The erect, disease resistant foliage makes this variety particularly suitable for organic cultivation.
12. **Tiessen - Guelph, Canada:** The variety of choice for growing asparagus in areas with cold winters and cool springs. In other areas it is out yielded by Millennium.

Propagation

Asparagus is grown by seeds or crown.

Seed Propagation

Seeds are sown in nursery beds but before sowing seeds are soaked in water for about 3-4 days. Three to four kilogram of seeds are required to raise the seedlings for planting one hectare field. The optimum temperature for seed germination is 25-30°C. Seedlings 10-12 weeks old are transplanted in the field at a distance of 20-50 cm in a row and 1.0-1.2 m between rows. Planting can be done in the double rows also but at slightly wider spacing. Seeds are also sown directly at the bottom of 15-20 cm deep furrows and covered with 3-5 cm of soil. Later when the seedlings develop, soil is gradually filled to cover them. This method of seed sowing is practiced for production of white or blanched spears. However, in the case of green spears, planting is done in the rows bed. About 15,000-25,000 plants/ha are required to use crowns from one year old plants for propagation of asparagus(Nath *et al.*, 2008).

Crown Propagation

Well grown, one year old crowns are considered to be most suitable for commercial planting (Shukla *et al.*, 1992). The crown are planted in shallow trenches (30 cm wide and 25 cm deep). The crowns should be covered in full depth of furrow. When blanching is done, it is necessary to mound the soil over the rows, to obtain white spears.

Sowing and Planting Time

Seed sowing or planting can be done during July to November in the plains. In Himachal Pradesh, in mid hills (Zone-II) seeds are sown in the month of February to April whereas in high hills (Zone-III and IV seed sowing is completed by April to May. Crown planting in mid hills (Zone -II October to mid November is ideal time and in high hills (Zone-III and IV) planting of crown may be completed by March to April (Shukla *et al.*, 1992). For a hectare planting 600 g seeds are required.

Manures and Fertilizers

About 20 tonnes of FYM/ha is mixed in the soil at the time of field preparation. Among the chemical fertilizers, N:P:K@100:120:160 kg/ha may be given (Shukla *et al.*, 1992). The optimum way of fertilizer application in asparagus is to apply in split dosages. The first dose may be given in early spring before start of growth and second after harvesting is completed.

Irrigation and Weed Management

The first irrigation is given at the time of planting and thereafter at 10-15 days intervals. The water is allowed to stand in furrows for percolating in the root zones of the plants. Heavy watering shoud be avoided and proper provision of drainage should be given. For managing the weeds herbicides are available to use, however, first, determine the weed problem, considering their nature of emergence, annuals or perennials, grasses or broadleaves. Then selection should be made for use of the herbicides that best fit the problem. Some herbicides control only germinating or sprouting weed seeds and provide little, if any, control of weeds already emerged. Some herbicides are effective only when applied postemergence, that is, applied to the leaves of growing weeds. Some herbicides are systemic in action, that is, after contact with leaves they are translocated throughout the plant including the underground parts.

Harvesting

Asparagus spears are cut by hand, as they emerge through the soil from the underground crowns. Spears are usually harvested at a height of 20-25 cm, with the tips slightly closed. Harvested spears are prepared for market by grading, sizing, and bunching, based on freshness, stalk diameter and length, spear color, tip tightness, and extent of bruising. After trimming the butt end, bunches are packed upright in boxes containing moisture pads or water. High quality fresh asparagus spears are dark green and firm, with tightly closed and compact tips. Stems are straight, tender and glossy in appearance. Spears with green butts are preferred over those with white butts, because the latter are associated with increased toughness. However, it has been shown that a small amount of white tissue at the butt will delay decay development under typical commercial handling practices.

Post harvest Management

Asparagus produces little ethylene, <0.1 μL·kg^{-1}·hr^{-1} at 20°C. Exposure to ethylene will accelerate lignification (toughening) of the spears, while low temperatures will minimize ethylene-induced toughening. Freshly harvested asparagus has one of the highest respiration rates among produce. However, rates decline after harvest and the apical tips have higher rates than basal portions of the stalks. Both sugars and vitamin C decrease rapidly, resulting in a loss of flavor and nutritional value. In addition, if the storage temperature is above 36°F, fibers develop in the spears, resulting in a tough, stringy texture, and the tip buds begin to open. Harvested spears that are to be sold to consumers within a few hours should be stored in a cool place away from sunlight to prevent wilting. Spears that must be held for 24 hours or longer or are to be shipped to wholesale markets should be washed and chilled with cold water. If hydrocooling is available, a water temperature of 32°F for 5 minutes will lower the spear temperature to about 40°F. Once asparagus is chilled, it can be stored up to 3 weeks at a temperature of 36°F and 95% relative humidity. High relative humidity can be obtained by packaging spears in perforated film. Non-perforated film is not recommended because of injury from increased carbon dioxide and decreased oxygen within the package. An increase of ethylene in non-perforated film also has been reported to cause toughening of spears. Storage at 32°F for 10 days will result in chilling injury. High temperatures, in addition to causing a more-rapid decline in quality, may result in mold development.

Pre-cooling and Storage Conditions

Asparagus has a high metabolic rate after harvest and thus is one of the most perishable crops. Rapid cooling to 0-2°C soon after harvest is essential for optimal postharvest keeping quality. It has been shown that a 4-hour delay in cooling can cause 40% increase in shear force due to tissue toughening. Hydro-cooling is the most effective method for quickly cooling asparagus. Hydro-coolers use either an immersion or a shower system to bring products in contact with cold water. Recommended commercial storage conditions for asparagus are 0-2°C with 95-99% relative humidity. Low temperature is essential to delay senescence, tissue toughening, and flavor loss, while high humidity is critical to prevent desiccation and loss of glossiness. With rapid cooling and these storage conditions, asparagus can be stored for up to 3 weeks. However, extended storage (>10 days) at 0°C may result in some chilling injury. Storage in controlled atmosphere with 5-10% CO_2 is beneficial in preventing decay and reducing the rate of spear lignification. These effects are more pronounced if the temperatures cannot be maintained below 5°C.

Postharvest Disease

Decay is an important source of postharvest loss in asparagus. The most prominent disease is bacterial soft rot, caused by *Erwinia carotovora* or *Pseudomonas ssp.* Decay, characterized by soft rot pits, may initiate anywhere on the spears, but is most frequently found on the tips or butts. Spears that are re-cut above the white portion of the butt have been reported to be more susceptible to this type of rot. Some fungi (e.g. *Fusarium*, *Penicillium* and *Phytopthora*) are also associated with postharvest decay and spoilage of asparagus.

Physiological Disorders

1) Elongation and Tip Bending : Asparagus will continue to grow and elongate after harvest if not cooled promptly and stored in low temperatures. Spear growth and elongation are also promoted if the butt is in contact with water. Tip bending occurs as a response to gravity, such upward growth when the spears are horizontal or when the tips are deflected upon reaching the package top.

2) **Spear Toughening:** This disorder results from tissue lignification and fiber development, and progresses from butt to tip. Spear toughening occurs rapidly at temperatures >10°C and is accelerated by the presence of ethylene. Bruising and tip breakage are often the results of rough handling, and can also promote toughening due to wound ethylene.

3) **Feathering:** Expansion and opening of the tips is a sign of senescence. This is often observed when asparagus is harvested over-mature or after an extended period of storage, especially at higher than optimal temperatures.

4) **Hollow Stem:** This problem tends to be more prevalent during periods of warm, wet weather resulting in rapid spear growth and is found in crops that are ready to be harvested. You may have a problem in specific years with certain cultivars. Plant vigour and spacing may also play a role.

5) **Hooking:** This results in bent spears. This problem has many causes but is always because of some sort of injury to the spear either above or below the ground. Spears will hook into prevailing winds where sandblasting or cold air is a problem. Under the soil, cutworm feeding, cutting knives, stones, etc. can cause severe hooking.

6) **Chilling Injury:** Asparagus can develop chilling injury after extended storage (>10 days) at 0°C. Symptoms include loss of glossiness and graying of tips, with possibly a limp, wilted appearance. Severe chilling injury may result in darkened spots or streaks near the spear tips.

7) **Freezing Injury:** Damage due to freezing will be initiated at 0.5°C in asparagus. Symptoms include water-soaked appearance and extreme tissue softening.

Major Diseases

The main diseases of asparagus are *Fusarium* stem and crown rot caused by *Fusarium moniliforme*, and (or) *F. oxysporum* f. sp. *asparagi*, *Fusarium* wilt and root rot caused by *F. oxysporum* f. sp. asparagi, and rust caused by *Puccinia asparagi*. One or both *Fusarium* spp. may be present as soilborne pathogens. Other common diseases are seedling blight, decline, or replant problems. Both pathogens can enter the plant through young feeder roots, spread throughout the root and crown regions and eventually weaken and kill the plant. *F. oxysporum* f. sp. *asparagi* causes reddish brown vascular discoloration and decay of feeder and storage roots and occasionally discoloration and rot of crowns and stems. By contrast, *F. moniliforme* is associated primarily with stem and crown lesions and rarely with root tissue. As both fungi increase in the soil over a period of years, yields progressively decline, and the disease can be intensified by nutrient stress, drought, and insect damage. *Fusarium* can be introduced into the field from infected transplants. Asparagus rust is caused by *Puccinia asparagi*. It is considered one of the most devastating diseases of asparagus. Symptoms of the disease first appear as small orange patches on spears and on fern branches. High humidity and warm temperatures influence repeated spore production and germination. Windblown spores are spread to adjoining plants, and with sufficient moisture new infections are initiated. Individual fern needles are shed, and the entire field may turn brown. The entire life cycle of the rust fungus occurs on asparagus, with the summer orange spores giving way to black overwintering spores.

Management

i) Select only disease-free material.

ii) Dip the roots in a recommended fungicide solution.
iii) Seed may be surface contaminated with *Fusarium* spores. Treat the seed for 2 minutes in a solution of 1 part commercial 5.25% sodium hypochlorite to 4 parts water. Remove the seed from the solution, rinse thoroughly with cold water, and spread out to surface dry, and then plant immediately unless presoaking is planned.
iv) Select well-drained fields where asparagus has never been grown.
v) Adopt resistant varieties Some varieties selected out of Washington strains have some rust resistance. Several New Jersey selections have good to excellent resistance.
vi) Destroy wild asparagus plants and volunteer seedlings in the vicinity of the crop.

Major Pests

1. Asparagus beetle (*Crioceris asparagi*)

It is a very important pest of asparagus.The adult beetles, which overwinter under plant refuse and debris along field borders, begin to move to asparagus as the spears first emerge in the spring. The beetles feed on the spears and lay eggs singly in vertical rows, usually near the tip of the spear . The eggs hatch in approximately 1 week, and the numerous fleshy, grey larvae move to foliage where they feed. The larval stage lasts 2-3 weeks, after which the larvae leave the plant, burrow into the soil, and pupate.

Management :

i) Apply neem based formulation.
ii) Follow clean culture and long term rotation

2. Spotted asparagus beetle (*Crioceris duodecimpunctata*)

Extensive spear damage may result from beetle feeding, which causes a distorted spear growth with a distinct 'shepherd's crook' shape.

Management :

i) Apply neem based formulation.
ii) Follow clean culture and long term rotation

3. Asparagus aphid (*Brachycolus asparagus*)

This fairly small, blue green aphid feeds on the fern. The feeding results in a 'witches'broom' growth of the plants.The damage results in short, bushy plants with a distinctive silvery, blue green color. Crowns may be affected, resulting in thin spears and poor yields.

Management:

i) Apply neem based formulation

Genetic Resources and Improvement

Important germplasm collection of *Asparagus officinalis* cultivars are maintained at the crops and Food Central research Institute, Lincoln, Canterbury, New Zealand and by USDA, United States (Nicholas, 1993). Somaclonal variation in plants regenerated by organogenesis from long-term cultured calluses of two diploid staminate genotypes of *Asparagus officinalis* cv. Argenteuil was characterized by plant phenotype, ploidy, meiotic behavior, pollen viability, fruit and seed set, and AFLP profiles. Phenotypic deviations from the donors were detected in foliage color, flower size, and cladode and flower morphology. Ploidy changes were observed in 37.8% of the 37 regenerants studied. Meiotic alterations in 12 out of 21 regenerants included laggards, dicentric bridges, micronuclei, restitution nuclei and polyads. Of the 408 AFLP markers screened in 43 regenerants and the donors, 2.94% showed polymorphism. High pollen viability was observed in the 22 regenerants analyzed. All crosses between one pistillate plant and 35 regenerants, as well as the controls, produced fruits and seeds; however, no plump seeds resulted in 35.3% of the crosses with regenerants, and no seeds germinated in 12.5% of those with apparently normal seeds. Fruit and seed set was similar in crosses with diploid regenerants with normal meiosis and the controls but was lower in crosses with diploid and polyploid regenerants with abnormal meiosis. This show that the regenerated plants exhibited conspicuous somaclonal variation that could be eventually exploited for *in vitro* selection systems (Pontaroli and Camadro, 2005)

Tissue culture

While practicing the tissue culture technique in regeneration of asparagus, two methods have been tried with reasonable success. One has been to culture the lateral buds excised from elongating spears. A new plant is obtained from each bud and as many plants can be obtained by this method as there are lateral buds. In the second method, slices of a spear (with buds removed) are stimulated to form callus and subsequently plantlets are regenerated from the callus. This method is more complex than that of lateral bud culture, but can provide an unlimited number of plants from each piece of excised tissue (Takatori *et al.*, 1967). A highly reproducible *in vitro* regeneration system in *Asparagus racemosus* using nodal segments has been developed. Multiple shoots were induced *in vitro* from nodal segments through adventitious shoot bud regeneration as well as indirectly through callus. The ability of nodal segments to produce shoot buds varies depending upon their position on the main shoot of the field grown plants. Nodal segments obtained from 7th node to 9th node was proved to be the most suitable for the induction of bud initials. Explants showed the best proliferation and growth on the medium fortified with BA 3.0 mgl^{-1} and NAA 0.5 $mg.l^{-1}$ with an average of 40-45 shoot buds / explant and an average 3-3.5 cm length. *In vitro* rooting of regenerated shoots was achieved on half strength MS medium fortified with NAA. The plants were successfully hardened for about 5 weeks thereafter transferred to the field. The acclimatized plantlets were successfully established in the field with approximately more than 75% survival (Vijay and Kumar, 2009). Shoot tips isolated from spears were cultured on basal MS medium. Multiple bud clusters were obtained on MS supplemented with 1mg/l ben-zyladenine . Addition of kinetin to multiplication medium was more effective at increasing the number of initiated shoots than BA. Supplementation of the culture medium with adenine sulfate generally improved the vegetative growth of shoots and shoot proliferation, with a concentration of 80 mg/litre producing better results than 40 or 120 mg/litre. The best rooting percentage as well as number of roots was obtained by adding 1 mg/litre IAA to the culture medium (half-strength MS). Transplanting to outdoor conditions was successful when plantlets were cultured in pots containing perlite and peat moss (1:1) (Bekheet, 1999).

Indirect organogenesis and somatic embryogenesis were used for the *in vitro* regeneration of *Asparagus officinalis*. For the organogenesis pathway, nodal explants were treated with different NAA and BAP concentrations for both callus induction and shoot development. The treatment inducing the formation of the highest number of shoots (mean shoot number per explant = 20.5) was the one containing 0.015 mgl^{-1} NAA and 0.5 mgl^{-1} BAP. The effect of two IBA concentrations (1.0 and 1.25 mgl^{-1}) on rooting of regenerated shoots was also investigated, with 1.25 mgl^{-1} giving a higher rooting percentage (50%). Shoot tips were found to generate the highest number of plantlets at NAA concentration of 12 mgl-^{1}, while for nodal section a concentration of 3 mgl^{-1} of NAA was required. The percentage regeneration of plantlets was highest (83.3 %) when nodal explants were treated with 0.015 mgl^{-1} NAA and 0.5 mgl^{-1} BAP (Bojnauth *et al.*, 2003).

References

Bekheet, S.A. (1999). *In vitro* propagation of *Asparagus officinalis*. Egyptian Journal of

Horticulture, 26(2): 237-248.

Bojnauth, G., Puchooa, S. and Bahorun, T (2003). *In vitro* regeneration of *Asparagus officinalis*-preliminary results. AMAS 2003. pp7-15.

Kim, B.Y., Cui, Z.G., Lee, S.R., Kim, S.J., Kang, H.K., Lee, Y.K. and Park, D.B. (2009). Effects of *Asparagus officinalis* extracts on liver cell toxicity and ethanol Metabolism Journal of Food Science, 74 I(7): H204 - H. 208.

Nath, P., Srivastava, V.K., Dutta, O.P. and Swamy, K.R.M. (2008). Vegetable Crops. Dr. Prem Nath. Agricultural Science Foundation, Bangalore.

Nicholas, N.A. (1993). *Asparagus officinalis* In : PROSEA, Plant Resorces of South East Asia. Vegetables 8. Sienosmma, J.S. and Piluek, K. (Eds). Pudoc Scientific Publishers, Wageningen, pp. 91-93.

Pontaroli, A.C. and Camadro, E.L. (2005). Somaclonal variation in *Asparagus officinalis* plants regenerated by organogenesis from long-term callus cultures. Genet. Mol. Biol. 28(3):423-430.

Shukla, Y.R., Jain, Y.C. and Sharma, O.P. (1992). Cultural practices for vegetable Crops in Himachal pradesh, Directorate of Extension Education, Dr.Y.S. Parmar University of Horticulture and Forestry, Solan, HP. 173-230.

Spence, M. (2006). Asparagus: The king of vegetables. German Agricultural Marketing Board. Retrieved Feb. 26, 2007.

Takatori, F.H., Murashige T. and Stillman, J.I (1967). Tissue culture of asparagus plants. California Agriculture, August, 1967, p. 8.

Vaughan, J. G., Geissler, C. A. and Nicholson, B.(1997). The New Oxford Book of Food Plants. Oxford University Press.

Vijay, N. and Kumar, A. (2009). *In vitro* plantlet regeneration in *Asparagus racemosus* through shoot bud differentiation on nodal segments. Science 2.0.

38

WINTER SQUASH

The flower and fruit of winter squash are eaten as a vegetable, the fruit is kept in proper conditions of light, temperature and humidity, up to six months in good condition. It is rich in beta carotene and glucose. Carotenoids found in winter squash include alpha-carotene, beta-carotene, beta-cryptoxanthin, lutein, and zeaxanthin. Pectin-containing cell wall polysaccharides found in winter squash are important anti-inflammatory nutrients provided by this food, as are its cucurbitacins (triterpene molecules). Winter squash is an excellent source of immune-supportive vitamin A (in its 'previtamin' carotenoid forms) and free radical-scavenging vitamin C. It is also a very good source of enzyme-promoting manganese and digestion-promoting dietary fiber. In addition, winter squash is a good source of heart-healthy folate, omega-3 fatty acids, vitamin B_2, vitamin B_6, magnesium, and potassium; and bone-building copper and vitamin K. Nutritional Composition of the fruit is given in table (38.1).

Table (38.1) Nutritional composition of the winter squash (per 100 gof edible portion)

Constituents	Amount
Water	85 per cent
Food energy	50 Kcal
Protein	1.4g
Fat	0.3 g
Carbohydrate total	12.4 g
Fibre	0.6 g
Ash	0.5 g
Calcium	22 mg
Phosphorus	38mg
Iron	0.6mg
Sodium	1 mg
Potassium	369 mg
Vitamin 'A'	3700 I.U.
Thiamine	0.05 mg
Riboflavin	0.11 mg
Niacin	0.6 mg
Ascorbic acid	3 mg

Source : Watt and Merrill (1963)

The winter squash seeds contained 39.25% crude protein, 27.83% crude oil, 4.59% ash, and 16.84% crude fiber; the corresponding values for the kernels were 39.22, 43.69, 5.14, and 2.13%, respectively. Pumpkin seed kernels contained moderate concentrations of minerals, especially P, Mg, and K. The amino acid profiles indicate that methionine and tryptophan were the most limiting amino acids, while arginine, glutamic, and aspartic acids were the most plentiful amino acids. The saturated fatty acid content was 27.73% and comprises of 16.41% palmitic acid and 11.14% stearic acid. The unsaturated fatty acid value was 73.03% and consisting mainly of 18.14% oleic acid and 52.69% linoleic acid. The oil obtained from the pumpkin seed kernels had a refractive index of 1.4656, specific gravity of 0.913, iodine value of 105.12 (g /100g oil), saponification value of 185.20 (mg KOH/g oil), acid value of 0.53 (mg KOH/g oil), and peroxide value of 0.85 (meq peroxide/kg oil). Considering the lipid and protein content in the kernels, and their fatty and amino acid compositions, the pumpkin seed kernels appear to be quite promising for commercial exploitation (Alfawaz, 2004).

Table (38.2). Proximate composition (%DM) of the whole seeds and kernels)

Components	Seed	Kernel
Crude protein	39.25	39.22
Crude oil	27.83	43.69
Total ash	4.59	5.14
Crude fiber	16.84	2.13
Carbohydrate	11.48	9.82

Source: Alfawaz (2004)

Table (38.3). Elemental nutrient composition (mg/100 g DM) of winter squash seed

Elements	Mean value
Calcium	139.70
Copper	0.30
Iron	13.66
Magnesium	364.43
Phosphorus	1036.82
Potassium	753.11
Sodium	68.58
Zinc	1.09

Source: Alfawaz (2004)

Table (38.4). Amino acids composition (g/100 g protein) of winter squash seed kernels

Amino Acids	Mean value
Methionine	1.83
Tyrosine	3.26
Phenylalanine	5.29
Leucine	7.25
Isoleucine	3.59
Lysine	3.71
Threonine	3.04
Tryptophan	1.10
Valine	4.45
Total essential AA	33.52
Histidine	2.66
Arginine	15.80
Aspartic acid	9.56
Glutamic acid	23.23
Serine	5.85
Glycine	6.01
Alanine	5.12
Total nonessential AA	68.23

Source: Alfawaz (2004)

Table (38.5). Fatty acids composition (%) of winter squash seed kernels

Fatty acid (FA)	Mean value
Myristic (C14:0)	0.18
Palmitic (C16:0)	16.41
Stearic (C18:0)	11.14
Palmitoleic (C16:1)	0.16
Oleic (C18:1)	18.14
Erucic (C22:1)	0.76
Linoleic (C18:2)	52.69
Linolenic (C18:3)	1.27
Total saturated FA (TSFA)	27.73
Total unsaturated FA (TUSFA)	73.03
Monounsaturated FA (MUSFA)	19.06
Polyunsaturated FA (PUSFA)	53.97

Source: Alfawaz (2004)

Origin and Distribution

All of the *Cucurbita* species originated in the New World. The four species of genus viz., *C. maxima, C. mixta, C. moschata* and *C. pepo* are important food plants to Native Americans and have been cultivated for millennia. *C. maxima* was developed from a wild ancestor in central South America. *C. mixta* was originally native to Guatemala and Mexico. *C. moschata* was domesticated more than 5000 years ago from wild ancestors in tropical America. *C. pepo* no longer occurs in the wild, but probably was derived from ancestors in North America.

Botanical Description

Taxonomic position of the genus *Cucurbita maxima* is given as under:

Kingdom	Plantae
Subkingdom	Tracheobionta
Superdivision	Spermatophyta
Division	Magnoliophyta
Class	Magnoliopsida
Subclass	Dilleniidae
Order	Violales
Family	Cucurbitaceae
Genus	*Cucurbita* L.
Species	*maxima*

It is often not easy to distinguish *Cucurbita maxima* plants or fruits from the related *Cucurbita moschata* and *Cucurbita pepo* L. The plant habit is similar and the fruit shape and size are variable. Distinction is easiest by observing differences of the fruit stalk, stems and leaves. *Cucurbita maxima* has a soft, rounded fruit stalk not enlarged at the apex, soft, rounded stems and soft, usually unlobed leaves. *Cucurbita moschata* has a hard, smoothly angled fruit stalk widened at the apex, hard, smoothly grooved stems and soft, moderately lobed leaves. *Cucurbita pepo* has an angular fruit stalk sometimes slightly widened at the apex, hard, angular, grooved, prickly stems and palmately lobed, often deeply cut and prickly leaves. *Cucurbita andreana,* a species related to *Cucurbita maxima* and sometimes considered a subspecies of the latter, occurs wild in Argentina and Bolivia; it hybridizes readily with *Cucurbita maxima*. It may be the ancestor of *Cucurbita maxima* (Chigumira and Grubben, 2004.)

Description of species

There are about 27 species of *Cucurbita*, but most of the squashes familiar to Europeans, Asians and North Americans belong to one of four species, all of which are annual, tendril-bearing vines (or vine-like "bush" types derived from viny relatives). They have coarse, rather large simple leaves, usually lobed and often prickly. Squash vines may run for 2-10 m and often take root at their nodes. The plants are monoecious, having both staminate (male) and pistillate (female) flowers on the same vines. The flowers are yellow, bell-shaped and 7.6-15.2 cm across. Squash plants usually produce staminate flowers for several days before they even begin producing pistillate flowers. The flowers are easily distinguished because the female flower has a swollen ovary at its bases. The fruits of squashes, pumpkins and gourds are technically called berries, defined as fruits that develop from single pistils and contain seeds that are not encased in stones. Winter squash possesses long shelf life, hard skins and dry flesh. A brief description of the all four species is given as under:

1.Cucurbita maxima

Plants have soft, spongy stems that are hairy and round in cross section. They have very large hairy leaves that are quite rounded and usually not lobed. The peduncle (fruit stalk) is cylindrical and rather soft. The seeds have lighter colored margins and thin papery coverings. Cultivars include all of the Hubbard, banana, buttercup, and turban squashes,

2. Cucurbita mixta

Under this group, cultivars have hard, angled stems with soft hairs. The leaves are shallowly lobed and glabrous (without hairs) or only slightly hairy. The leaves are usually marked with white blotches. The peduncle is hard and sharply 5-angled with a corky swelling (not flared) where it attaches to the fruit. The seeds are white or tan with a lighter colored margin.

3.Cucurbita moschata

Plants have hard, angular stems with soft hairs. The leaves have shallow rounded lobes, white markings, and are quite hairy. The peduncle flares out where it attaches to the fruit. The seeds are white to brown with darker margins.

4.Cucurbita pepo

Plants of this group have hard, 5-angled stems with prickly hairs. The leaves are deeply lobed and covered with bristly hairs. The peduncle is deeply grooved and 5-angled and very hard where it attaches to the fruit. The seeds have distinctively colored margins. Important varieties under this group are 'Patty Pan', 'Sweet Dumpling', 'Delicata',

Morphological Description of Plant

Annual, scandent herb, climbing by lateral, 2-5-branched tendrils, strongly branched; stems rounded, long running, softly pubescent, often rooting at nodes. Leaves alternate, simple, without stipules; petiole (5-)10-20 cm long; blade usually reniform, not lobed to shallowly 5-7-lobed, 7-25(-30) cm in diameter, deeply cordate at base, margins finely toothed, softly hairy, occasionally with white blotches, 3-veined from the base. Flowers solitary, unisexual, regular, 5-merous, large, 10-20 cm in diameter, lemon yellow to orange-yellow; sepals free, subulate to linear, 0.5-2 cm long; corolla campanulate, with widely spreading lobes; male flowers long-pedicelled (up to 23 cm), with 3 stamens, filaments free, anthers connivent into a long twisted body; female flowers shortly pedicelled (up to 5.5 cm), with inferior, ellipsoid, 1-celled ovary, style thick, stigmas 3-5, 2-lobed. Fruit a large, globose to ovoid or obovoid berry, weighing up to 50 kg, with a wide range of colours; flesh yellow-orange, many-seeded; fruit stalk cylindrical, not enlarged at apex. Seeds obovoid, flattened, 1.5-2.5 cm × 1-1.5 cm, white to pale brown, surface smooth to somewhat rough, margin prominent. Seedling with epigeal germination; hypocotyl 3-3.5 cm long; cotyledons elliptical, 2.5-4 cm long, cuneate at base, obtuse at apex, entire (Chigumira and Grubben, 2004). Its chromosome number(2n) is =40.

Climate and Soil

Winter squash are grown in the tropics from the lowlands up to 2000 m altitude. Optimum mean daily temperatures for growth are 18-27°C. It is more tolerant to low temperatures than *Cucurbita moschata*, but less tolerant to high temperatures. It is almost insensitive to photoperiod and tolerates some shading. It can be produced year round in frost-free areas, although excessive humidity during the rainy season stimulates the development of fungal and bacterial diseases causing leaf decay, wilting and fruit rot. It can be grown in almost all kinds of soils, however, well-drained soil with a neutral or slightly acid reaction (pH 5.5-6.8) should be preferred.

Sowing and Seed Rate

Seed germinates within 5-7 days after sowing. A common planting spacing is 2 m × 2 m, giving a plant population of 2500 plants/ha. The 1000-seed weight is about 200 g. The seed requirement for a hectare sowing is 2-3 kg. Because of the branching and creeping plant habit, the optimal planting density is flexible, varying from 600-4000 plants/ha; the number of plants per hill is 1-4.The plants form an extensive fibrous root system. The growth habit is indeterminate; under suitable conditions, the trailing stems continue to grow indefinitely when they are permitted to root at the nodes. The stems may reach a length of more than 20 m, but they are usually no longer than about 5 m. Bushy cultivars with short internodes and semi-erect stems also exist.

Flowering and Fruit Set

It is a cross pollinated crop, flowering starts 35-60 days after germination and is more or less continuous. The ratio of male to female flowers is about 20:1. This ratio is influenced by the growing conditions, long days and high temperatures favouring male sex expression. Production of the sticky pollen is abundant. Anthesis and pollination take place early in the morning. The fruits mature 30-40 days after pollination. One or two fruits develop per stem. The fruit-harvesting period extends from 2-6 months after sowing.

Manures and Fertilizers

The crop responds well to farm yard manure. At the time of field preparation apply FYM @ 15-20 t/ha for obtaining a good yield a doze of NPK@ 100:60:60/ ha shoud be given . Apply one third N and full doze of P an K while remaining doze of N may be given in two split dozes i.e., 30 and 60 days after sowing.

Major Disease

1. Downy mildew

It is caused by *Pseudoperonospora cubensis*. It is most serious disease of humid regions. Its first symptoms appear on older leaves with pale green angular spots, bounded by leaf veins. The lower surface becomes covered with a faint, purplish layer and the entire leaf dies.

Control :

1. Follow wide spacing, good drain age and aeration .

2. Adopt long term crop rotation.

2. Powdery mildew

It is caused by *Erysiphe cichorace-arum*. The upper side of affected leaves is covered with white powdery fungus growth; the leaves become yellowish and dry out.

Control :

1.Apply sulphur based fungicide

2. Use resistant variety.

3.*Alternaria* leaf spot

It is a fungal disease caused by *Al-*

ternaria cucumerina. Infested plants become defoliated and die within weeks. This fungus also causes fruit rot. The symptoms first appear as small round whitish necrotic spots, later with concentric circles, appear on the oldest leaves; the leaves turn grey or yellow and dry out.

Control :

1.Use healthy seed.

2.Remove plant debris and follow crop rotation

4. Gummy stem blight

It is caused by *Didymella bryoniae*. First the nodes with nearly ripe fruits are attacked; they become oily-green and sap exudes forming drops of gum, while the stems beyond the affected area wilt. This fungus also causes seedling death, fruit stalk canker and fruit rot.

Control:

1.Use healthy seed.

2.Remove plant debris and follow long term crop rotation.

5. *Fusarium* wilt

It is caused by *Fusarium oxysporum.* Symptoms appear as yellowing of the leaves followed by wilting of stems or the whole plant. It can be controlled by using healthy seed, removing debris, crop rotation and avoiding excessive nitrogen fertilization.

Control :

1.Use healthy seed.

2.Remove plant debris and follow long term crop rotation.

6. Anthracnose

It is caused by *Colletotrichum lagenarium.* It is most serious disease of cucurbits worldwide. It causes defoliation and lesions on the fruits.

Control :

1. Use certified disease free seed.
2 .Remove plant debris and follow long term crop rotation.

Virus diseases

Many viruses are reported on *Cucurbita maxima* the most important ones are cucumber mosaic virus (CMV), papaya ring spot virus (PRSV-W), squash leaf curl virus (SLCV), watermelon mosaic virus (WMV-2) and zucchini yellow mosaic virus (ZYMV). These viruses often occur in combination. They are spread by mechanical infection and insects (mainly aphids, whitefly and thrips), and probably some by seed.

Control :

1. Use disease free seed materials
2. Remove weeds, debris and follow crop rotation
3. Judicious use of pesticide to check popolation of vectors.

Major Pests

Squash bug

Adult squash bugs are about 5/8" long and 1/4" wide. They are grey or black and have orange and brown stripes on the edges of their abdomen. Their eggs are a yellow to bronze in color and are deposited on underside of leaves along leaf veins in groups Squash bugs feed on cucumbers, summer and winter squash and pumpkins

Control :

1.Follow good agricultural pactices.

2. Apply neem based formulation.

Harvesting and Yield

Fruits of winter squash are picked when nearly or fully mature, 4-6 weeks after flowering, and are harvested in several rounds until the crop stops producing, 90-120 days after planting. The yield largely depends on cultivar and growing conditions. The number of mature fruits harvested per plant is low, and the weight of individual fruits, mainly depending on cultivar, varies widely from 1-50 kg. With good care, a yield of 15-20 t/ha is reasonable, and with improved cultivars a yield of 30 t/ha is possible.

Genetic Resources and Improvement

The National Seed Storage Laboratory (NSSL), Fort Collins, Colorado, United

States, and the Vavilov Institute of Plant Industry (VIR), Petersburg, Russia maintains important base collections. With the introduction of improved cultivars, ancient landraces are in danger of disappearing .The Centre for Conservation and Breeding of Agricultural Biodiversity (COMAV) holds a collection of more than 900 accessions of different species under *Cucurbita*. A large number of these accessions belong to *C. maxima,* collected from almost all the Spanish provinces. Most of them are traditional landraces adapted to diverse agro-ecological conditions, from mountainous dry lands to plain irrigated areas. Since *C. maxima* is partially allogamous, a great deal of heterogeneity is expected within landraces. However, the number of plants included in traditional Spanish gardens is usually low. This small effective population size could influence the structure of the variation. Determining the degree of variability within an accession is necessary as a preliminary step for studying the genetic diversity among pumpkin accessions. In addition to the morphological characterization, a molecular screening is essential for determining this variability (Ferriol *et al.*, 2001). Populations (115) of winter squash, were collected from different provinces of the Black Sea region and phenotypic diversity in their fruit characters was assessed. The collection showed appreciable phenotypic variation in fruit shape, fruit colour, fruit brightness, fruit dimension and fruit weight. Cluster analysis based on 14 quantitative and 7 qualitative variables identified 10 different groups. The first five principal component axes accounted for 65.0 per cent of the total multivariate variation among the populations. The greater part of variance was accounted for by fruit weight, fruit diameter, fruit length, length of seed cavity and flesh thickness. This evaluation of fruit trait variability can assist geneticists and breeders to identify populations with desirable characterisics for inclusion in variety breeding programs (Balkaya *et al.*, 2010).

The six SSR markers tested performed poorly over a range of conditions: annealing temperature from 45°C to 60°C in 2° increments and 1.5 to 2.5 mM MgCl2. Marker CMCT168 amplified, but this was not highly reproducible. None of the other five markers amplified Banding patterns obtained using SBAPs markers appeared to be highly reproducible. Polymorphism among individuals, and within and among accessions. The SSR molecular genetic markers developed in melon and cucumber species were not promising. It is possible the primer binding sites have diverged so much between species that the primer did not effectively bind any longer. It may be possible to further optimize them in *C. maxima.* Preliminary analysis of SBAPs banding patterns showed no evidence of population structure. Identical genotypes were spread among the continents. Due to their high reproducibility and high levels of polymorphism, SBAPs appear to be promising markers for studying genetic diversity was observed (Kanellis and Sheffer, 2011). *Cucurbita maxima* is cross-pollinated but self-compatible, inbreeding causing little loss of vigour. A considerable degree of heterosis of inbred lines has been observed. Hand pollination of the large flowers is easy. Seed companies are less interested in open-pollinated cultivars, because farmers can easily produce their own seed. Western seed companies have commercialized many F_1 hybrids. A Japanese seed company has developed 'Shintoza', an interspecific hybrid of *Cucurbita maxima* and *Cucurbita moschata*, which combines high vigour and resistances to soil borne diseases and nematodes.

References

Alfawaz, M.A. (2004). Chemical composition and oil characteristics of pumpkin (*Cucurbita maxima*) seed kernels. Res. Bult., 129: 5-18.

Balkaya, A., Özbakir, M. and Kurtar, E.S. (2010). The phenotypic diversity and fruit characterization of winter squash (*Cucurbita maxima*) populations from the black Sea Region of Turkey . African Journal of Biotechnology, 9 (2):152-162.

Chigumira Ngwerume, F. and Grubben, G.J.H. (2004). *Cucurbita maxima* Duchesne In: Grubben, G.J.H. & Denton, O.A. (Editors). PROTA 2: Vegetables/Légumes. PROTA, Wageningen, Netherlands.

Ferriol , M, Picó, B. and Nuez , F. (2001). Genetic variability in pumpkin (*Cucurbita maxima*) using RAPD markers. Cucurbit Genetics Cooperative Report , 24:94-96 .

Kanellis, M.M. and Sheffer, S.M. (2011). Molecular genetic marker comparison in winter squash (*Cucurbita maxima*). Hobart and William Smith Colleges, 2. USDA, ARS Plant Genetic Resources Unit, Geneva, NY 14456

Watt, B.K. and Merrill, A.L. (1963). Composition of Food Raw, Processed, Prepared, USDA Agricultural Handbook.

D

E

F

G

H

I

Cherry Tomato

Paprika

Snow Pea

Baby Corn

Gherkin

Sweet Corn

Broccoli

Romanesco Cauliflower

Red Cabbage

Savoy Cabbage

Chinese Cabbage

Brussels Sprout

Chive

Welsh Onion

Leek

Rakkyo

Parsnip

Chinese Chive

Horse Radish

Jerusalem Artichoke

Endive

Lettuce

Globe Artichoke

Sugar Loaf Chicory

Red Chicory

Rhubarb

Parsley

Celery

Celeriac

Kale

Collard

Winter Squash

Wintercress

Florence fennel

Watercress

Swiss Chard

Asparagus